Springer Proceedings in Complexity

Springer Proceedings in Complexity publishes proceedings from scholarly meetings on all topics relating to the interdisciplinary studies of complex systems science. This series is indexed by DBLP.

Springer welcomes book ideas from authors. Potential authors who wish to submit a book proposal should contact Christopher Coughlin, Publishing Editor, Springer.

Proposals must include the following:

- name, place and date of the scientific meeting
- a link to the committees (local organization, international advisors etc.)
- scientific description of the meeting
- list of invited/plenary speakers
- an estimate of the planned proceedings book parameters (number of pages/articles, requested number of bulk copies, submission deadline).

submit your proposals to: christopher.coughlin@springer.com

More information about this series at http://www.springer.com/series/11637

Gian Italo Bischi · Anastasiia Panchuk
Davide Radi
Editors

Qualitative Theory of Dynamical Systems, Tools and Applications for Economic Modelling

Lectures Given at the COST Training School on New Economic Complex Geography at Urbino, Italy, 17–19 September 2015

Springer

Editors
Gian Italo Bischi
Università di Urbino "Carlo Bo"
Urbino
Italy

Davide Radi
LIUC - Università Cattaneo
Castellanza
Italy

Anastasiia Panchuk
National Academy of Sciences of Ukraine
Kiev
Ukraine

ISSN 2213-8684 ISSN 2213-8692 (electronic)
Springer Proceedings in Complexity
ISBN 978-3-319-33274-1 ISBN 978-3-319-33276-5 (eBook)
DOI 10.1007/978-3-319-33276-5

Library of Congress Control Number: 2016939062

Printed on acid-free paper

This Springer imprint is published by Springer Nature
The registered company is Springer International Publishing AG Switzerland

Preface

This volume contains the lessons delivered during the "Training School on qualitative theory of dynamical systems, tools and applications" held at the University of Urbino (Italy) from 17 September to 19 September 2015 in the framework of the European COST Action "The EU in the new complex geography of economic systems: models, tools and policy evaluation" (Gecomplexity). Gecomplexity is a European research network, inspired by the New Economic Geography approach, initiated by P. Krugman in the early 1990s, which describes economic systems as multilayered and interconnected spatial structures. At each layer, different types of decisions and interactions are considered: interactions between international or regional trading partners at the macrolevel; the functioning of (financial, labour, goods) markets as social network structures at mesolevel; and finally, the location choices of single firms at the microlevel. Within these structures, spatial inequalities are evolving through time following complex patterns determined by economic, geographical, institutional and social factors. In order to study these structures, the Action aims to build an interdisciplinary approach to develop advanced mathematical and computational methods and tools for analysing complex nonlinear systems, ranging from social networks to game theoretical models, with the formalism of the qualitative theory of dynamical systems and the related concepts of attractors, stability, basins of attraction, local and global bifurcations.

Following the same spirit, this book should provide an introduction to the study of dynamic models in economics and social sciences, both in discrete and in continuous time, by the methods of the qualitative theory of dynamical systems. At the same time, the students should also practice (and, hopefully, appreciate) the interdisciplinary "art of mathematical modelling" of real-world systems and time-evolving processes. Indeed, the set-up of a dynamic model of a real evolving system (physical, biological, social, economic, etc.) starts from a rigorous and critical analysis of the system, its main features and basic principles. Measurable quantities (i.e. quantities that can be expressed by numbers) that characterize its state and its behaviour must be identified in order to describe the system mathematically. This leads to a schematic description of the system, generally a simplified representation, expressed by words, diagrams and symbols. This task is, commonly, carried out by specialists of the real system, such as economists and social scientists. The following stage consists in the translation of the schematic model into a mathematical model, expressed by mathematical symbols and operators. This leads us to the mathematical study of the model by using mathematical tools, theorems, proofs, mathematical expressions and/or numerical methods. Then, these mathematical results must be translated into the natural language and terms typical of the system described, that is economic or biologic or physical terms, in order to obtain laws or statements useful for the application considered. This closes the path of mathematical modelling, but often it is not the end of the modelization process. In fact, if the results obtained are not satisfactory, in the sense that they do not agree with the observations or experimental data, then one needs to re-examine the model, by adding some details or by changing some basic assumptions, and start again the whole procedure. The chapters of this volume are mainly devoted to the mathematical methods for the analysis of dynamical models by using the qualitative theory of dynamical systems, developed through a continuous and fruitful interaction among analytical, geometric and numerical methods. However, several examples of model building are given as well, because this is the most creative stage, leading from reality to its formalization in the form of a mathematical model. This requires competence and fantasy, the reason why we used the expression "art of mathematical modelling".

The simulation of the time evolution of economic systems by using the language and the formalism of dynamical systems (i.e. differential or difference equations according to the assumption of continuous or discrete time) dates back to the early steps of the mathematical formalization of models in economics and social sciences, mainly in the nineteenth century. However, in the last decades, the importance of dynamic modelling increased because of the parallel trends in mathematics on one side and economics and social sciences on the other side. The two developments are not independent, as new issues in mathematics favoured the enhancement of understanding of economic systems, and the needs of more and more complex mathematical models in economics and social studies stimulated the creation of new branches in mathematics and the development of existing ones. Indeed, in recent mathematical research, a flourishing literature in the field of qualitative theory of nonlinear dynamical systems, with the related concepts of attractors, bifurcations, dynamic complexity, deterministic chaos, has attracted the attention of many scholars of different fields, from physics to biology, from chemistry to economics and sociology, etc. These mathematical topics become more and more popular even outside the restricted set of academic specialists. Concepts such as bifurcations (also called catastrophes in the Eighties), fractals and chaos entered and deeply modified several research fields.

On the other side, during the last decades, also economic modelling has been witnessing a paradigm shift in methodology. Indeed, despite its notable achievements, the standard approach based on the paradigm of the rational and representative agent (endowed with unlimited computational ability and perfect information) as well as the underlying assumption of efficient markets failed to explain many important features of economic systems and has been criticized on a number of grounds. At the same time, a growing interest has emerged in alternative approaches to economic agents' decision-making, which allow for factors such as bounded rationality and heterogeneity of agents, social interaction and learning, where agents' behaviour is governed by simpler "rules of thumb" (or "heuristics") or "trial and error" or even "imitations mechanisms". Adaptive system, governed by local (or myopic) decision rules of boundedly rational and heterogeneous agents, may converge in the long run to a rational equilibrium, i.e. the same equilibrium forecasted (and instantaneously reached) under the assumption of full rationality and full information of all economic agents. This may be seen as an evolutionary interpretation of a rational equilibrium, and some authors say that in this case, the boundedly rational agents are able to learn, in the long run, what rational agents already know under very pretentious rationality assumptions. However, it may happen that under different starting conditions, or as a consequence of exogenous perturbations, the same adaptive process leads to non-rational equilibria as well, i.e. equilibrium situations which are different from the ones forecasted under the assumption of full rationality, as well as to dynamic attractors characterized by endless asymptotic fluctuations that never settle to a steady state. The coexistence of several attracting sets, each with its own basin of attraction, gives rise to path dependence, irreversibility, hysteresis and other nonlinear and complex phenomena commonly observed in real systems in economics, finance and social sciences, as well as in laboratory experiments.

From the description given above, it is evident that the analysis of adaptive systems can be formulated in the framework of the theory of dynamical systems, i.e. systems of ordinary differential equations (continuous time) or difference equations (discrete time); the qualitative theory of nonlinear dynamical systems, with the related concepts of stability, bifurcations, attractors and basins of attraction, is a major tool for the analysis of their long-run (or asymptotic) properties. Not only in economics and social sciences, but also in physics, biology and chemical sciences, such models are a privileged instrument for the description of systems that change over time, often described as "nonlinear evolving systems", and their long-run aggregate outcomes can be interpreted as "emerging properties", sometimes difficult to be forecasted on the basis of the local (or step by step) laws of motion. As we will see in this book, a very important role in this theory is played by graphical analysis, and a fruitful trade-off between analytic, geometric and numerical methods. However, these methods built up a solid mathematical theory based on general theorems that can be found in the textbooks indicated in the references.

Chapter 1, by Gian Italo Bischi, Fabio Lamantia and Davide Radi, is the largest one, as it contains the basic lessons delivered during the Training School. It introduces some general concepts, notations and a minimal vocabulary about the mathematical theory of dynamical systems both in continuous time and in discrete time, as well as optimal control.

Chapter 2, by Anastasiia Panchuk, points out several aspects related to global analysis of discrete time dynamical systems, covering homoclinic bifurcations as well as inner and boundary crises of attracting sets.

Chapter 3, by Anna Agliari, Nicolò Pecora and Alina Szuz, describes some properties of the nonlinear dynamics emerging from two oligopoly models in discrete time. The target of this chapter is the investigation of some local and global bifurcations which are responsible for the changes in the qualitative behaviours of the trajectories of discrete dynamical systems. Two different kinds of oligopoly models are considered: the first one deals with the presence of differentiated goods and gradient adjustment mechanism, while the second considers the demand function of the producers to be dependent on advertising expenditures and adaptive adjustment of the moves. In both models, the standard local stability analysis of the Cournot-Nash equilibrium points is performed, as well as the global bifurcations of both attractors and (their) basins of attraction are investigated.

Chapter 4, by Ingrid Kubin, Pasquale Commendatore and Iryna Sushko, acquaints the reader with the use of dynamic models in regional economics. The focus is on the New Economic Geography (NEG) approach. This chapter briefly compares NEG with other economic approaches to investigation of regional inequalities. The analytic structure of a general multiregional model is described, and some simple examples are presented where the number of regions assumed to be small to obtain more easily analytic and numerical results. Tools from the mathematical theory of dynamical systems are drawn to study the qualitative properties of such multiregional model.

In Chap. 5, Fabio Lamantia, Davide Radi and Lucia Sbragia review some fundamental models related to the exploitation of a renewable resource, an important topic when dealing with regional economics. The chapter starts by considering the growth models of an unexploited population and then introduces commercial harvesting. Still maintaining a dynamic perspective, an analysis of equilibrium situations is proposed for a natural resource under various market structures (monopoly, oligopoly and open access). The essential dynamic properties of these models are explained, as well as their main economic insights. Moreover, some key assumptions and tools of intertemporal optimal harvesting are recalled, thus providing an interesting application of the theory of optimal growth.

In Chap. 6, Fabio Tramontana considers the qualitative theory of discrete time dynamical systems to describe the time evolution of financial markets populated by heterogeneous and boundedly rational traders. By using these assumptions, he is able to show some well-known stylized facts observed in financial markets that can be replicated even by using small-scale models.

Finally, in Chap. 7, Ugo Merlone and Paul van Geert consider some dynamical systems which are quite important in psychological research. They show how to implement a dynamical model of proximal development using a spreadsheet, statistical software such as R or programming languages such as C++. They discuss strengths and weaknesses of each tool. Using a spreadsheet or a subject-oriented statistical software is rather easy to start, hence being likely palatable for people with background in both economics and psychology. On the other hand, employing C++ provides better efficiency at the cost of requiring some more competencies. All the approaches proposed in this chapter use free and open-source software.

Urbino	Gian Italo Bischi
Kiev	Anastasiia Panchuk
Castellanza	Davide Radi

Acknowledgements

We wish to thank the 30 participants of the Training School, mainly Ph.D. students, but also young researchers as well as some undergraduate students, coming from many different European countries. The continuous and fruitful interactions with them helped the teachers to improve their lessons and, consequently, greatly contributed to the quality of this book. A special thanks to Prof. Pasquale Commendatore, Chair of the COST Action, and Ingrid Kubin, vice-chair, who encouraged the project of the Training School and collaborated for its realization. We are deeply indebted to Laura Gardini for her efforts to increase the scientific quality of the School, as well as her help in the organization process. Of course the publication of this book would not have been possible without the high quality of the lessons delivered by the teachers, and we want to thank them for sending so accurate written versions of their lessons. We would also like to express special thanks to Mrs. Sabine Lehr, the Springer-Verlag Associate Editor who facilitated the book's publication and carefully guided the entire editorial process. The usual disclaimers apply.

Contents

Contributors

Anna Agliari Department of Economics and Social Sciences, Catholic University, Piacenza, Italy

Gian Italo Bischi DESP-Department of Economics, Society, Politics, Università di Urbino "Carlo Bo", Urbino, PU, Italy

Pasquale Commendatore Department of Law, University of Naples Federico II, Naples, NA, Italy

Ingrid Kubin Department of Economics, Institute for International Economics and Development (WU Vienna University of Economics and Business), Vienna, Austria

Fabio Lamantia Department of Economics, Statistics and Finance, University of Calabria, Rende, CS, Italy; Economics—School of Social Sciences, The University of Manchester, Manchester, UK

Ugo Merlone Department of Psychology, Center for Cognitive Science, University of Torino, Torino, Italy

Anastasiia Panchuk Institute of Mathematics, National Academy of Sciences of Ukraine, Kiev, Ukraine

Nicolò Pecora Department of Economics and Social Sciences, Catholic University, Piacenza, Italy

Davide Radi School of Economics and Management, LIUC - Università Cattaneo, Castellanza, VA, Italy

Lucia Sbragia Department of Economics, Durham University Business School, Durham, UK

Iryna Sushko Institute of Mathematics, National Academy of Sciences of Ukraine, Kiev, Ukraine

Alina Szuz Independent Researcher, Cluj-Napoca, Romania

Fabio Tramontana Department of Mathematical Sciences, Mathematical Finance and Econometrics, Catholic University, Milan, MI, Italy

Paul van Geert Heymans Institute, Groningen, The Netherlands

Chapter 1
Qualitative Methods in Continuous and Discrete Dynamical Systems

Gian Italo Bischi, Fabio Lamantia and Davide Radi

Abstract This chapter gives a general and friendly overview to the qualitative theory of continuous and discrete dynamical systems, as well as some applications to simple dynamic economic models, and is concluded by a section on basic principles and results of optimal control in continuous time, with some simple applications. The chapter aims to introduce some general concepts, notations and a minimal vocabulary concerning the study of the mathematical theory of dynamical systems that are used in the other chapters of the book. In particular, concepts like stability, bifurcations (local and global), basins of attraction, chaotic dynamics, noninvertible maps and critical sets are defined, and their applications are presented in the following sections both in continuous time and discrete time, as well as a brief introduction to optimal control together with some connections to the qualitative theory of dynamical systems and applications in economics.

G.I. Bischi (✉)
DESP-Department of Economics, Society, Politics, Università di Urbino "Carlo Bo",
42 Via Saffi, 61029 Urbino, PU, Italy
e-mail: gian.bischi@uniurb.it

F. Lamantia
Department of Economics, Statistics and Finance, University of Calabria,
3C Via P. Bucci, 87036 Rende, CS, Italy
e-mail: fabio.lamantia@unical.it

F. Lamantia
Economics—School of Social Sciences, The University of Manchester,
Arthur Lewis Building, Manchester, UK

D. Radi
School of Economics and Management, LIUC - Università Cattaneo,
22 C.so Matteotti, 21053 Castellanza, VA, Italy
e-mail: dradi@liuc.it

G.I. Bischi et al. (eds.), *Qualitative Theory of Dynamical Systems, Tools and Applications for Economic Modelling*, Springer Proceedings in Complexity,
DOI 10.1007/978-3-319-33276-5_1

1.1 Some General Definitions

In this section we introduce some general concepts, notations and a minimal vocabulary about the mathematical theory of dynamical systems. A *dynamical system* is a mathematical model, i.e., a formal, mathematical description, of a system evolving as time goes on. This includes, as a particular case, systems whose state remains constant, that will be denoted as systems at equilibrium.

The first step to describe such systems in mathematical terms is the characterization of their "state" by a finite number, say n, of measurable quantities, denoted as "state variables", expressed by real numbers $x_i \in \mathbb{R}$, $i = 1, \ldots, n$. For example in an economic system these numbers may be the prices of n commodities in a market, or the respective quantities, or they can represent other measurable indicators, like level of occupation, or salaries, or inflation. In an ecologic system these n numbers used to characterize its state may be the numbers (or densities) of individuals of each species, or concentration of inorganic nutrients or chemicals in the environment. In a physical system[1] the state variables may be the positions and velocities of the particles, or generalized coordinates and related momenta of a mechanical system, or temperature, pressure etc. in a thermodynamic system.

This ordered set of real numbers can be seen as a vector $\mathbf{x} = (x_1, \ldots, x_n) \in \mathbb{R}^n$, i.e., a "point" in an n-dimensional space, and this allows us to introduce a "geometric language", in the sense that a 1-dimensional dynamical system is represented by point along a line, a 2-dimensional one by a point in a Cartesian plane and so on.

Sometimes only the values of the state variables included in a subset of $\mathbb{R}^n$ are suitable to represent the real system. For example only non-negative values of x_i are meaningful if x_i represents a price in an economic system or the density of a species in an ecologic one, or it can be that in the equations that define the system a state variable x_i is the argument of a mathematical function that is defined in a given domain, like a logarithm, a square root or a rational function. As a consequence, only the points in a subset of $\mathbb{R}^n$ are admissible states for the dynamical system considered, and this leads to the following definition.

Definition 1.1 The *state space* (or *phase space*) $M \subseteq \mathbb{R}^n$ is the set of admissible values of the state variables.

As a dynamical system is assumed to evolve with time, these numbers are not fixed but are functions of time $x_i = x_i(t)$, $i = 1, \ldots, n$, where t may be a real number (*continuous time*) or a natural number (*discrete time*). The latter assumption may sound quite strange, whereas it represents a common assumption in systems where changes of the state variables are only observed as a consequence of events occurring at given time steps (event-driven time). For example, it is quite common in economic and social sciences where in many systems the state variables can change as a consequence of human decisions that cannot be continuously revised, e.g., after

[1] Physics is the discipline where the formalism of dynamical system has been first introduced, since 17th century, even if the modern approach, often denoted as qualitative theory of dynamics systems, has been introduced in the early years of the 20th century.

production periods (the typical example is output of agricultural products) or after the meetings of an administration council or after the conclusions of contracts etc. (decision-driven time).

So, in the following we will distinguish these two cases, according to the domain of the state functions: $x_i : \mathbb{R} \to \mathbb{R}$ or $x_i : \mathbb{N} \to \mathbb{R}$, i.e., the continuous or discrete nature of time. In any case, the purpose of dynamical systems is the following: given the state of the system at a certain time t_0, compute the state of the system at time $t \neq t_0$. This is equivalent to the knowledge of an operator

$$\mathbf{x}(t) = \mathbf{G}\left(t, \mathbf{x}(t_0)\right) \ , \tag{1.1}$$

where boldface symbols represent vectors, i.e., $\mathbf{x}(t) = (x_1(t), \ldots, x_n(t)) \in M \subseteq \mathbb{R}^n$ and $\mathbf{G}(\cdot) = (G_1(\cdot), \ldots, G_n(\cdot)) : M \to M$. If one knows the *evolution operator* $\mathbf{G}$ then from the knowledge of the *initial condition* (or *initial state*) $\mathbf{x}(t_0)$ the state of the system at any future time $t > t_0$ can be computed, as well as at any time of the past $t < t_0$. Generally we are interested in the forecasting of future states, especially in the asymptotic (or long-run) evolution of the system as $t \to +\infty$, i.e., the fate, or the destiny of the system. However, even the flashback may be useful in some cases, like in detective stories when the investigators from the knowledge of the present state want to know what happened in the past.

The vector function $\mathbf{x}(t)$, i.e., the set of n functions $x_i(t)$, $i = 1, \ldots, n$ obtained by (1.1), represents the parametric equations of a *trajectory*, as t varies. In the case of continuous time $t \in \mathbb{R}$ the trajectory is a curve in the space $\mathbb{R}^n$, that can be represented in the $n+1$-dimensional space $(\mathbb{R}^n, t)$, and denoted as *integral curve*, or in the state space (also denoted as "phase space") $\mathbb{R}^n$, see Fig. 1.1. In the latter case the direction of increasing time is represented by arrows, and the curve is denoted as *phase curve*.

In the case of discrete time a trajectory is a sequence (i.e., a countable set) of points, and the time evolution of the system jumps from one point to the successive one in the sequence. Sometimes line segments can be used to join graphically the points, moving in the direction of increasing time, thus getting an ideal piecewise smooth curve by which the time evolution of the system is graphically represented.

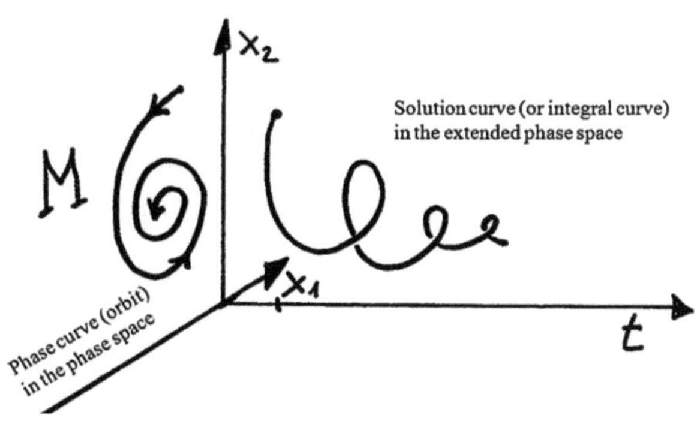

Fig. 1.1 Solution curve and its projection in the phase space

An *equilibrium* (*stationary state* or *fixed point*) $\mathbf{x}^* = \left(x_1^*, \ldots, x_n^*\right)$ is a particular trajectory such that all the state variables are constant

$$\mathbf{x}(t) = \mathbf{G}\left(t, \mathbf{x}^*\right) = \mathbf{x}^* \text{ for each } t > t_0 \ .$$

An equilibrium is a trapping point, i.e., any trajectory through it remains in it for each successive time: $\mathbf{x}(t_0) = \mathbf{x}^*$ implies $\mathbf{x}(t) = \mathbf{x}^*$ for $t \geq t_0$. This definition can be extended to any subset of the phase space:

Definition 1.2 A set $A \subseteq M$ is *trapping* if $\mathbf{x}(t_0) \in A$ implies $\mathbf{x}(\mathbf{t}) = \mathbf{G}\,(t, \mathbf{x}(t_0)) \in A$ for each $t > t_0$.

This can also be expressed by the notation $\mathbf{G}\,(t, A) \subseteq A$, where

$$\mathbf{G}\,(t, A) = \{\mathbf{x}(t) \in M : \exists t \geq t_0 \text{ and } \mathbf{x}(t_0) \in A \text{ so that } \mathbf{x}(t) = \mathbf{G}\,(t, \mathbf{x}(t_0))\} \ .$$

So, any trajectory starting inside a trapping set cannot escape from it. We now define a stronger property, in the sense that it concerns particular kinds of trapping sets.

Definition 1.3 A closed set $A \subseteq M$ is *invariant* if $\mathbf{G}\,(t, A) = A$, i.e., each subset $A' \subset A$ is not trapping.

In other words, any trajectory starting inside an invariant set remains there, and all the points of the invariant set can be reached by a trajectory starting inside it. Notice that an equilibrium point is a particular kind of invariant set (let's say the simplest). However, we will see many other kinds of invariant sets, where interesting cases of nonconstant trajectories are included.

We now wonder what happens if we start a trajectory from an initial condition close to an invariant set, i.e., in a neighborhood of it. The trajectory may enter the invariant set (and then it remains trapped inside it) or it may move around it or it may go elsewhere, far from it. This leads us to the concept of stability of an invariant set (Fig. 1.2).

Definition 1.4 (*Lyapunov stability*) An invariant set A is *stable* if for each neighborhood U of A there exists another neighborhood V of A with $V \subseteq U$ such that any trajectory starting from V remains inside U.

In other words, Lyapunov stability means that all the trajectories starting from initial conditions outside A and sufficiently close to it remain around it. Instability is the negation of stability, i.e., an invariant set A is unstable if a neighborhood $U \supset A$

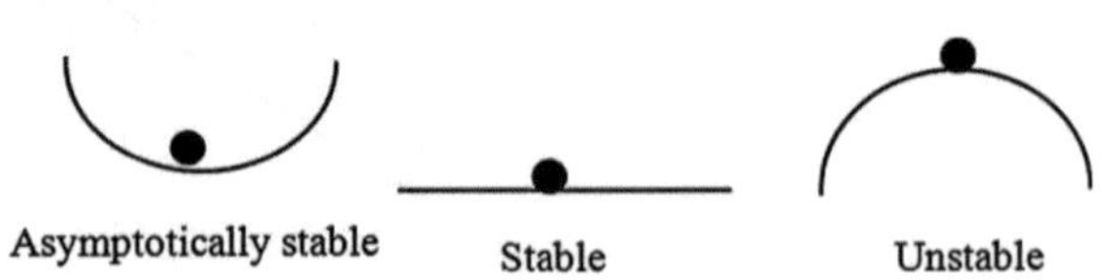

Fig. 1.2 Analogies with the gravity field

exists such that initial conditions taken arbitrarily close to A exist that generate trajectories that exit U. The following definition is stronger.

Definition 1.5 (*Asymptotic stability*) An invariant set A is *asymptotically stable* (and it is often called *attractor*) if:

(i) A is stable (according to the definition given above);
(ii) $\lim_{t\to+\infty} \mathbf{G}(t, \mathbf{x}) \in A$ for each initial condition $\mathbf{x} \in V$.

In other words, asymptotic stability means that the trajectories starting from initial conditions outside A and sufficiently close to it not only remain around it, but tend to it in the long run (i.e., asymptotically), see the schematic pictures in Fig. 1.3. At a first sight, the condition (ii) in the definition of asymptotic stability seems to be stronger than (i), hence (i) seems to be superfluous. However it may happen that a neighborhood $U \supset A$ exists such that initial conditions taken arbitrarily close to A generate trajectories that exit U and then go back to A in the long run (see the last picture in Fig. 1.3).

Of course, all these definitions expressed in terms of neighborhoods can be restated by using a norm (and consequently a distance) in $\mathbb{R}^n$, for example the euclidean norm $\|\mathbf{x}\| = \sqrt{\sum_{i=1}^n x_i^2}$ from which the distance between two points $\mathbf{x} = (x_1, \ldots, x_n)$ and

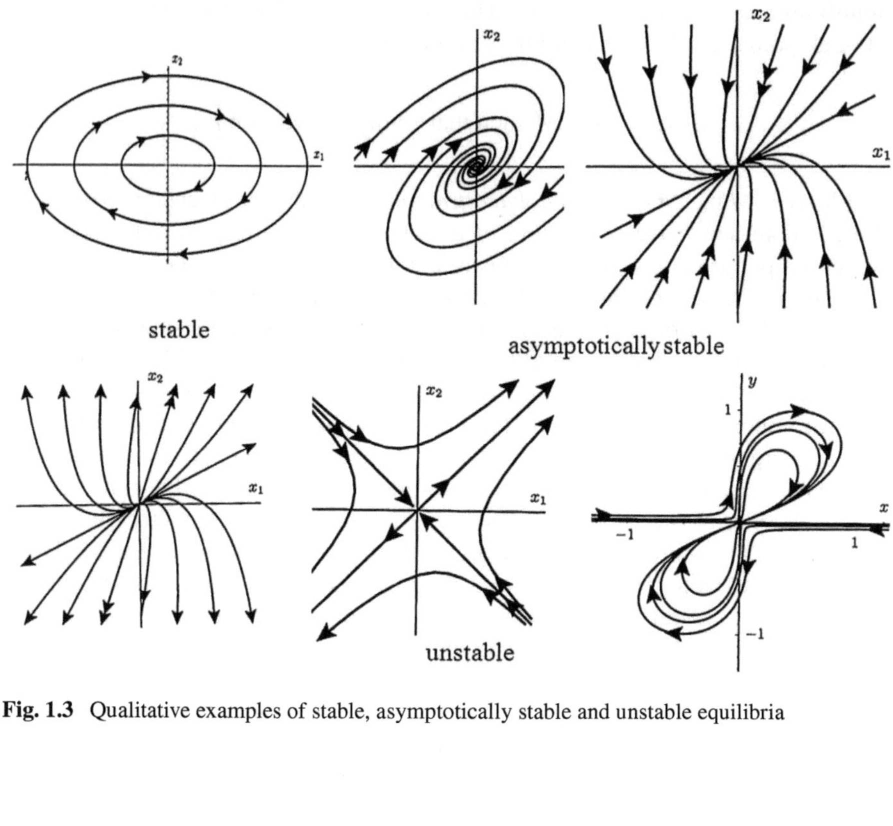

Fig. 1.3 Qualitative examples of stable, asymptotically stable and unstable equilibria

$\mathbf{y} = (y_1, \ldots, y_n)$ can be defined as $\|\mathbf{x} - \mathbf{y}\| = \sqrt{\sum_{i=1}^{n}(x_i - y_i)^2}$. As an example we can restate the definitions given above for the particular case of an equilibrium point.

Let $\mathbf{x}(t) = \mathbf{G}(t, \mathbf{x}(t_0))$, $t \geq 0$, a trajectory starting from the initial condition $\mathbf{x}(t_0) = \mathbf{G}(t_0, \mathbf{x}(t_0))$ and $\mathbf{x}^*$ an equilibrium point $\mathbf{x}^* = \mathbf{G}(t, \mathbf{x}^*)$ for $t \geq 0$. The equilibrium $\mathbf{x}^*$ is stable if for each $\varepsilon > 0$ there exists $\delta_\varepsilon > 0$ such that $\|\mathbf{x}(t_0) - \mathbf{x}^*\| < \delta_\varepsilon \implies \|\mathbf{x}(t) - \mathbf{x}^*\| < \varepsilon$ for $t \geq 0$. If in addition $\lim_{t\to\infty} \|\mathbf{x}(t) - \mathbf{x}^*\| = 0$ then $\mathbf{x}^*$ is asymptotically stable. Instead, if an $\varepsilon > 0$ exists such that for each $\delta > 0$ we have $\|\mathbf{x}(t) - \mathbf{x}^*\| > \varepsilon$ for some $t > 0$ even if $\|\mathbf{x}(t_0) - \mathbf{x}^*\| < \delta$, then $\mathbf{x}^*$ is unstable.

These definitions are local, i.e., concern the future behavior of a dynamical system when its initial state is in an arbitrarily small neighborhood of an invariant set. So, they can be used to characterize the behavior of the system under the influence of small perturbation from an equilibrium or another invariant set. In other words, they give an answer to the question: given a system at equilibrium, what happens when small exogenous perturbation move its state slightly outside the equilibrium state? However, in the study of real systems we are also interested in their global behavior, i.e., far from equilibria (or more generally from invariant sets) in order to consider the effect of finite perturbations and to answer questions like: how far can an exogenous perturbation shift the state of a system from an equilibrium remaining sure that it will spontaneously go back to the originary equilibrium? This kind of questions leads to the concept of basin of attraction.

Definition 1.6 (*Basin of attraction*) The *basin of attraction* of an attractor A is the set of all points $\mathbf{x} \in M$ such that $\lim_{t\to+\infty} \mathbf{G}(t, \mathbf{x}) \in A$, i.e.,

$$B(A) = \left\{ \mathbf{x} \in M \text{ such that } \lim_{t\to+\infty} \mathbf{G}(t, \mathbf{x}) \in A \right\}.$$

If $B(A) = M$ then A is called *global attractor*. Generally the extension of the basin of a given attractor gives a measure of its robustness with respect to the action of exogenous perturbations. However this is a quite rough argument, because a greater extension of the basin of an attractor may does not imply greater robustness if the attractor is close to a basin boundary. Moreover, when basins are considered, one realizes that in some cases stable equilibria may be even more vulnerable than unstable ones (see Fig. 1.4).

Other important indicators should be critically considered. For example, how fast is the convergence towards an attractor? Even if an invariant set is asymptotically stable and it has a large basin, an important question concerns the speed of convergence, i.e., the amount of time which is necessary to reduce the extent of a perturbation. In some cases this time interval may be too much for any practical purpose. These arguments lead us to the necessity of a deep understanding of the global behavior of a dynamical system in order to give useful indications about the performance of the real system modeled. The main problem is that, generally, the operator $\mathbf{G}$ that allows to get an explicit representation of the trajectories of the dynamical system for any initial condition in the phase space, is not known, or cannot be expressed in terms of elementary functions, or its expression is so complicated that it cannot be

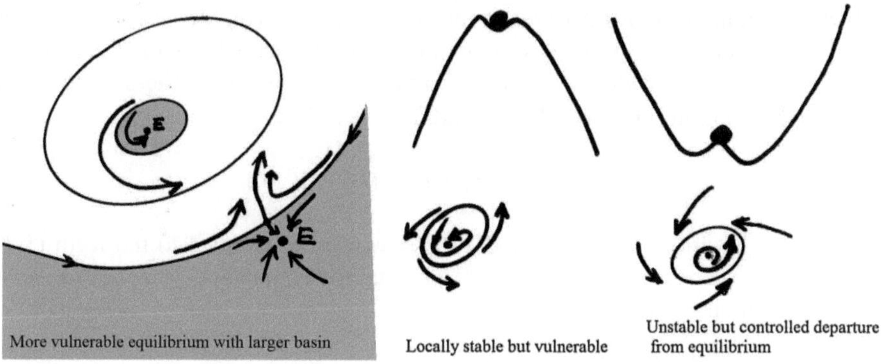

Fig. 1.4 Stability and vulnerability

used for any practical purpose. In general a dynamical system is expressed in terms of *local evolution equations*, also denoted as *dynamic equations* or *laws of motion*, that state how the dynamical system changes as a consequence of small time steps. In the case of continuous time the evolution equations are expressed by the following set of *ordinary differential equations* (ODE) involving the time derivative, i.e., the speeds of change, of each state variable

$$\begin{aligned} \frac{dx_i(t)}{dt} &= f_i(x_1(t), \ldots, x_n(t); \alpha) , \quad i = 1, \ldots, n , \\ x_i(t_0) &= \overline{x}_i , \end{aligned} \tag{1.2}$$

where the time derivative at the left hand side represents, as usual, the speed of change of the state variable $x_i(t)$ with respect to time variations, the functional relations give information about the influence of the same state variable x_i (self-control) and of the other state variables x_j, $j \neq i$ (cross-control) on such rate of change, and $\alpha = (\alpha_1, \ldots \alpha_m)$, $\alpha_i \in \mathbb{R}$, represents m real parameters, fixed along a trajectory, which can assume different numerical values in order to represent exogenous influences on the dynamical systems, e.g., different policies or effects of the outside environment. The modifications induced in the model after a variation of some parameters α_i are called *structural modifications*, as such changes modify the shape of the functions f_i, and consequently the properties of the trajectory.

The set of equations (1.2) are "differential equations" because their "unknowns" are functions $x_i(t)$ and they involve not only $x_i(t)$ but also their derivatives. In the theory of dynamical systems it is usual to replace the Leibniz notation $\frac{dx}{dt}$ of the derivative with the more compact "dot" notation $\dot{x}$ introduced by Newton. With this notation, the dynamical system (1.2) is indicated as

$$\dot{x}_i = f_i(x_1, \ldots, x_n; \alpha), \quad i = 1, \ldots, n , \tag{1.3}$$

Differential equations of order greater than one, i.e., involving derivatives of higher order, can be easily reduced to systems of differential equations of order one in the form (1.2) by introducing auxiliary variables. For example the second order differential equation (involving the second derivative $\ddot{x} = \frac{d^2x}{dt^2}$)

$$\ddot{x}(t) + a\dot{x}(t) + bx(t) = 0 \tag{1.4}$$

with initial conditions $x(0) = x_0$ and $\dot{x}(0) = v_0$ can be reduced to the form (1.3) by defining $x_1(t) = x(t)$ and $x_2(t) = \dot{x}(t)$, so that the equivalent system of two first order differential equations becomes

$$\begin{aligned} \dot{x}_1 &= x_2 , \\ \dot{x}_2 &= -bx_1 - ax_2 \end{aligned}$$

with $x_1(0) = x_0, x_2(0) = v_0$. If along a trajectory the parameters explicitly vary with respect to time, i.e., some $\alpha_i = \alpha_i(t)$ are functions of time, then the model is called *nonautonomous*. Also a nonautonomous model can be reduced to an equivalent autonomous one in the normal form (1.2) of dimension $n + 1$ by introducing the dynamic variable $x_{n+1} = t$ whose time evolution is governed by the added first order differential equation $\dot{x}_{n+1} = 1$.

In the case of discrete time, the evolution equations are expressed by the following set of *difference equations* that inductively define the time evolution as a sequence of discrete points starting from a given initial condition

$$\begin{aligned} x_i(t+1) &= f_i(x_1(t), \ldots, x_n(t); \alpha), \quad i = 1, \ldots, n , \\ x_i(0) &= \overline{x}_i \end{aligned} \tag{1.5}$$

Also in this case a higher order difference equation, as well as a nonautonomous difference equation, can be reduced to an expanded system of first order difference equations. For example, the second order difference equations

$$x(t+1) + ax(t) + bx(t-1) = 0$$

starting from the initial conditions $x(-1) = x_0$, $x(0) = x_1$ can be equivalently rewritten as

$$\begin{aligned} x(t+1) &= -ax(t) - by(t) , \\ y(t+1) &= x(t) , \end{aligned}$$

where $y(t) = x(t-1)$, with initial conditions being $x(0) = x_1$, $y(0) = x_0$. Analogously, a nonautonomous difference equation

$$x(t+1) = f(x(t), t)$$

becomes

$$x(t+1) = f(x(t), y(t)),$$
$$y(t+1) = y(t) + 1,$$

where $y(t) = t$.

So, the study of (1.2) and (1.5) constitutes a quite general approach to dynamical systems in continuous and discrete time respectively. They are local representations of the evolution of systems that change with time. Their qualitative analysis consists in the study of existence and main properties of attracting sets, their basins, and their qualitative changes as the control parameters are let to vary. We refer the reader to standard textbooks and the huge literature about difference and differential equations in order to study their general properties and methods of solutions. The aim of this lecture note is just to give a general overview of the basic elements for a qualitative understanding of the long run behavior of some dynamic models. We will first consider the case of continuous time, then the case of discrete time by stressing the analogies and differences between these two time scales, and finally we shall give some concepts and results about optimal control analysis.

1.2 Continuous-Time Dynamical Systems

In this section we consider dynamic equations in the form (1.2), starting from problems with $n = 1$, i.e., 1-dimensional models where the state of the system is identified by a single dynamical variable, then we move to $n = 2$ and finally some comments on $n > 2$. For each case, we will first consider linear models, for which an explicit expression of the solution can be obtained, and then we move to nonlinear models for which we will only give a qualitative description of the equilibrium points, their stability properties and the long-run (or asymptotic) properties of the solutions without giving their explicit expression. We will see that such qualitative study (also denoted as qualitative or topological theory of dynamical systems, a modern point of view developed in the 20th century) essentially reduces to the solution of algebraic equations and inequalities, without the necessity to use advanced methods for solving integrals. We start with a sufficiently general (for the goals of these lecture notes) theorem of existence and uniqueness of solutions of ordinary differential equations.

Theorem 1.1 (Existence and Uniqueness) *If the functions f_i have continuous partial derivatives in M and $x(t_0) \in M$, then there exists a unique solution $x_i(t)$, $i = 1, \ldots, n$, of the system (1.2) such that $x(t_0) = \overline{\mathbf{x}}$, and each $x_i(t)$ is a continuous function.*

Indeed, the assumptions of this theorem may be weakened, by asking for bounded variations of the functions f_i in the equations of motion (1.2), such as the so called Lipschitz conditions. However the assumptions of the previous Theorem are suitable for our purposes. Moreover, other general theorems are usually stated to define the conditions under which the solutions of the differential equations have a regular behavior. We refer the interested reader to more rigorous books, see the bibliography for details.

1.2.1 *One-Dimensional Dynamical Systems in Continuous Time*

1.2.1.1 The Simplest One: A Linear Dynamical System

Let us consider the following dynamic equation

$$\dot{x} = \alpha x \quad \text{with initial condition} \quad x(t_0) = x_0 \ . \tag{1.6}$$

It states that the rate of growth of the dynamic variable $x(t)$ is proportional to itself, with proportionality constant α (a parameter). If $\alpha > 0$ then whenever x is positive it will increase (positive derivative means increasing). Moreover, as x increases also the derivative increases, so it increases faster and so on. This is what, even in the common language, is called "exponential growth", i.e., "the more we are, the more we increase". Instead, whenever x is negative it will decrease (negative derivative) so it will become even more negative and so on. This is a typical unstable behavior.

On the contrary, if $\alpha < 0$ then whenever x is positive it will decrease (and will tend to zero) whereas when x is negative the derivative is positive, so that x will increase (and tend to zero). A stabilizing behavior.

In this case an explicit solution can be easily obtained to confirm these arguments. In fact, it is well known, from elementary calculus, that the only function whose derivative is proportional to the function itself is the exponential, so $x(t)$ will be in the general form $x(t) = ke^{\alpha t}$, where k is an arbitrary constant that can be determined by imposing the initial condition $x(t_0) = x_o$, hence $ke^{\alpha t_0} = x_0$, from which $k = x_0 e^{-\alpha t_0}$. After replacing k in the general form we finally get the (unique) solution

$$x(t) = x_0 e^{\alpha(t-t_0)} \ . \tag{1.7}$$

The same solution can be obtained by a more standard integration method, denoted as separation of the variables: from $\frac{dx}{dt} = \alpha x$ we get $\frac{dx}{x} = \alpha dt$ and then, integrating both terms we get

$$\int_{x_0}^{x(t)} \frac{dx}{x} = \int_{t_0}^{t} \frac{dx}{x} \Longrightarrow \ln x(t) - \ln x_0 = \alpha(t - t_0) \Longrightarrow \ln \frac{x(t)}{x_0} = \alpha(t - t_0) \ ,$$

from which (1.7) is obtained by taking exponential of both members. Some graphical representations of (1.7), with different values of the parameter α and different initial conditions, are shown in Fig. 1.5 in the form of integral curves, with time t represented along the horizontal axis and the state variable along the vertical one. Among all the possible solutions there is also an equilibrium solution, corresponding to the case of vanishing time derivative $\dot{x} = 0$ (equilibrium condition). In fact, from (1.6) we can see that the equilibrium condition corresponds to the equation $\alpha x = 0$ which, for $\alpha \neq 0$, gives the unique solution $x^* = 0$. Indeed, the trajectory starting from the

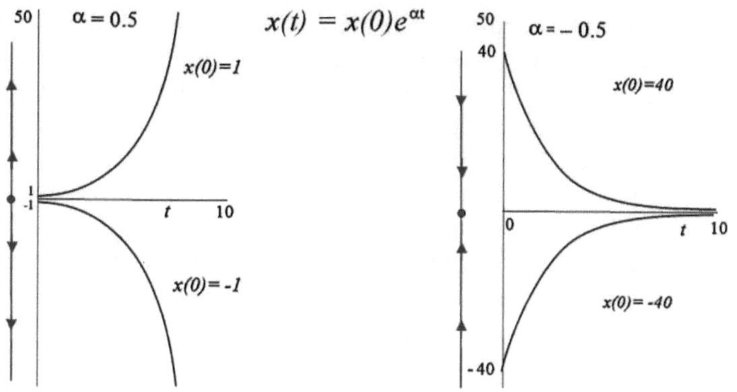

Fig. 1.5 Integral curves and phase portraits of $\dot{x} = \alpha x$

initial condition $x_0 = 0$ is given by $x(t) = 0$ for each t, i.e., starting from $x_0 = 0$ the system remains there forever. However, as shown in Fig. 1.5, different behaviors of the system can be observed if the initial condition is slightly shifted from the equilibrium point, according to the sign of the parameter α. In fact if $\alpha > 0$ (left panel) then the system amplifies this slight perturbation and exponentially departs from the equilibrium (unstable, or repelling, equilibrium) whereas if $\alpha < 0$ (right panel) then the system recovers from the perturbation going back to the equilibrium after a given return time (asymptotically stable, or attracting, equilibrium).

This qualitative analysis of existence and stability of the equilibrium can be obtained even without any computation of the explicit analytic solution (1.7), by solving the equilibrium equation $\alpha x = 0$ and by a simple algebraic study of the sign of the right hand side of the dynamic equation (1.6) around the equilibrium, as shown in Fig. 1.6. This method simply states that if the right hand side of the dynamic equation (hence $\dot{x}$) is positive then the state variable increases (arrow towards positive direction of the axis), if $\dot{x} < 0$ then x decreases (arrow towards negative direction).

This 1-dimensional representation (i.e., along the line) is the so called phase diagram of the dynamical system, where the invariant sets are represented (the equilibrium in this case) together with the arrows that denote tendencies associated with any point of the phase space (and consequently stability properties). Of course, the knowledge of the explicit analytic solution gives more information, for example the time required to move from one point to another. For example, in the case $\alpha < 0$, corresponding to stability of the equilibrium $x^* = 0$, we can state that after a

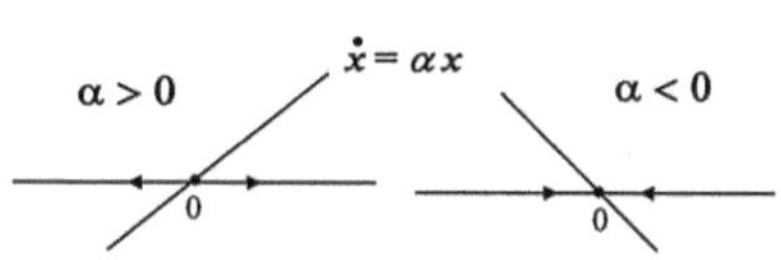

Fig. 1.6 Graphic of the the line $y = \alpha x$ and the corresponding one-dimensional phase diagram

displacement of the initial condition at distance $d = \|x_0 - x^*\|$ from the equilibrium, the time required to reduce such a perturbation at the fraction d/e (where e is the Neper constant $e \simeq 2.7$) is $T_r = -1/a$, an important stability indicator known as *return time*. As it can be seen, as the parameter α goes to 0 the return time tends to infinity. In fact, if $\alpha = 0$ all the points are equilibrium points, i.e., any initial condition generates a constant trajectory that remains in the same position forever.

As an example, let us consider the dynamic equation that describes the growth of a natural population. If $x(t)$ represents the number of individuals in a population (of insects, or bacteria, or fishes or humans), $n > 0$ represents the natality (or birth) rate and $m > 0$ represents the mortality (or death) rate then a basic balance equation used in any population model states that

$$\dot{x} = nx - mx = (n - m)\,x$$

which is of the form (1.6) with $\alpha = n - m$. Of course in this case, due to the meaning of the model, only non-negative values of the state variable x are admissible. The qualitative analysis of this model states that if natality is greater than mortality then the population exponentially increases, if the two rates are identical the population remains constant and if mortality exceeds natality the population exponentially goes to extinction. A quite reasonable result. We now introduce a modification in the simple population growth model by introducing a constant immigration (emigration) term $b > 0\ (< 0)$

$$\dot{x} = \alpha x + b\,. \tag{1.8}$$

Now the equilibrium condition $\dot{x} = 0$ becomes $\alpha x + b = 0$ from which the equilibrium is $x^* = -b/a$. If $\alpha < 0$ and $b > 0$ (endogenously decreasing population with constant immigration) then the equilibrium is positive and stable (as $\dot{x} < 0$ for $x > x^*$ and $\dot{x} > 0$ for $x < x^*$). Instead, for $\alpha > 0$ and $b < 0$ (endogenously increasing population with constant emigration) the equilibrium is positive and unstable. We conclude by noticing that the dynamic model (1.8) is called linear nonhomogeneous (or affine) and can be reduced into the form (1.6) by a change of variable (a translation). In fact, let us define the new dynamic variable $X = x - x^* = x + b/a$. This change of variable corresponds to a translation that brings the new zero coordinate into the equilibrium point. If we replace $x = X - b/a$ into (1.8) we get $\dot{X} = \alpha X$. Then we have the linear model (1.6) in the dynamic variable $X(t)$, with initial condition $X(t_0) = x_0 + b/a$, whose solution is $X(t) = X(t_0)\mathrm{e}^{\alpha(t-t_0)}$. Going back to the original variable, by using the transformation $X = x + b/a$, we obtain

$$x(t) = \left(x_0 + \frac{b}{\alpha}\right)\mathrm{e}^{\alpha(t-t_0)} - \frac{b}{\alpha}\,.$$

This is a first example of conjugate dynamical systems, as the models (1.6) and (1.8) can be transformed one into the other by an invertible change of coordinates. We will give a more formal definition of conjugate (or qualitative equivalent) dynamic models in the following chapters.

As an example, let us now consider a dynamic formalization of a partial market of a single commodity, under the Walrasian assumption that the price of the good increases (decreases) if the demand is higher (lower) than supply. The simplest dynamic equation to represent this assumption is given by

$$\dot{p} = f(p) = k\left[D(p) - S(p)\right] , \tag{1.9}$$

where $q = D(p)$ represents the demand function, i.e., the quantity demanded by consumers when the price of the good considered is p, $q = S(p)$ represents the supply function, i.e., the quantity of the good that producers send to the market when the price is p, $k > 0$ is a constant that gives the speed by which the price reacts to a disequilibrium between supply and demand. The standard occurrence is that supply function $S(p)$ is increasing and demand function is decreasing, as shown in Fig. 1.7. The equilibrium point p^* is located at the intersection of demand and supply curves, and it is stable because the derivative of p is positive on the left and negative on the right of p^*, so that p^* is always reached in the long run even if the initial price $p(0)$ is not an equilibrium one (or equivalently if the price has been displaced from the equilibrium price). An analytic solution of the dynamic equations can be obtained under the assumption that demand and supply functions are linear

$$D(p) = a - bp , \quad S(p) = a_1 + b_1 p ,$$

where all the parameters a, b, a_1, b_1 are positive. In fact, in this case the dynamic equation is a linear differential equation with constant coefficients

$$\dot{p} = -k\,(b + b_1)\,p + k\,(a - a_1)$$

which is in the form (1.8) and has equilibrium point $p^* = (a - a_1)/(b + b_1)$. As we will see in the next sections, a similar analysis, based on the linearization of the model around the equilibrium point, is possible by computing the slopes of the functions (i.e., their derivatives) at the equilibrium point.

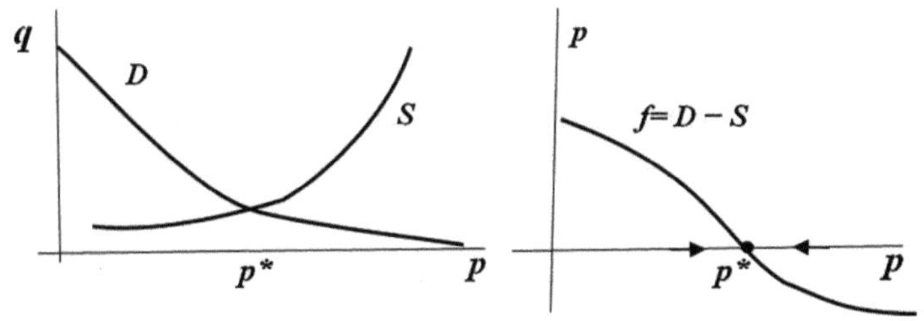

Fig. 1.7 Qualitative graphical analysis of price dynamics with standard demand and supply functions

Fig. 1.8 Qualitative analysis of (1.9) with bimodal demand function

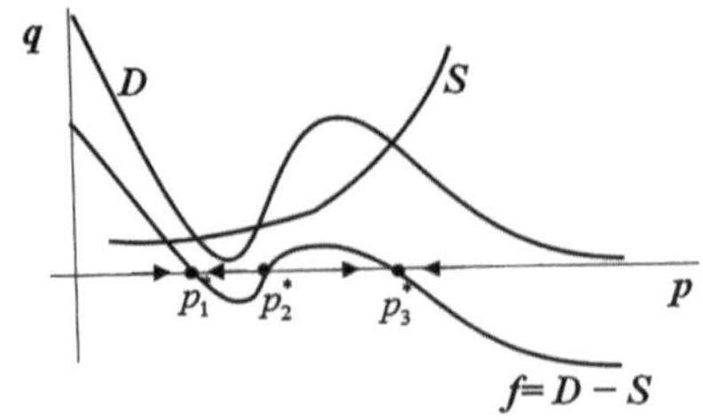

Let us now consider a different demand curve, obtained by assuming that consumers exhibit a nonstandard behavior for intermediate prices. In the situation shown in Fig. 1.8, even if demanded quantity is high for low prices and low for high prices, like in the standard case, we assume that for intermediate prices consumers prefer to buy the good at higher price because they use price as a quality indicator. Such assumption leads to a"bimodal" shape of the demand function (i.e., with two inversion points, a relative minimum and relative maximum) that may intersect the supply curve in three points, like in Fig. 1.8, and consequently three coexisting equilibrium prices, say $p_1^* < p_2^* < p_3^*$. By using the qualitative analysis, we can see that the time derivative of the price $p(t)$ is positive whenever $p < p_1^*$ or $p_2^* < p < p_3^*$, i.e., where $D(p) > S(p)$. This leads to a situation of *bistability* as both the lowest equilibrium price p_1^* and the highest one p_3^* are asymptotically stable, each with its own basin of attraction, whereas the intermediate unstable equilibrium price p_2^* separates the basins, i.e., it acts as a watershed located on the boundary between the two basins.

1.2.1.2 Qualitative Analysis and Linearization Procedure for the Logistic Model

The population model described in the Sect. 1.2.1.1 is quite unrealistic as it admits unbounded population growth, which is impossible in a finite world. As already noticed by Malthus [27], when the population density becomes too high, scarcity of food or space (overcrowding effect) causes mortality, proportional to the population density. So an extra mortality term, say sx, should be added to the natural mortality m, and thus the model becomes

$$\dot{x} = f(x) = nx - (m + sx)x = \alpha x - sx^2 \tag{1.10}$$

which is a nonlinear dynamic model. Also in this case, after separation of the variables, an analytic solution can be found by integrating a rational function. In fact, after some algebraic transformations of the rational function the following solution is obtained

$$x(t) = \frac{\alpha x_0 e^{\alpha t}}{\alpha + s x_0 \left(e^{\alpha t} - 1\right)}, \tag{1.11}$$

whose graph (for different initial conditions) is shown in Fig. 1.9.

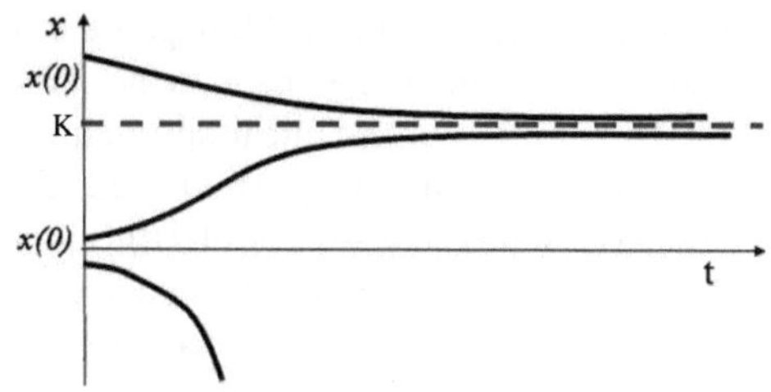

Fig. 1.9 Graph of function (1.11) with three different initial conditions $x(0)$

As it can be seen from the graph of $x(t)$ in (1.11), all solutions starting from a positive initial condition asymptotically converge to the attracting equilibrium $K = \alpha/s$ (usually called carrying capacity in the language of ecology) represented by the horizontal asymptote. Another equilibrium point exists, given by the extinction equilibrium $Q = 0$, which is repelling.

However, the possibility to find an analytic solution by integrating a nonlinear differential equation is a rare event, so we now try to infer the same conclusions without finding the explicit solution, i.e., by using qualitative methods. As usual, the first step is the localization of the equilibrium points, solutions of the equilibrium condition $\dot{x} = 0$, i.e., $f(x) = x(\alpha - sx) = 0$, from which the two solutions $x_0^* = 0$ and $x_1^* = \alpha/s$ are easily computed. In order to determine their local stability properties, it is sufficient to notice that the graph of the right hand side of (1.10), see Fig. 1.10, has negative slope around the equilibrium x_1^*, so that $\dot{x}$ is positive on the left and negative on the right, and vice versa at the equilibrium x_0^*, as indicated by the arrows along the x axis (the 1-dimensional state space of the system). This can be analytically determined even without the knowledge of the whole graph of the function, as it is sufficient to compute the sign of the x-derivative of the right hand side at each equilibrium point. In fact, it is well known that the derivative computed in a given point of the graph represents the slope of the graph (i.e., of the line tangent to the graph) at that point. So, the local behavior of the dynamical system in a neighborhood of an equilibrium point, hence its local stability as well, is generally the same as the one of the linear approximation (i.e., the tangent). This rough argument will be explained more formally in the next sections. In the particular case of the logistic model (1.10) the derivative is $\frac{df}{dx} = f'(x) = \alpha - 2sx$, and computed at the two equilibrium points becomes $f'(0) = \alpha > 0, f'(\alpha/s) = -\alpha < 0$, hence $Q = 0$ is

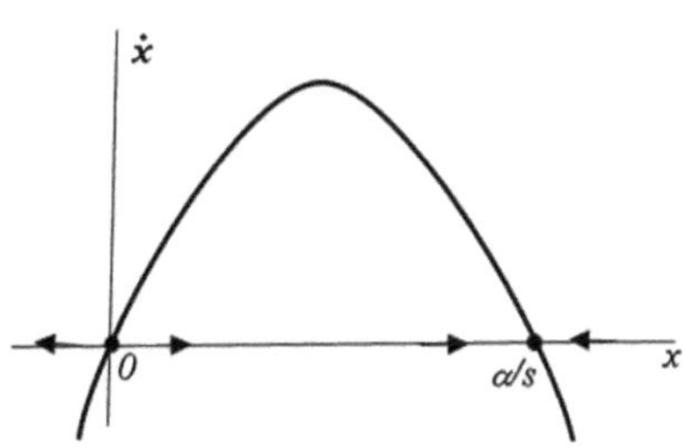

Fig. 1.10 Qualitative dynamic analysis of logistic equation (1.10)

unstable, $K = \alpha/s$ is stable. Moreover the parameter α can be seen as an indicator of how fast the system will go back to the stable equilibrium after a small displacement, as the *return time* for the linear approximation is $T_r = 1/\alpha$.

Before ending this part, we notice that the equilibrium points $x_0^* = 0$ and $x_1^* = \alpha/s$ are two (constant) solutions of (1.10), whose graphs in the plane (t, x) are horizontal lines. Thus, by the theorem of existence and uniqueness of a solution stated above, any other (nonconstant) solution $x(t)$ of (1.10) cannot cross these two horizontal lines. From (1.10) by a simple second-degree inequality, it is easy to see that $\dot{x} > 0$ occurs whenever $x \in (0, \alpha/s)$. Moreover, being $\frac{d^2x}{dt^2} = \frac{d\dot{x}}{dt} = \alpha\dot{x} - 2sx\dot{x} = \dot{x}\,(\alpha - 2sx)$, we deduce that $x(t)$ is strictly decreasing and concave whenever $x(0) \in (-\infty, 0)$ and that $x(t)$ is strictly decreasing and convex whenever $x(0) \in (\alpha/s, +\infty)$. Finally, when $x(0) \in (0, \alpha/s)$, $x(t)$ is strictly increasing and from convex becomes concave when $x(t) = \alpha/(2s)$, see again Fig. 1.9.

1.2.1.3 Qualitative Analysis of One-Dimensional Nonlinear Models in Continuous Time

The qualitative method used to understand the dynamic properties of the logistic equation can be generalized to any one-dimensional dynamic equation in continuous time

$$\dot{x} = f(x) \tag{1.12}$$

It consists, first of all, in the localization of the equilibrium points according to the equilibrium condition $\dot{x} = 0$, i.e., the solutions of the equation $f(x) = 0$. As a consequence of the Theorem of uniqueness, oscillations are not possible for a 1-dimensional dynamical system in continuous time, hence for a system starting from any initial condition which is not an equilibrium, only increasing or decreasing solutions can be obtained. Hence just four different phase portraits characterize the dynamic behavior of the 1-dimensional system around an equilibrium, as shown in Fig. 1.11.

Of course, if an initial condition coincides with an equilibrium point, i.e., $x(0) = x^*$ and $f(x^*) = 0$, then the unique solution is $x(t) = x^*$ for $t \geq 0$. In other words, starting from an equilibrium point, the system remains there forever. The natural question arising is what happens if the initial condition is taken close to an equilibrium point, i.e., if the system is slightly perturbed from the equilibrium considered. Will the distance from the equilibrium increase or will the perturbation be reduced so that the system spontaneously goes back to the originary equilibrium? An answer to this question is easy in the case of hyperbolic equilibria, defined as equilibrium points with nonvanishing derivative, i.e., $f'(x^*) \neq 0$. In fact, if x^* is one of such solutions

Fig. 1.11 The four different phase diagrams around an equilibrium point

and $f'(x^*) \neq 0$, then the right hand side of (1.12) can be approximated by the first order Taylor expansion (linear approximation)

$$f(x) = f(x^*) + f'(x^*)(x - x^*) + o(x - x^*) = f'(x^*)(x - x^*) + o(x - x^*)$$

being $f(x^*) = 0$. So, if $f'(x^*) \neq 0$ and we neglect the higher order terms then we obtain a *linear approximation* of the dynamical system (1.12). In fact, if we translate the origin of the x coordinate into the equilibrium point by the change of variable $X = x - x^*$, that represents the displacement between $x(t)$ and the equilibrium points x^*, then (1.12) becomes

$$\dot{X} = \alpha X$$

with $\alpha = f'(x^*)$, i.e., a linear differential equation in the form (1.6), that governs the time evolution of the system in a neighborhood of the equilibrium point x^*. Of course, this linear differential equation constitutes only a local approximation, i.e., for initial conditions taken in a sufficiently small neighborhood of the equilibrium point considered. This leads to the following result:

Proposition 1.1 (1D Local Asymptotic Stability in Continuous Time) *Let x^* be an equilibrium point of (1.12), i.e., $f(x^*) = 0$. If $f'(x^*) < 0$, then x^* is a locally asymptotically stable equilibrium; if $f'(x^*) > 0$, then x^* is unstable.*

This gives a simple method to classify the stability of a hyperbolic equilibrium. Instead, for a nonhyperbolic equilibrium, i.e., a point x^* such that $f(x^*) = 0$ and $f'(x^*) = 0$, nothing can be said about the stability of x^*, and further investigations are necessary, involving higher order derivatives or, equivalently, the knowledge of the shape of the function $f(x)$ around x^*. In Fig. 1.12 we can see, through four simple examples, that all possible phase portraits can be obtained around a nonhyperbolic equilibrium.

These situations characterized by a nonhyperbolic equilibrium point have been denoted as *structurally unstable*, in the sense that a slight (i.e., arbitrarily small) modification of the shape of the function $f(x)$ generally leads to a modification in the stability property as well as in the number of equilibrium points. Such a modification may be caused by the presence of parameters that may be used as devices (or policies) to modify the shape of the function f. Such slight modifications leading to qualitatively different dynamic scenarios are denoted as *bifurcations*, and are described in Sect. 1.2.1.4. To end this section we stress that the notion of structural stability should not be confused with that of dynamic stability: The latter deals with the effect on the trajectories of a small displacement of the initial condition (i.e., of the phase point), whereas the former deals with the effect, on the phase portrait (i.e., the dynamic scenario) of a slight modification of the function f due to a slight change of the value of a parameter.

Before giving a complete classifications of the bifurcations, we give some examples. Let us consider the case of a fishery with constant harvesting, i.e., a fish population $x(t)$ characterized by a logistic growth equation, which is exploited for commercial purposes. Let us assume that in each time period a constant quota h is harvested.

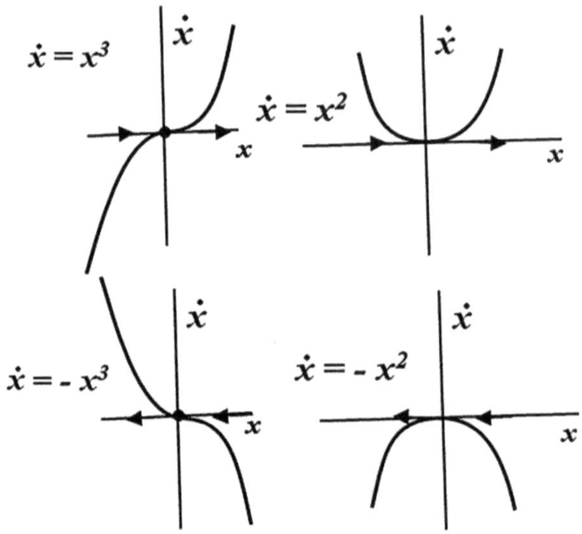

Fig. 1.12 Four examples of different phase portraits around the nonhyperbolic equlibrium $x^* = 0$ such that $f(x^*) = 0$ and $f'(x^*) = 0$

This leads to the following dynamic model

$$\dot{x} = x(\alpha - sx) - h \tag{1.13}$$

where the quota h is a parameter that indicates the policy imposed by an authority to regulate the fishing activity. The right hand side of the dynamic equation is a vertically translated parabola, and the equilibrium points, determined by imposing the equilibrium condition $x(\alpha - sx) - h = 0$, are given by $x_0^* = \left(\alpha - \sqrt{\alpha^2 - 4hs}\right)/(2s)$ and $x_1^* = \left(\alpha + \sqrt{\alpha^2 - 4hs}\right)/(2s)$ provided that $h < \alpha^2/(4s)$. The qualitative analysis shows that the higher equilibrium x_1^* is stable, and gives the equilibrium value at which the harvested population settles, whereas the lower is unstable, and constitutes the boundary that separates the basin of attraction of x_1^* and the set of initial conditions leading to extinction. A sort of "survival threshold": If, due to some accident, the initial condition falls below x_0^* then the dynamics of the system will lead it to extinction. Moreover, if the harvesting quota exceeds the value $\alpha^2/(4s)$, then the two equilibrium points merge and then disappear. This occurs when the graph of the right hand side of (1.13) is tangent to the horizontal axis: the two equilibria merge into a unique (nonhyperbolic) equilibrium. This is a bifurcation, after which no equilibrium exists and the only possible evolution is a decrease of population towards extinction. This sequence of dynamic situations can be summarized by a *bifurcation diagram*, see Fig. 1.13, where in the horizontal axis is represented the bifurcation parameter h and in the vertical axis are reported the equilibrium values, represented by a continuous line when stable and by a dashed line when unstable. As it can be seen, as far as $h < \alpha^2/(4s)$ we observe only quantitative modifications, i.e., the stable equilibrium decreases and the unstable one increases (thus causing the

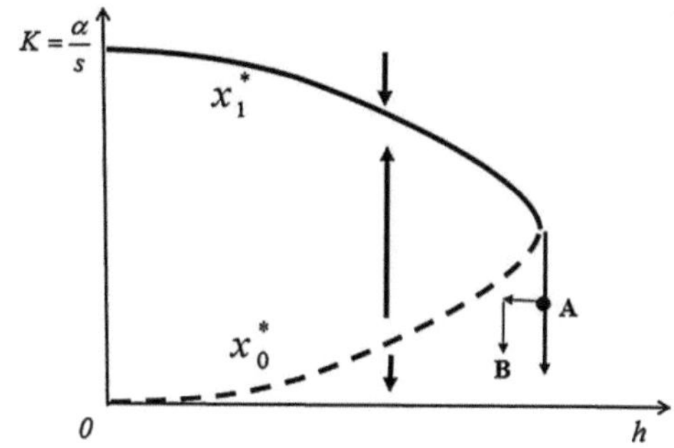

Fig. 1.13 Bifurcation diagram of the harvesting model (1.13) with constant harvesting h as bifurcation parameter

shrinking of the basin of attraction), whereas at the bifurcation value an important qualitative change occurs, leading to the disappearance of the two equilibrium points and consequently to a completely different dynamic scenario. This is the essence of the concept of bifurcation, related to slight modifications of a parameter leading to a qualitatively different phase diagram. It is worth noting that in this case the bifurcation occurring for increasing values of the "policy parameter" h is characterized by irreversibility (or hysteresis effect). In fact, if the harvesting quota h is gradually increased until it crosses the bifurcation point, then the fish population will decrease, see point A in Fig. 1.13. At this stage, even if the parameter h is decreased to reach a pre-bifurcation value $h < \alpha^2/(4s)$, it may be not sufficient to bring the system back to the stable equilibrium, because the phase point is trapped below the survival threshold x_0^*.

1.2.1.4 Local Bifurcations in One-Dimensional Nonlinear Models in Continuous Time

Two one-dimensional dynamical systems $\dot{x} = f(x)$ and $\dot{x} = g(x)$ are qualitatively equivalent if they have the same number of equilibrium points that orderly have, along the phase line, the same stability properties. This equivalence relation defines classes of equivalent dynamical systems on the line, see, e.g., the sketch represented in Fig. 1.14. One of these dynamical systems is structurally stable if after a slight modification of the graph of the function at the right hand side, for example a small variation of a parameter, it remains in the same equivalence class. In other words, such small variation only causes quantitative modifications of the equilibrium points. Instead if an arbitrarily small modification causes a qualitative change in the number and/or in the stability properties of the equilibria, so that the system enters a different equivalence class, then a bifurcation occurs at the boundary between two equivalence classes, and the system is said structurally unstable when it is along the boundary. These bifurcation situations, i.e., these situations of structural instability, are characterized by the presence of one or more nonhyperbolic equilibrium points.

The kinds of bifurcations through which such qualitative changes occur can be classified into a quite limited number of categories.

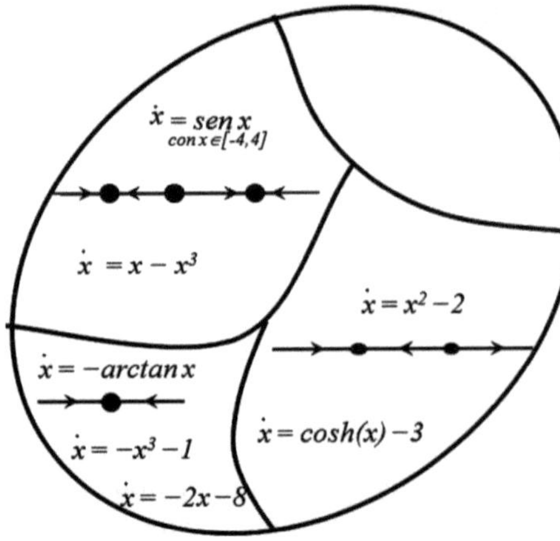

Fig. 1.14 Equivalence classes of dynamical systems of the line

Fold Bifurcation This bifurcation is characterized by the creation of two equilibrium points, one stable and one unstable, as a parameter varies. Of course, if the same parameter varies in the opposite direction, at the bifurcation point two equilibrium points, one stable and one unstable, merge and then disappear. A canonical example is given by the dynamical system $\dot{x} = f(x) = \mu - x^2$ as the parameter μ varies through the bifurcation value $\mu_0 = 0$ (see Fig. 1.15, where the bifurcation diagram is shown as well). Notice that two equilibrium points $x^*_{1,2} = \pm\sqrt{\mu}$ only exist for $\mu \geq 0$, and they are coincident $x^*_{1,2} = 0$ for $\mu = 0$ and nonhyperbolic, as $f'(x) = -2x$ vanishes for $x = 0$. Instead, for $\mu > 0$ the two equilibrium points are one stable and one unstable being $f'(x^*_1) = f'(-\sqrt{\mu}) = 2\sqrt{\mu} > 0$ and $f'(x^*_2) = f'(\sqrt{\mu}) = -2\sqrt{\mu} < 0$. Of course, if we start our analysis from a positive value of the parameter μ and decrease it until it reaches and crosses the bifurcation value $\mu = 0$, we observe two equilibrium points, one stable and one unstable that join at $\mu = 0$ and then disappear. It is worth noticing that the unstable equilibrium represents the boundary of the basin of attraction of the stable one, so we may describe this bifurcation by saying that a stable equilibrium collides with the boundary of its basin and then disappears.

Transcritical (or Stability Exchange) Bifurcation This bifurcation is characterized by the existence of two equilibrium points, one stable and one unstable, that merge at the bifurcation point and after the bifurcation they still exist but both with opposite stability property, i.e., the once stable becomes unstable whereas the once unstable becomes stable. A canonical example is given by the dynamical system $\dot{x} = f(x) = \mu x - x^2$ as the parameter μ varies through the bifurcation value $\mu_0 = 0$ (see Fig. 1.16, where the bifurcation diagram is shown as well). Notice that two equilibrium points $x^*_1 = 0$ and $x^*_2 = \mu$ always exist: they are coincident $x^*_{1,2} = 0$ for $\mu = 0$ and nonhyperbolic, as $f'(x) = \mu - 2x$ vanishes for $\mu = 0$ and $x = 0$. As $f'(x^*_1) = f'(0) = \mu$ and $f'(x^*_2) = f'(\mu) = -\mu$, $x^*_1 = 0$ is stable for $\mu < 0$ and unstable for $\mu > 0$ whereas $x^*_2 = \mu$ is unstable for $\mu < 0$ and stable for $\mu > 0$. So we can say that they merge at the bifurcation point and exchange their stability.

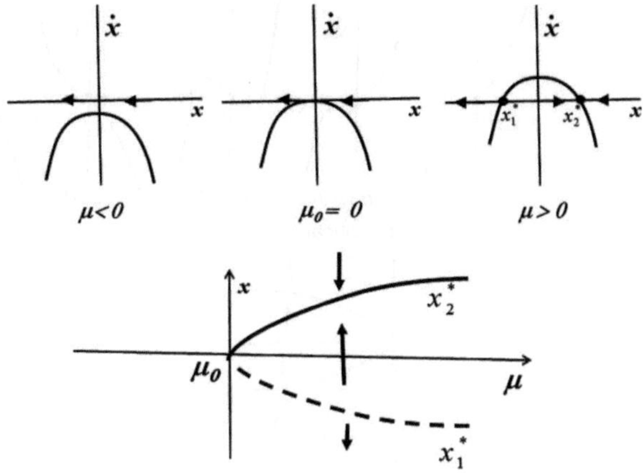

Fig. 1.15 Fold bifurcation

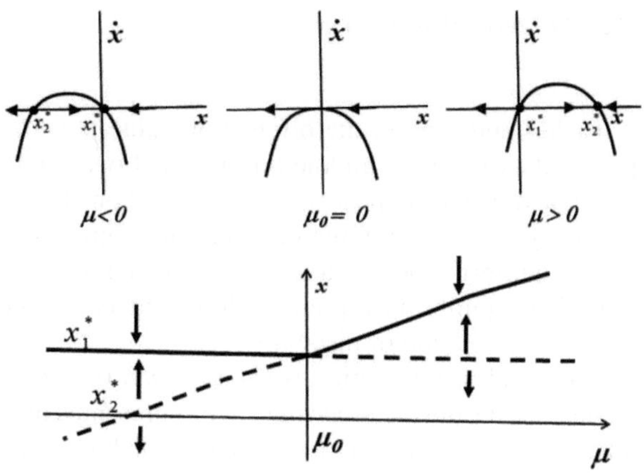

Fig. 1.16 Transcritical bifurcation

Pitchfork Bifurcation This bifurcation is characterized by a transition from a single equilibrium point to three equilibria: the one already existing changes its stability property as the bifurcation parameter crosses the bifurcation point, and this leads to the simultaneous creation of two further equilibria. Of course, if the same parameter varies in the opposite direction, at the bifurcation point two equilibrium points merge and disappear and only the central one survives, even if it changes its stability property. A canonical example is given by the dynamical system $\dot{x} = f(x) = \mu x - x^3$ as the parameter μ varies through the bifurcation value $\mu_0 = 0$ (see Fig. 1.17, where the bifurcation diagram is shown as well). Notice that the equilibrium points $x_0^* = 0$ always exists, and two further ones, $x_{1,2}^* = \pm\sqrt{\mu}$ for $\mu \geq 0$. All three are coincident $x_0^* = x_{1,2}^* = 0$ for $\mu = 0$, thus giving a unique nonhyperbolic equilibrium at the bifurcation point. In fact, from $f'(x) = \mu - 3x^2$ follows that $f'(0) = \mu$, hence x_0^* is stable for $\mu < 0$ and unstable for $\mu > 0$. Instead, for $\mu > 0$ the two newly born equilibrium points $x_{1,2}^*$ are both stable being $f'(\pm\sqrt{\mu}) = -\mu < 0$. Of course, if we start our analysis from a positive value of the parameter μ and decrease it until

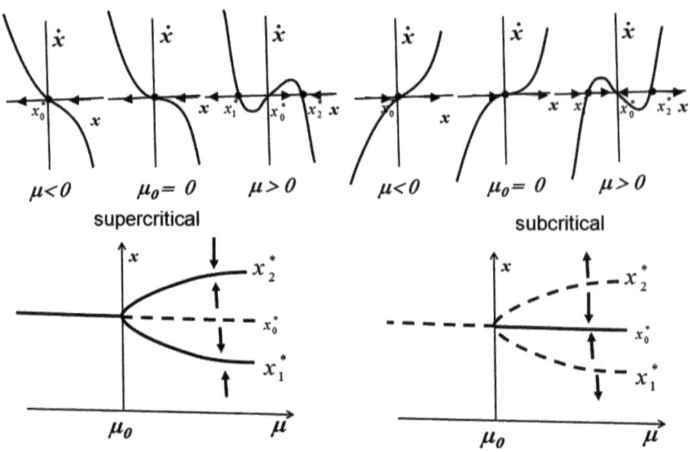

Fig. 1.17 Pitchfork bifurcation

it reaches and crosses the bifurcation value $\mu = 0$, we observe three equilibrium points, the one in the middle unstable and two stable at opposite sides, that join at $\mu = 0$ and then disappear while the central one becomes stable. It is worth noticing that for $\mu > 0$, when three equilibrium points exist, a situation of two coexisting stable equilibria, each with its own basin of attraction, occurs. Moreover, the central (unstable) equilibrium represents the boundary that separates the two basins of attraction in this situation of bistability.

This kind of bifurcation is called *supercritical pitchfork bifurcation* in order to distinguish it from the *subcritical* one, represented in the same picture, where a unique unstable equilibrium becomes stable at the bifurcation value with the simultaneous creation of two unstable equilibrium points located at opposite sides, and constitutes the upper and lower boundary of the basin of attraction of the central stable one. The canonical dynamical system that gives rise to a subcritical pitchfork bifurcation is $\dot{x} = f(x) = x^3 - \mu x$, as the parameter μ is increased through the bifurcation value $\mu_0 = 0$.

1.2.2 Two-Dimensional Dynamical Systems in Continuous Time

We now consider dynamic models of systems whose state is described by two variables, say $x_1(t)$ and $x_2(t)$, which are interdependent, i.e., the time evolution of $x_1(t)$, expressed by its time derivative $\dot{x_1}$, can be influenced by itself and by $x_2(t)$, and the same holds for $\dot{x_2}$:

$$\begin{aligned} \dot{x_1} &= f_1(x_1(t), x_2(t)) \,, \\ \dot{x_2} &= f_2(x_1(t), x_2(t)) \,. \end{aligned} \tag{1.14}$$

A general method to get a qualitative global view of the phase portrait of a model in the form (1.14) is obtained by a representation, in the phase space (x_1, x_2), of the two curves of equation $f_1(x_1, x_2) = 0$ and $f_2(x_1, x_2) = 0$, usually called *nullclines*. The points of intersection of these curves are the equilibrium points, solutions of the following system of two equations with two unknowns

$$\begin{aligned} f_1(x_1, x_2) &= 0 \,, \\ f_2(x_1, x_2) &= 0 \,. \end{aligned} \tag{1.15}$$

Moreover, the two curves subdivide the phase plane into zones characterized by different signs of the time derivatives $(\dot{x_1}, \dot{x_2})$. The resulting directions (obtained by the usual graphical rule of vector sum) give a qualitative idea of the dynamics of the model in each region of the phase plane. As an example of this method, let us consider the prey-predator model, also known as Lotka-Volterra model

$$\begin{aligned} \dot{x}_1 &= \alpha x_1 - s x_1^2 - b x_1 x_2 = x_1(\alpha - s x_1 - b x_2) \\ \dot{x}_2 &= -d x_2 + c x_1 x_2 = x_2(c x_1 - d) \end{aligned} \tag{1.16}$$

where $x_1 = x_1(t)$ represents the numerosity (or the density) in a given region of a species (the prey) that feeds from the environment, and $x_2 = x_2(t)$ represents the numerosity (or density) of predators that can only take nourishment from the prey population x_1. In the absence of predators ($x_2 = 0$) the prey population evolves according to the usual logistic growth function, whereas in the absence of preys ($x_1 = 0$) predators exhibit an exponential decay at rate d (mortality for starvation). The interaction term, proportional to the product xy under the assumption of random motion of prey and predators in the region considered (like in gas kinetics) has a negative effect on preys and positive on predators. This simple ecological model was proposed by the Italian mathematician Vito Volterra to explain the endogenous mechanism leading to oscillations in the fish harvesting observed in the Adriatic Sea.

Let us first consider the simpler case obtained by assuming $s = 0$ (like in the first model proposed by Volterra). In this case, the nullcline $\dot{x}_1 = 0$ is given by $x_1 = 0$, i.e., the vertical axis, or the horizontal line $x_2 = \alpha/b$, and the nullcline $\dot{x}_2 = 0$ is given by $x_2 = 0$, i.e., the horizontal axis, or the vertical line $x_1 = d/c$ (see Fig. 1.18). The coordinate axes are trapping sets, i.e., any trajectory starting from an initial condition

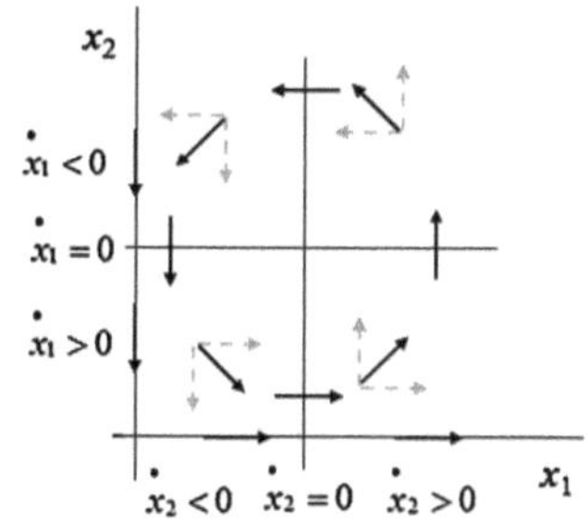

Fig. 1.18 Phase plane analysis of the Lotka-Volterra model (1.16) with $s = 0$

taken on the vertical axis $x_1 = 0$ remains there (as the rate of change of x_1 is $\dot{x}_1 = 0$ on it) and the corresponding trajectory goes to 0, the exponential decline of predators in the absence of preys. Instead along the trapping horizontal axis $x_2 = 0$ the prey population increases without any bound, as the term of overcrowding is neglected being in this case $s = 0$. In order to understand what happens starting from initial conditions interior to the positive quadrant, i.e., from initial situations of coexistence of preys and predators, we represent the horizontal and vertical arrows with orientations according to the signs of $\dot{x}_1$ and $\dot{x}_2$ (see Fig. 1.18). The directions of the phase vectors (also called *phasors*) clearly indicate a counterclockwise cyclic motion. This represents an oscillatory motion of both $x_1(t)$ and $x_2(t)$, hence endogenous or self-sustained oscillations. This is an important result, because it states that a dynamic system can exhibit autonomous oscillations, without any oscillatory forcing term. In other words a system with interacting components can oscillate even if nobody shakes it from outside.

An intuitive explanation of this dynamic behavior can be easily provided in the case of the prey-predator system modeled by Volterra. In fact, let us assume that at the initial time a few preys and a lot of predators are present, i.e., a small x_1 value and a large x_2 value, an initial state located in the upper-left quadrant of the phase space. In this case predators suffer for scarcity of food, and their number will decline. After this decline a few predators remain and preys will increase because of low predatory pressure. After this preys' population increase predators will have plenty of available food and consequently their population will increase, and this will lead to severe predatory pressure and thus a decay in preys' population. So, we again find the system in a situation with a few preys and a lot of predators, and the same process will be repeated, thus giving the cyclic time evolution.

Some trajectories starting from different initial conditions are shown in the left panel of Fig. 1.19, whereas the versus time representation of a typical trajectory can be seen in the right panel. Of course, the trajectory starting from the positive equilibrium point $E = (d/c, \alpha/b)$, located at the intersection of two nullclines, will remain there forever. However, if a perturbation causes a shift of the phase point from E then endless oscillations will start, with greater amplitude according to the distance of the initial condition (i.e., the entity of the shift) from the equilibrium

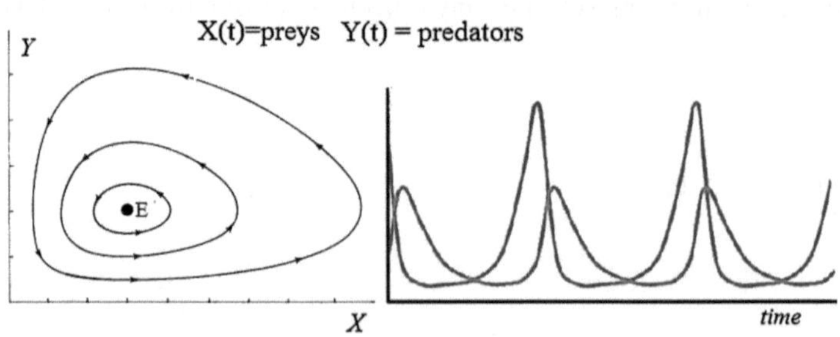

Fig. 1.19 Phase portrait (*left*) and versus time (*right*) representation of the trajectories of the model (1.16) with $s = 0$

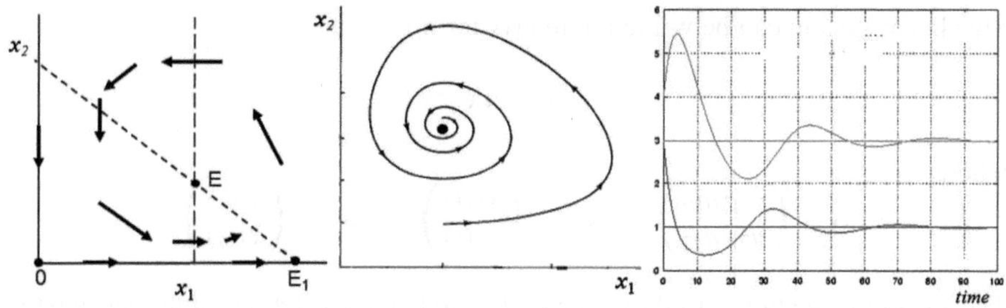

Fig. 1.20 Phase portrait and versus time representation of the trajectories of the model (1.16) with $s > 0$

point. Of course also $O = (0, 0)$ is an equilibrium point, located at the intersections of the nullclines that coincide with the coordinate axes. A classification of these equilibrium points will be proposed in Sect. 1.2.2.1.

One may wonder what happens if the overcrowding parameter $s > 0$, i.e., the prey population alone follows a logistic growth. In this case the prey nullcline has equation $\alpha - sx_1 - bx_2 = 0$, i.e., it is a tilted line with negative slope (see Fig. 1.20). It is not easy to understand how the trajectories change by the qualitative method of nullclines and phasors. A numerical representation of a typical trajectory in the phase plane as well as the corresponding time paths $x_1(t)$ and $x_2(t)$ are shown in Fig. 1.20; however a more detailed analysis will be possible with the methods described in the next sections.

We end this section by stressing the fact that endogenous oscillations are a well known phenomenon in a capitalistic economy, where up and down patterns have been (and currently are) observed in the main macroeconomic indicators. As we will see in more details later in these lecture notes, the same dynamic equations proposed by Volterra to describe the time evolution of preys' and predators' populations have been used (with founded motivations) by the economist Richard Goodwin in [14] to represent endogenous business cycles, by using salaries and occupation as dynamic variables. This is an example of how dynamic models can be usefully applied in different fields.

1.2.2.1 Linear Systems

Following the same path as for the one-dimensional case, let us first of all consider a linear homogeneous system of two differential equations of first order (i.e., involving only the first derivative of the dynamic variables) with constant coefficients in the (normal) form:

$$\begin{cases} \dot{x_1} = a_{11}x_1(t) + a_{12}x_2(t) \ , \\ \dot{x_2} = a_{21}x_1(t) + a_{22}x_2(t) \ . \end{cases} \tag{1.17}$$

This linear system can be written in matrix form

$$\dot{\mathbf{x}} = \mathbf{A}\mathbf{x} , \tag{1.18}$$

where

$$\mathbf{A} = \begin{pmatrix} a_{11} & a_{12} \\ a_{21} & a_{22} \end{pmatrix} , \quad \mathbf{x}(t) = \begin{pmatrix} x_1(t) \\ x_2(t) \end{pmatrix} , \quad \dot{\mathbf{x}}(t) = \begin{pmatrix} \dot{x}_1(t) \\ \dot{x}_2(t) \end{pmatrix} .$$

The aim of this section is to show a procedure to find the solutions of this system that reduces to a very simple algebraic method that essentially consists in the solution of a second degree algebraic equation (in the field of complex numbers).

However, before stating this result, we outline the arguments at the basis of the proof.

An important general property of a linear system of ordinary differential equations is that given two solutions, say

$$\varphi(t) = \begin{pmatrix} \varphi_1(t) \\ \varphi_2(t) \end{pmatrix} \quad \text{and} \quad \psi(t) = \begin{pmatrix} \psi_1(t) \\ \psi_2(t) \end{pmatrix}$$

any linear combination of them

$$\mathbf{y}(t) = \alpha\varphi(t) + \beta\psi(t) \quad \text{with } \alpha, \beta \in \mathbb{R} \tag{1.19}$$

is a solution of (1.17) as well. In fact, assuming that both $\varphi(t)$ and $\psi(t)$ satisfy (1.18) being then solutions, we obtain

$$\dot{\mathbf{y}} = \alpha\dot{\varphi} + \beta\dot{\psi} = \alpha\mathbf{A}\varphi + \beta\mathbf{A}\psi = \mathbf{A}\left(\alpha\varphi + \beta\psi\right) = \mathbf{A}\mathbf{y} ,$$

so that also $\mathbf{y}(t)$ is a solution. This means that the set of all the solutions, obtained with different "weights" α and β in the linear combination, is a vector space. Moreover, it is possible to prove that it has dimension 2, i.e., all the solutions can be generated as linear combinations of just two independent solutions, that form a base of the vector space. The definition of independent solutions is the usual one: given two solutions, say again $\varphi(t)$ and $\psi(t)$, they are independent if $\alpha\varphi(t) + \beta\psi(t) = \mathbf{0}\ \forall t$ implies $\alpha = \beta = 0$. In order to check such independence it is possible to use the Wronskian determinant

$$W(t) = \text{Det}\begin{pmatrix} \varphi_1(t) & \psi_1(t) \\ \varphi_2(t) & \psi_2(t) \end{pmatrix} = \varphi_1(t)\psi_2(t) - \psi_1(t)\varphi_2(t) .$$

If $W(t) \neq 0$ for at least a t value, then the two solutions $\varphi(t)$ and $\psi(t)$ are independent. In fact, it is possible to prove that only one of the following is true: $W(t) = 0\ \forall t$ or $W(t) \neq 0\ \forall t$. Hence it is sufficient to check it for just one value of t, for example $W(0) \neq 0$.

So, in order to find the general solution (i.e., all the possible solutions) of (1.17) it is sufficient to find just two of them which are independent. Then the general solution will be in the form (1.19) and, by imposing an initial conditions $x_1(0), x_2(0)$ the two constants α and β can be uniquely determined by solving the following linear algebraic system

$$\alpha\begin{pmatrix}\varphi_1(0)\\ \varphi_2(0)\end{pmatrix}+\beta\begin{pmatrix}\psi_1(0)\\ \psi_2(0)\end{pmatrix}=\begin{pmatrix}\varphi_1(0) & \psi_1(0)\\ \varphi_2(0) & \psi_2(0)\end{pmatrix}\begin{pmatrix}\alpha\\ \beta\end{pmatrix}=\begin{pmatrix}x_1(0)\\ x_2(0)\end{pmatrix},$$

which is a linear system whose coefficient matrix is nonsingular, being $W(0) \neq 0$ its determinant.

In the following we show a direct method to find two independent solutions. Following again the same arguments as in the one-dimensional case, let us propose a "trial solution" in exponential form, i.e.,

$$x_i(t) = v_i e^{\lambda t}, \; i = 1, 2 . \tag{1.20}$$

As $\dot{x}_1(t) = \lambda v_1 e^{\lambda t}$ and $\dot{x}_2(t) = \lambda v_2 e^{\lambda t}$, after replacing this trial solution into (1.17) we get

$$\begin{cases}\lambda v_1 e^{\lambda t} = a_{11} v_1 e^{\lambda t} + a_{12} v_2 e^{\lambda t}\\ \lambda v_2 e^{\lambda t} = a_{21} v_1 e^{\lambda t} + a_{22} v_2 e^{\lambda t}\end{cases}$$

and after simplification of all the identical exponential it becomes

$$\begin{cases}(a_{11}-\lambda)v_1 + a_{12}v_2 = 0\\ a_{21}v_1 + (a_{22}-\lambda)v_2 = 0\end{cases} \quad \text{or} \quad (\mathbf{A}-\lambda\mathbf{I})\,\mathbf{v} = \mathbf{0} \tag{1.21}$$

an algebraic linear homogeneous system with unknowns v_1 and v_2 and parameter λ. This homogeneous systems has nontrivial solutions (i.e., solutions different from $(0, 0)$) provided that

$$\text{Det}\left(\begin{bmatrix}a_{11}-\lambda & a_{12}\\ a_{21} & a_{22}-\lambda\end{bmatrix}\right) = 0 \quad \text{or} \quad \text{Det}\,(\mathbf{A}-\lambda\mathbf{I}) = \mathbf{0}$$

This condition can be expressed in the form of the "characteristic equation"

$$\lambda^2 - \text{Tr}(A)\lambda + \text{Det}(A) = 0 , \tag{1.22}$$

where $\text{Tr}(A) = a_{11} + a_{22}$ (the sum of diagonal elements of the matrix $\mathbf{A}$) and $\text{Det}(A) = a_{11}a_{22} - a_{12}a_{21}$ (the determinant of the matrix $\mathbf{A}$).

This is a standard problem of linear algebra, known as eigenvalue problem. In fact, in matrix form it can be expressed as

$$\mathbf{A}\mathbf{v} = \lambda\mathbf{v} \quad \text{with } \mathbf{v} \neq \mathbf{0} \quad \text{and} \quad \lambda \in \mathbb{C} ,$$

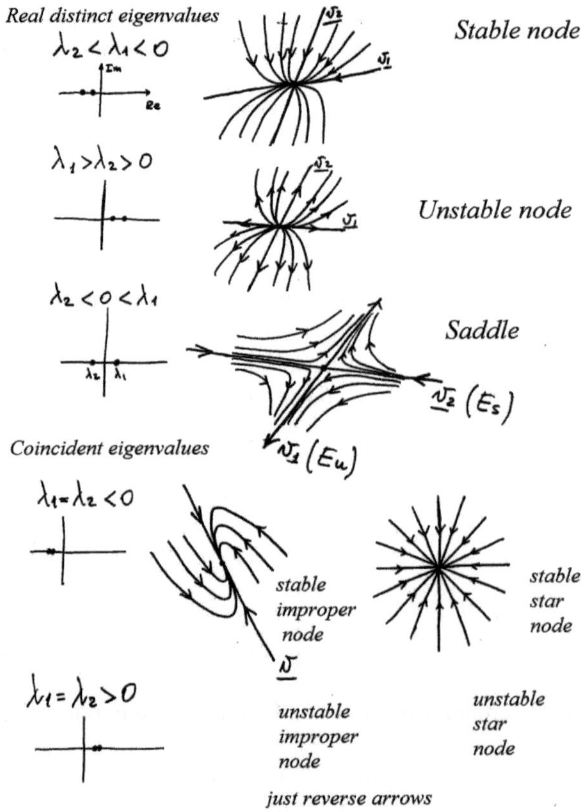

Fig. 1.21 Some phase diagrams of a linear system of the plane with real eigenvalues

i.e., the linear operator **A**, applied to the vector **v**, gives a vector $\lambda\mathbf{v}$ proportional to it, let's say in the same direction. The real number λ is called *eigenvalue* and the solution vector **v** *eigenvector*.

To sum up, $\mathbf{v}e^{\lambda t}$ is a solution of (1.17) if and only if $[\mathbf{A} - \lambda\mathbf{I}]\,\mathbf{v} = 0$, i.e., λ is an eigenvalue of **A** with corresponding eigenvector $\mathbf{v} \neq \mathbf{0}$, i.e., $\text{Det}\,[\mathbf{A} - \lambda\mathbf{I}] = 0$ or, equivalently, if λ satisfies the characteristic equation (1.22). This reduces the problem of finding two independent solutions of (1.17) to the computation of two solutions of the second degree algebraic equation (1.22). According to the sign of the discriminant of the characteristic equation, $\text{Tr}(A)^2 - 4\text{Det}(A)$, we can have two real distinct, two real coincident or two complex conjugate eigenvalues. This will give rise to different kinds of phase portraits, as explained below.

1. If we have two real, distinct and negative eigenvalues $\lambda_2 < \lambda_1 < 0$, i.e., $\text{Tr}\,(A)^2 - 4\text{Det}\,(A) > 0$ with $\text{Tr}(A) < 0$ and $\text{Det}(A) > 0$, then the two independent solutions are $\mathbf{v}_1e^{\lambda_1 t}$ and $\mathbf{v}_2e^{\lambda_2 t}$, both decreasing to 0 as $t \to \infty$. The general solution is

$$\mathbf{x}(t) = c_1\mathbf{v}_1e^{\lambda_1 t} + c_2\mathbf{v}_2e^{\lambda_2 t}\ , \tag{1.23}$$

where the two constants c_1 and c_2 are uniquely determined according to the initial condition $x(0) = (x_1(0), x_2(0))$. The corresponding phase diagram is represented in Fig. 1.21 and the asymptotically stable equilibrium is called *stable node* (or *sink*).

2. If we have two real, distinct and positive eigenvalues $\lambda_1 > \lambda_2 > 0$, i.e., $\text{Tr}\,(A)^2 - 4\text{Det}\,(A) > 0$ with $\text{Tr}(A) > 0$ and $\text{Det} > 0$, then the two independent solutions are $\mathbf{v}_1 e^{\lambda_1 t}$ and $\mathbf{v}_2 e^{\lambda_2 t}$, both increasing to ∞ as $t \to \infty$. The general solution is again of the form (1.23), the corresponding phase diagram is represented in Fig. 1.21 and the unstable equilibrium $(0, 0)$ is called *unstable node* (or *source*).
3. If we have two real distinct eigenvalues of opposite sign, $\lambda_2 < 0 < \lambda_1$, i.e., $\text{Tr}\,(A)^2 - 4\text{Det}\,(A) > 0$ with $\text{Det} < 0$, then the two independent solutions $\mathbf{v}_1 e^{\lambda_1 t}$ and $\mathbf{v}_2 e^{\lambda_2 t}$ are one increasing to ∞ and one decreasing to 0 as $t \to \infty$. The corresponding phase diagram is represented in Fig. 1.21 and the unstable equilibrium $(0, 0)$ is called *saddle*. Notice that an invariant line exists, called *stable manifold* (along the direction indicated by the eigenvector $\mathbf{v}_2$ associated with the negative eigenvalue, whereas the line along the eigenvector $\mathbf{v}_1$, associated to the positive eigenvalue, is referred to as the *unstable manifold*), on which the dynamics is asymptotically convergent to the equilibrium $(0, 0)$. Nevertheless the equilibrium is unstable, and the kind of notion around it may even be misleading, as the generic trajectory first moves towards it (so that it may look as a convergent trajectory) whereas it then turns away from the equilibrium.
4. If we have two coincident and negative eigenvalues $\lambda_1 = \lambda_2 = \lambda < 0$ (and consequently $\mathbf{v}_1 = \mathbf{v}_2 = \mathbf{v}$), i.e., $\text{Tr}\,(A)^2 - 4\text{Det}\,(A) = 0$ with $\text{Tr} < 0$, then two independent solutions are $\mathbf{v}e^{\lambda t}$ and $\mathbf{v}te^{\lambda t}$, both converging to 0 as $t \to \infty$, and the general solution becomes

$$\mathbf{x}(t) = c_1 \mathbf{v} e^{\lambda t} + c_2 \mathbf{v} t e^{\lambda t} \,. \tag{1.24}$$

The corresponding phase diagram is represented in Fig. 1.21 and the stable equilibrium $(0, 0)$ is called *stable improper node* (or *stable star node* in particular symmetric situations).

5. If we have two coincident and positive eigenvalues $\lambda_1 = \lambda_2 = \lambda > 0$ (and consequently $\mathbf{v}_1 = \mathbf{v}_2 = \mathbf{v}$), i.e., $\text{Tr}\,(A)^2 - 4\text{Det}\,(A) = 0$ with $\text{Tr} > 0$, then we have the same general solution and the corresponding phase diagram is obtained from the previous one just reversing the arrows and is called *unstable improper node* (or *unstable star node* in particular symmetric situations).
6. If $\text{Tr}\,(A)^2 - 4\text{Det}\,(A) < 0$ we have two complex conjugate eigenvalues $\lambda_1 = a + ib$ and $\lambda_2 = a - ib$, where $a = \text{Re}(\lambda) = \text{Tr}(A)/2$ is the real part and $b = \text{Im}\,(\lambda) = \sqrt{4\text{Det}(A) - \text{Tr}(A)^2}/2$ is the imaginary part. Again two independent solutions are $\varphi(t) = \mathbf{v}_1 e^{\lambda_1 t}$ and $\bar{\varphi}(t) = \mathbf{v}_2 e^{\lambda_2 t}$, with both the eigenvalues as well as the corresponding eigenvectors complex. However, $\bar{\varphi}(t)$ is the complex conjugate of $\varphi(t)$, and we can write them in trigonometric form

$$\varphi(t) = \mathbf{v}_1 e^{at} \left(\cos(bt) + \imath \sin(bt)\right) \text{ and } \bar{\varphi}(t) = \bar{\mathbf{v}}_1 e^{at} \left(\cos(bt) - \imath \sin(bt)\right).$$

As any linear combination of two solutions is again a solution of (1.17) we can obtain two independent real solutions in the form

$$\frac{1}{2}(\varphi(t)+\bar{\varphi}(t)) = \text{Re } \varphi(t) = (\text{Re } \mathbf{v}_1)e^{at}\cos(bt)$$
$$\frac{1}{2i}(\varphi(t)-\bar{\varphi}(t)) = \text{Im } \varphi(t) = (\text{Im } \mathbf{v}_1)e^{at}\sin(bt)$$

So, the general solution can be written in the form

$$\mathbf{x}(t) = e^{at}\left[c_1(\text{Re } \mathbf{v}_1)\cos(bt) + c_2(\text{Im } \mathbf{v}_1)\sin(bt)\right] \tag{1.25}$$

from which we can see that the part inside square brackets causes oscillations around the equilibrium $(0, 0)$ whereas the exponential term outside the square brackets determines the expanding or contracting nature of the oscillations: if $a < 0$, i.e., $\text{Tr}(A) < 0$, then the oscillations exhibit decreasing amplitude and converge to the equilibrium $(0, 0)$, if $a > 0$, i.e., $\text{Tr}(A) > 0$, then the oscillations increase in amplitude and diverge. Finally, if $a = 0$, i.e., $\text{Tr}(A) = 0$, then the oscillations are of constant amplitude. The corresponding phase diagram are represented in Fig. 1.22, and the corresponding phase diagrams are denoted as *stable*

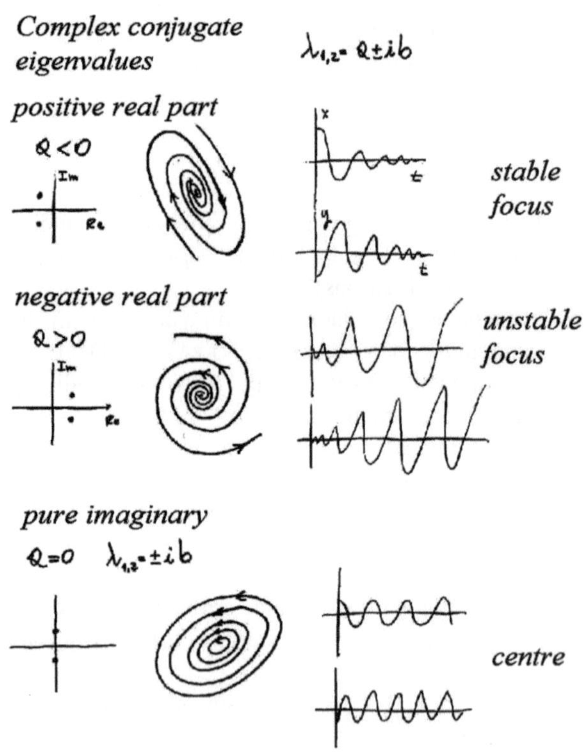

Fig. 1.22 Some phase portraits of a linear system of the plane with complex eigenvalues

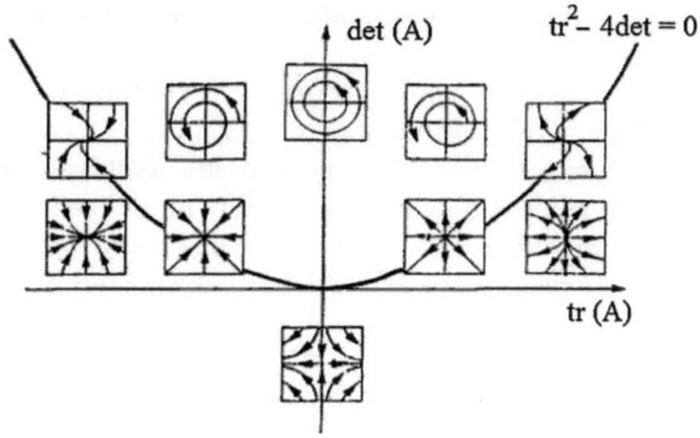

Fig. 1.23 The trace-determinant plane for a linear two-dimensional continuous time system

focus (or *stable spiral*), *unstable focus* (or *unstable spiral*) and *center* respectively. Notice that the case of complex eigenvalues is the first one giving oscillations, and the imaginary part of the eigenvalues $b = 0.5\sqrt{4\text{Det}(A) - \text{Tr}(A)^2}$ determines the time required to complete a whole oscillation, given by $T = 2\pi/b$.

To summarize all these cases it is useful to represent the trace $\text{Tr}(A)$ and the determinant $\text{Det}(A)$ on the coordinate axes of a Cartesian plane (see Fig. 1.23), together with the curve of equation $\text{Tr}(A)^2 - 4\text{Det}(A) = 0$ (a parabola with vertex in the origin of the axes). Above the parabola we have $\text{Tr}(A)^2 - 4\text{Det}(A) < 0$, hence oscillatory behavior, below it we have $\text{Tr}(A)^2 - 4\text{Det}(A) > 0$, so we have nodes and saddles according to the sign of $\text{Det}(A)$.

Remark 1.1 It is worth noting that asymptotic stability of the unique equilibrium occurs only in the quadrant with $\text{Tr}(A) < 0$ and $\text{Det}(A) > 0$. Moreover, in this case of linear dynamic models the local asymptotic stability is equivalent to global asymptotic stability, i.e., if the equilibrium is stable it attracts all the initial conditions $(x_1(0), x_2(0)) \in \mathbb{R}^2$. Instead, when the equilibrium is unstable, then all the initial conditions starting outside the equilibrium generate diverging trajectories.

Some other particular cases can be noticed. For example, if $\text{Det}(A) \neq 0$ then the linear homogeneous algebraic system to obtain the equilibrium points $\mathbf{Ax} = \mathbf{0}$ has the unique solution $(0, 0)$, whereas if $\text{Det}(A) = 0$ then infinitely many equilibrium points exist, located along a line through the origin. These equilibria are nonhyperbolic being one of the eigenvalues equal to zero. Even in the case of $\text{Tr}(A) = 0$ and $\text{Det}(A) > 0$ the equilibrium point is denoted as nonhyperbolic as the real part of the eigenvalues vanishes. These will identify the bifurcation cases when dealing with 2-dimensional nonlinear dynamic models (in continuous time) that depend on a bifurcation parameter.

1.2.2.2 Nonlinear Dynamic Models in Two Dimensions

Let us consider a nonlinear model in the form (1.14) and let $E = \left(x_1^*, x_2^*\right)$ be an equilibrium point, solution of the system (1.15). Differently from the linear models, in the nonlinear case several equilibrium points can coexist. However, what we have seen for a linear system can be used to understand the local behavior of a nonlinear system around a single equilibrium, i.e., *locally*, in a neighborhood of the equilibrium point. We recall that a neighborhood of a point of $\mathbf{x}^* \in \mathbb{R}^n$ is a set $N_r(\mathbf{x}^*)$ defined as

$$N_r(\mathbf{x}^*) = \left\{\mathbf{x} \in \mathbb{R}^n | \left\|\mathbf{x} - \mathbf{x}^*\right\| < r\right\} \text{ for some } r > 0$$

where $\|\cdot\|$ is a norm, such as the Euclidean norm $\|\mathbf{x}\| = \sqrt{\sum_{i=1}^n x_i^2}$. Hence in $\mathbb{R}^n$ a neighborhood is an open disk of radius r and center $\mathbf{x}^*$. In the following we will characterize the local phase portrait in a neighborhood of an equilibrium point $E = \left(x_1^*, x_2^*\right)$ by using the linear approximation of the nonlinear system obtained by the first order Taylor expansion

$$f_1(x_1, x_2) = f_1\left(x_1^*, x_2^*\right) + \left.\frac{\partial f_1}{\partial x_1}\right|_E \left(x_1 - x_1^*\right) + \left.\frac{\partial f_1}{\partial x_2}\right|_E \left(x_2 - x_2^*\right) + o(\left\|\mathbf{x} - \mathbf{x}^*\right\|),$$
$$f_2(x_1, x_2) = f_2\left(x_1^*, x_2^*\right) + \left.\frac{\partial f_2}{\partial x_1}\right|_E \left(x_1 - x_1^*\right) + \left.\frac{\partial f_2}{\partial x_2}\right|_E \left(x_2 - x_2^*\right) + o(\left\|\mathbf{x} - \mathbf{x}^*\right\|),$$

where the symbol $o(\cdot)$ represents higher order infinitesimal terms as $\mathbf{x} \to \mathbf{x}^*$. Being E an equilibrium, $f_i\left(x_1^*, x_2^*\right) = 0$, $i = 1, 2$, so if we define the *Jacobian matrix* as the matrix that collects the four partial derivatives

$$\mathbf{J}(x_1, x_2) = \begin{bmatrix} \frac{\partial f_1}{\partial x_1}(x_1, x_2) & \frac{\partial f_1}{\partial x_2}(x_1, x_2) \\ \frac{\partial f_2}{\partial x_1}(x_1, x_2) & \frac{\partial f_2}{\partial x_2}(x_1, x_2) \end{bmatrix}$$

and we substitute the Taylor expansion, then (1.14) can be written as

$$\begin{bmatrix} \dot{X}_1 \\ \dot{X}_2 \end{bmatrix} = \mathbf{J}\left(x_1^*, x_2^*\right) \begin{bmatrix} X_1 \\ X_2 \end{bmatrix} + o\,\|\mathbf{X}\| ,$$

where $X_1 = x_1 - x_1^*$, $X_2 = x_2 - x_2^*$ are coordinates centered in E, i.e., that measure the displacement from the equilibrium. Under suitable conditions we can use the linear approximation around $\mathbf{x}^*$ to classify the local phase portrait according to the following result.

Theorem 1.2 (Linearization Theorem) *Let the nonlinear system (1.14) have an equilibrium $\mathbf{x}^*$ such that all the eigenvalues of $J(\mathbf{x}^*)$ have nonvanishing real part. Then in a neighborhood of $\mathbf{x}^*$ the local phase portrait of (1.14) is qualitatively equivalent to that of the linear approximation.*

This is a quite informal and intuitive version of a more general theorem known as Hartman-Grobman Theorem. Here below we give a more rigorous version of it, extended to n dimensions, where the definition of qualitative (or topological) equivalence is included as well.

Theorem 1.3 (Hartman-Grobman [15, 18]) *Given a nonlinear system of differential equations* $\dot{\mathbf{x}} = f(\mathbf{x})$, $\mathbf{x} \in \mathbb{R}^n$, *let* $\mathbf{x}^* \in \mathbb{R}^n$ *be an equilibrium point, i.e.,* $f(\mathbf{x}^*) = 0$. *If* $\mathbf{x}^*$ *is hyperbolic, i.e., all the eigenvalues of the Jacobian matrix* $J(\mathbf{x}^*)$ *have nonvanishing real part, then the general solution* $\mathbf{y}(t) \in \mathbb{R}^n$ *of the linear system* $\dot{\mathbf{y}} = \mathbf{J}(\mathbf{x}^*)\,\mathbf{y}$ *is such that a neighborhood* U *of* $\mathbf{x}^*$ *exists and homeomorphism* $\mathbf{y} = \mathbf{h}(\mathbf{x})$ *exists, with* $\mathbf{h}$ *defined in* U *and with values in a neighborhood of the equilibrium* $\mathbf{0}$ *of the linear system, such that* $\mathbf{y}(t) = \mathbf{h}(\mathbf{x}(t))\quad \forall t \in R$ *with* $\mathbf{x}(t) \in U$ *solution of* $\dot{\mathbf{x}} = f(\mathbf{x})$.

We recall that a homeomorphism is a continuous and invertible function.

The Hartman-Grobman theorem essentially states that the trajectories of a nonlinear dynamic model in a neighborhood of a hyperbolic equilibrium are similar to the ones of its linear approximation whose matrix of coefficients is given by the Jacobian matrix computed at the equilibrium. This implies that any hyperbolic equilibrium point of a nonlinear dynamical system can be classified as a stable (unstable) node, or a saddle, or a stable (unstable) focus as for the corresponding linear approximation. The corresponding phase portraits may be in some way distorted (stretched, rotated etc.) however they are topologically equivalent. In particular the stable and unstable invariant manifold of saddles still exist, even if they are no longer lines but smooth curves tangent to the eigenvectors of the corresponding linear approximation.

A corollary of the Hartman-Grobman Theorem is given by the following proposition about local asymptotic stability of an equilibrium.

Theorem 1.4 (Local Asymptotic Stability) *Let* $\mathbf{x}^*$ *be an equilibrium point of* $\dot{\mathbf{x}} = f(\mathbf{x})$, $\mathbf{x} \in \mathbb{R}^n$. *If all the eigenvalues of* $J(\mathbf{x}^*)$ *have negative real part then* $\mathbf{x}^*$ *is a locally asymptotically stable equilibrium. If at least one eigenvalue of the Jacobian matrix* $J(\mathbf{x}^*)$ *has positive real part then* $\mathbf{x}^*$ *is unstable.*

To sum up, given an equilibrium point we can analyze the local qualitative behavior in a neighborhood of an hyperbolic equilibrium (i.e., the qualitative structure of the phase portrait around it) just studying the eigenvalues of the Jacobian matrix computed in it, that follows immediately from the computation of the trace and the determinant according to the classification listed for linear systems. However this procedure only gives information about the local behavior around the equilibrium points, and nothing about the global behavior, even if this is usually a good starting point to have a global view as well.

We also stress that the Hartman-Grobman Theorem provides no information about the behavior of the dynamical system around nonhyperbolic equilibria, i.e., when the determinant or the trace of the Jacobian matrix vanish. Such nongeneric situations are often characterized by structural instability, and in the presence of a parameter they may give rise to bifurcations, as will be discussed in the next section.

A Simple Example Let us consider the system of nonlinear differential equations

$$\begin{cases} \dot{x}_1 = 2x_1 - x_1^2 - 2x_1x_2 \\ \dot{x}_2 = 2x_2 - x_2^2 - 2x_1x_2 \end{cases} \tag{1.26}$$

with a generic initial condition $(x_1(0), x_1(0)) \in \mathbb{R}^2$.

The equilibrium points are the solutions of

$$\begin{aligned} x_1 (2 - x_1 - 2x_2) = 0 \\ x_2 (2 - x_2 - 2x_1) = 0 \end{aligned}$$

given by $O = (0, 0)$, $A = (2, 0)$, $C = (0, 2)$, $E = (2/3, 2/3)$. Given the Jacobian matrix

$$\mathbf{J}(x_1, x_2) = \begin{bmatrix} 2 - 2x_1 - 2x_2 & -2x_1 \\ -2x_2 & 2 - 2x_1 - 2x_2 \end{bmatrix}$$

a classification of the equilibrium points is easily obtained by the method of linear approximation based on the computation of the Jacobian in each of them.

$$\mathbf{J}(O) = \begin{bmatrix} 2 & 0 \\ 0 & 2 \end{bmatrix}$$

is a diagonal matrix, hence the eigenvalues are readily computed being them the diagonal entries[2]: $\lambda_1 = \lambda_2 = 2 > 0$. Hence the equilibrium is a repelling node, and due to the particular symmetric structure of the model, the equilibrium point of the corresponding linear approximation is a star node.

$$\mathbf{J}(A) = \begin{bmatrix} -2 & 0 \\ -4 & -2 \end{bmatrix}$$

is a triangular matrix, so even in this case the eigenvalues are given by the diagonal entries: $\lambda_1 = \lambda_2 = -2$, and we have a stable improper node. Analogously for the equilibrium B. Finally,

$$\mathbf{J}(E) = \begin{bmatrix} -\frac{2}{3} & -\frac{4}{3} \\ -\frac{4}{3} & -\frac{2}{3} \end{bmatrix}$$

from which we can see that $\mathrm{Tr}(E) = -4/3 < 0$ and $\mathrm{Det}(E) = -4/3 < 0$. Hence, E is a saddle. It is easy to verify that the eigenvalues are $\lambda_1 = -2$ with corresponding eigenvalue $\mathbf{v}_1 = (1, 1)$ (tangent to the stable manifold) and are $\lambda_2 = 2/3$ with corresponding eigenvalue $\mathbf{v}_2 = (1, -1)$ (tangent to the unstable manifold), see Fig. 1.24. Notice that, due to the symmetric form of (1.26), in this case the line along the direction indicated by $\mathbf{v}_1 = (1, 1)$ is the invariant stable manifold. Moreover, as usual,

[2]The property that the diagonal entries coincide with the eigenvalues of the matrix holds for all triangular matrices.

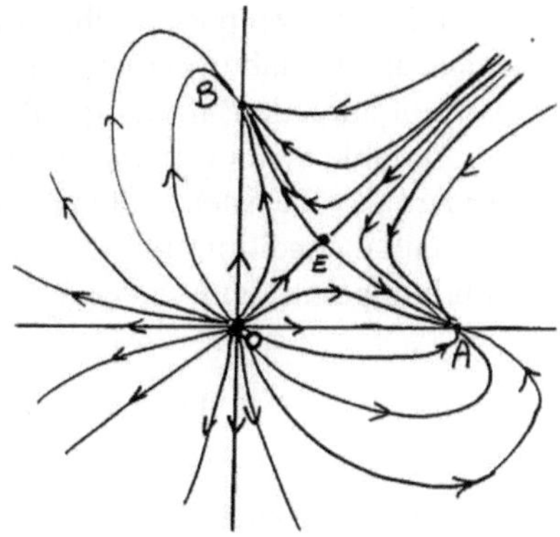

Fig. 1.24 Phase portrait for the competition model (1.26)

such stable manifold constitutes the boundary that separates the basins of the two stable equilibrium points A and B.

The restriction of the system (1.26) to the positive quadrant $\mathbb{R}^2_+ = \{(x_1, x_2) \in \mathbb{R}^2 | x_1 \geq 0; x_2 \geq 0\}$, is an example of Volterra model of competition between two species

$$\begin{cases} \dot{x}_1 = \alpha_1 x_1 - s_1 x_1^2 - b_1 x_1 x_2 \ , \\ \dot{x}_2 = \alpha_2 x_2 - s_2 x_2^2 - b_2 x_1 x_2 \ , \end{cases} \tag{1.27}$$

where each species alone grows according to a logistic law of motion, and the interaction has a negative effect on both, as each of them is assumed to subtract food from the other one. The study of this model led to the mathematical formulation of the principle of competitive exclusion: if two species need the same vital resources, then only one will survive. Which species will survive depends on the parameters that characterize each species behavior as well as on the initial advantage (i.e., the initial condition). This principle can be extended to the case of n species, and may have several applications even in social and economic systems. A good exercise is to generalize the results obtained in the particular case considered above to the more general model (1.27) in order to understand the role of each parameter on existence and stability of equilibria.

Another Ecological Example: the Prey-Predator Lotka-Volterra Model The prey-predator Lotka-Volterra model has been already described at the beginning of this section. Let us consider again the model (1.16) and compute its equilibrium points, solutions of the algebraic system

$$\begin{aligned} x_1(\alpha - s x_1 - b x_2) = 0 \ , \\ x_2(c x_1 - d) = 0 \ , \end{aligned} \tag{1.28}$$

given by

$$O = (0, 0) \ ; \quad A = \left(\frac{\alpha}{s}, 0\right) \ ; \quad E = \left(\frac{d}{c}, \frac{\alpha c - sd}{bc}\right) \ . \tag{1.29}$$

The equilibrium O represents the extinction of both species, so its stability indicates that there are no viability conditions for both species, at least for initial conditions in its basin of attraction; if the equilibrium A is stable, then the ecological conditions are not suitable to allow predators' survival. Only the stability of the equilibrium E can ensure the coexistence of the two species, provided it is positive, i.e., $\alpha c > sd$, and for initial conditions in its basin. In order to study the local stability of the three equilibrium points, let us consider, as usual, the Jacobian matrix

$$\mathbf{J}(x_1, x_2) = \begin{bmatrix} \alpha - 2sx_1 - bx_2 & -bx_1 \\ cx_2 & cx_1 - d \end{bmatrix}$$

and compute it in each equilibrium point. At the equilibrium

$$\mathbf{J}(0,0) = \begin{bmatrix} \alpha & 0 \\ 0 & -d \end{bmatrix}$$

we have a diagonal matrix with eigenvalues $\lambda_1 = \alpha > 0$ and $\lambda_2 = -d < 0$, so that the equilibrium O is a saddle. It is easy to check that, as usual with a diagonal matrix, the eigenvector associated with the first eigenvalue (the positive one in this case) is $\mathbf{v}_1 = [1, 0]$, hence the unstable manifold is along the horizontal axis. Instead, the eigenvector associated with the second eigenvalue (the negative one in this case) is $\mathbf{v}_2 = [0, 1]$, so the stable manifold of the saddle is along the vertical axis. Notice that both the coordinate axes are invariant lines. In fact, $x_1 = 0$ implies $\dot{x}_1 = 0$, so a trajectory starting from an initial condition on the vertical axis, i.e., $(0, x_2(0))$ with $x_2(0) > 0$, is trapped inside it and is governed by the one dimensional restriction $\dot{x}_2 = -dx_2$, hence exhibits an exponential decay (the decay of predator population in the absence of preys). The same holds for the horizontal axis $x_2 = 0$, and the dynamics on that trapping line is given by the logistic growth of preys' population in the absence of predators. As shown in Sect. 1.2.1.2, such dynamics converge to the equilibrium A. This is confirmed by the analysis of

$$\mathbf{J}(A) = \begin{bmatrix} -\alpha & -b\frac{\alpha}{s} \\ 0 & c\frac{\alpha}{s} - d \end{bmatrix},$$

which is a triangular matrix, hence the eigenvalues are given, again, by the diagonal entries, $\lambda_1 = -\alpha < 0$ and $\lambda_2 = c\alpha/s - d < 0$, negative for $\alpha c < sd$ and positive otherwise. Hence the equilibrium A is a stable node if $\alpha c < sd$, whereas it is a saddle if $\alpha c > sd$. Notice that the latter is also the condition for the positivity of the equilibrium E, thus confirming that stability of A is equivalent to the extinction of predators' population. When $\alpha c = sd$, equilibrium A is nonhyperbolic and merges with E, being in this case also $d/c = \alpha/s$ which implies $E = A$. This is typical example of a transcritical bifurcation. Notice that $\mathbf{v}_1 = [1, 0]$, hence the equilibrium A is always stable in the horizontal direction, whereas the eigenvector associated to λ_2, given by $\mathbf{v}_2 = [1, (sd - \alpha c - \alpha s)/(ab)]$, is transverse to the horizontal axis, and is tangent to the unstable manifold when A is a saddle.

The stability of the equilibrium of coexistence E is obtained through

$$\mathbf{J}(E) = \begin{bmatrix} -\frac{sd}{c} & \frac{-bd}{c} \\ \frac{\alpha c - sd}{b} & 0 \end{bmatrix},$$

from which $\text{Tr}(\mathbf{J}) = -sd/c \leq 0$ and $\text{Det}(\mathbf{J}) = d(\alpha c - sd)/c > 0$ whenever the equilibrium E has positive coordinates. If $s > 0$ then the equilibrium E is locally asymptotically stable whenever it is positive. Moreover, being $\text{Tr}(\mathbf{J})^2 - 4\text{Det}(\mathbf{J}) = d^2 s(s+4)/c^2 - 4d\alpha < 0$, if $\alpha > ds(s+4)/\left(4c^2\right)$ then E is a stable focus (see Fig. 1.20). A particular case occurs if $s = 0$, as $\text{Tr}(\mathbf{J}) = 0$ and consequently we have pure imaginary eigenvalues (i.e., eigenvalues with real part equal to zero). This implies that E is nonhyperbolic, hence the Hartman-Grobman Theorem cannot be applied. However, it can be shown that in this very particular case the trajectories are given by closed curves around E, which is marginally stable. This has been proved by Volterra in [34] and can be confirmed numerically as shown in Fig. 1.19.

We now propose another modification of the prey-predator model by introducing an effect of satiation of predators' appetite, i.e., we assume that predators cannot eat more than a given upper limit. This is expressed by the following model proposed in [30]

$$\begin{aligned} \dot{x}_1 &= \alpha x_1 - s x_1^2 - b\frac{x_1 x_2}{h + x_1}, \\ \dot{x}_2 &= -d x_2 + c\frac{x_1 x_2}{h + x_1}, \end{aligned} \tag{1.30}$$

where the function $g(x_1) = x_1/(h + x_1)$ is a typical "saturation function" with the following properties: $g(0) = 0$ (no eating without preys), $g'(x_1) > 0$, i.e., it increases with x_1 (more preys implies more food to eat) but saturates, i.e., $g(x_1) \to 1$ as $x_1 \to \infty$ (too many preys lead to appetite saturation). The constant h, called "half saturation constant", gives a measure of how fast is appetite satiation, as $g(h) = 1/2$. The dynamic behavior of this model is characterized by the presence of an invariant closed orbit on which trajectories move periodically, like in the classical Lotka-Volterra model. However, in this case the closed orbit is unique and attracts the trajectories around it. Such orbit is called limit cycle (Fig. 1.25).

1.2.2.3 Periodic Solutions and Limit Cycles

From the examples shown in the previous sections we have seen that with 2-dimensional dynamical systems in continuous time, differently from what happens for 1-dimensional systems, the invariant sets are not only given by equilibrium points. In fact, we can also have invariant closed orbits on which periodic trajectories exist, defined as solutions $\mathbf{x}(t) = \varphi(t)$ for which there is a $T > 0$ such that $\varphi(t + T) = \varphi(t)$ and for each $|t_1 - t_2| < T$ we have $\varphi(t_1) \neq \varphi(t_2)$. T is called *period* of the periodic trajectory. As usual for an invariant set, the question of stability arises (see Sect. 1.1):

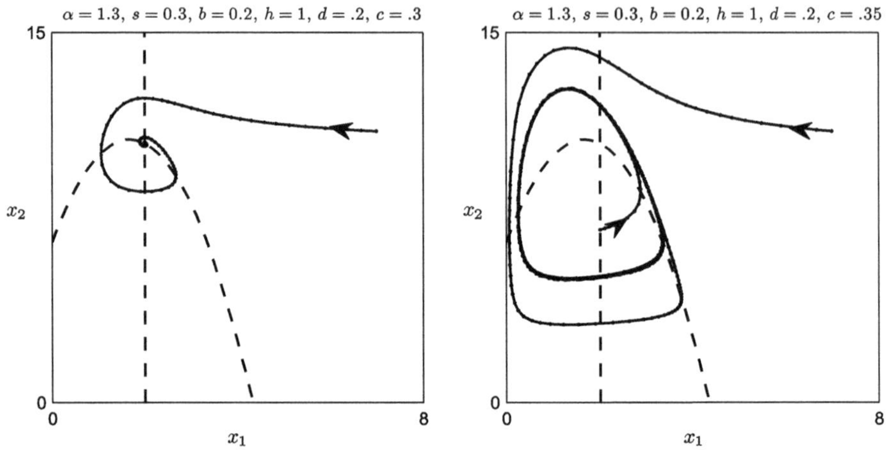

Fig. 1.25 Phase portrait for the prey-predator model with saturation of predators' appetite (1.30)

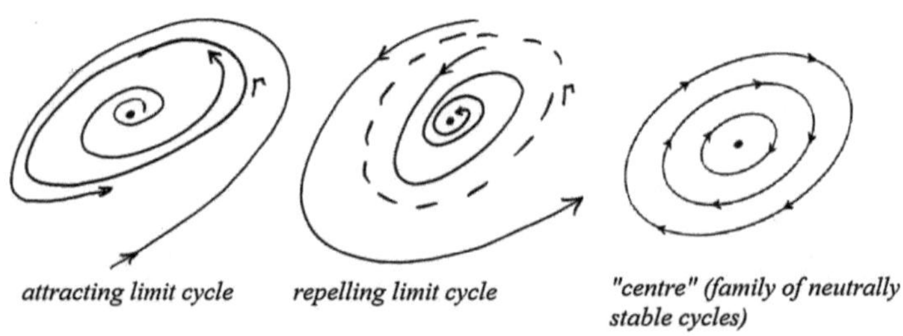

Fig. 1.26 Limit cycles

if a trajectory starts from an initial condition belonging to the invariant closed orbit Γ, then it remains trapped inside Γ by definition, but what about trajectories starting *around* it, i.e., from a neighborhood of Γ? Do they approach Γ asymptotically for $t \to \infty$? or, do they move away from it? or, do they remain distinct from Γ and close to it? These are the cases shown in Fig. 1.26, where a new kind of attractor or repellor existing in 2-dimensional dynamical systems is shown: the *limit cycle*. These kinds of solutions are very interesting in economic modeling, as they represent self-sustained cyclic behaviors, that every time go back to an already "visited" state, repeating the same path periodically.

Some general theorems and methods exist, for continuous time 2-dimensional dynamical systems, to detect the presence of limit cycles, as well as some results on bifurcations that create them, as we will see in the following.

First of all, in $\mathbb{R}^2$ the *Jordan curve lemma* states that any closed orbit Γ divides the plane into two connected and disjoint regions, one inside and one outside the closed curve, such that two points taken one in the inside region and one outside can only

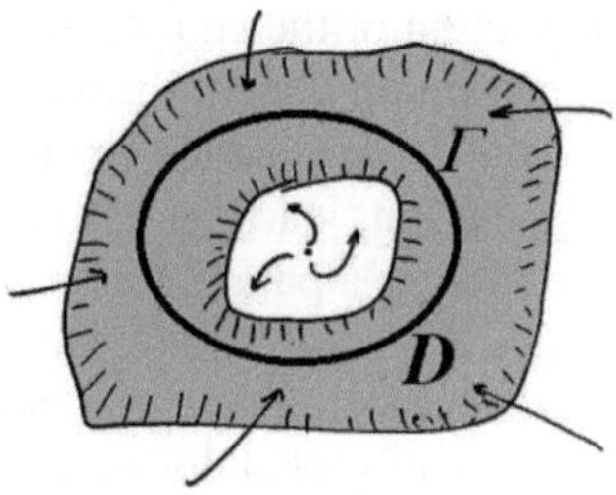

Fig. 1.27 Qualitative illustration of Poincaré-Bendixson Theorem

be connected by a trajectory crossing Γ. This implies that if a smooth[3] dynamical system of the plane has a closed invariant curve, then any trajectory starting from an initial condition inside it remains inside forever, and the same must hold for a trajectory starting outside. In other words, both regions are trapping. This is due to the fact that two trajectory cannot cross in an ordinary point due to the Theorem of uniqueness, hence a trajectory starting inside Γ cannot exit it because this cannot occur without crossing the orbit Γ. This lemma, which is quite intuitive for a system of the plane, is no longer true in more than two dimensions, as it is possible to connect any points without crossing the closed curve Γ if the third dimension is available. And the same holds with discrete time even in two dimensions as the trajectories in discrete time can jump from a point to another. So, the following Theorem, which is a consequence of the *Jordan curve lemma*, only holds for continuous-time two-dimensional dynamical systems.

Theorem 1.5 (Poincaré-Bendixson Theorem) *Let* $\dot{\mathbf{x}} = f(\mathbf{x})$ *be a set of two ordinary differential equations defined in an open set* $G \subseteq \mathbb{R}^2$*, and let* $D \subset G$ *be a compact (i.e., closed and bounded) trapping set that does not contain any equilibrium point. Then* D *must contain at least one closed invariant orbit of the dynamical system.*

Figure 1.27 illustrates the meaning of the theorem.

A corollary of this theorem states that if $K \subseteq G$ is a nonempty compact and trapping set then it must contain an equilibrium point or a closed invariant orbit. Moreover, if Γ is a closed orbit such that its interior region is entirely included into G (the set where the dynamical system is defined) then Γ must include at least one equilibrium point.

The Poincaré-Bendixson gives an existence result, that is it can be used to detect the presence of limit cycles, but gives no information about their stability or creation/destruction as a consequence of bifurcations as some parameters are varied.

1.2.2.4 Bifurcations of Two-Dimensional Dynamical Systems

In Sect. 1.2.2.1 we have seen that a topological classification of the unique equilibrium point of a 2-dimensional linear dynamical system is reduced to a simple

[3]By the term *smooth* we mean a $C^{(1)}$ dynamical system, i.e., expressed by equations of motion with continuous derivatives, so that the Theorem of existence and uniqueness apply.

inspection of the sign of the trace and the determinant of the matrix of coefficients. In particular, the equilibrium is asymptotically stable whenever the trace is negative and the determinant is positive. If the coefficients depend on some parameters it may happen that, starting from a configuration with a stable equilibrium, a continuous variation of a parameter leads to a change in sign of the trace or of the determinant, so that the equilibrium loses stability. In a linear system this implies that a dynamic scenario of global asymptotic convergence to the equilibrium is transformed into a situation of global divergence, i.e., any initial condition outside the equilibrium leads to an explosive trajectory going infinitely far from the equilibrium point. In other words, in a linear system the local behavior and the global behavior coincide.

According to the Hartman-Grobman Theorem, the topological classification of an (hyperbolic) equilibrium point of a nonlinear system can be obtained by the same procedure, provided that the matrix of coefficient of the linear approximation is obtained from the Jacobian matrix computed at the equilibrium considered. However, this equivalence is only local, i.e., it holds in a neighborhood (no information on the size) of the equilibrium point considered. So, in general nothing guarantees that such local classification can be extended globally, to the whole phase space. Moreover, a nonlinear system may have several equilibrium points (and even other invariant sets, such as the closed invariant orbits discussed in the previous section) so the global phase portrait may be quite complicated and cannot be, in general, deduced by a simple union of local phase portraits obtained around the hyperbolic equilibria. But the differences between linear and nonlinear models are not limited to these local/global considerations, as remarkable differences are related to the study of structural stability, i.e., what happens when, due to slight variations of some parameters, one or more equilibrium points change their stability properties, i.e., the trace and/or the determinant of a Jacobian matrix computed at an equilibrium point show a sign change. Indeed, in general such transitions of an equilibrium point from stable to unstable do not just imply a transition from stability to instability, but are associated with the creation/destruction of other equilibrium points around them or even to the creation/destruction of invariant closed orbits. The former occurrence will be described in terms of *fold*, *transcritical* or *pitchfork* bifurcations, as already seen in the case of one-dimensional dynamical systems, whereas the latter case will be described by a new kind of bifurcation, that has no one-dimensional analogue, denoted as *Andronov-Hopf bifurcation*. It will be characterized by the presence of complex conjugate eigenvalues crossing the imaginary axis, i.e., changing the sign of their real part, or equivalently situations with positive determinant and vanishing trace in the Jacobian matrix. All phenomena related to the presence of oscillatory behavior (due to focus, or spiral, equilibrium points) hence only occurring in dimension greater than one.

Therefore, while for linear systems a loss of stability leads to uninteresting dynamic scenarios, as loss of stability implies global divergence, in the case of nonlinear models the bifurcations leading to the loss of stability of an equilibrium may open new interesting dynamic scenarios, characterized by new equilibria and even new kinds of attracting sets. This means that the regions of the space of parameters characterized by instability of one or more equilibrium points may indicate the outcome of more interesting and even intriguing global phase portraits, characterized by coexistence of several attracting sets each with its own basin of attraction separated by basin boundaries on which unstable equilibria are located. In the following part of this section we recall in a more formal way some definitions and local bifurcations already described in an intuitive (mainly graphical) way in Sect. 1.2.1.4, and then we will introduce the Andronov-Hopf bifurcation.

As already intuitively stated, the notion of structural stability is strictly related to the definition of topological equivalence between two dynamical systems.

Definition 1.7 A dynamical system $\dot{\mathbf{x}} = f(\mathbf{x})$, $\mathbf{x} \in \mathbb{R}^n$, is topologically equivalent (or conjugate) to the dynamical system $\dot{\mathbf{y}} = g(\mathbf{y})$, $\mathbf{y} \in \mathbb{R}^n$, if a homeomorphism $\mathbf{h} : \mathbb{R}^n \to \mathbb{R}^n$, $\mathbf{y} = \mathbf{h}(\mathbf{x})$, exists that transforms the phase portrait (i.e., all the orbits) of the $\mathbf{x}$ of the former into the phase portrait in the $\mathbf{y}$ space of the latter, preserving the direction of time.

We recall that a homeomorphism is an invertible function $\mathbf{h}$ such that both $\mathbf{h}$ and $\mathbf{h}^{-1}$ are continuous.

Given a dynamical system that depends on a parameter

$$\dot{\mathbf{x}}(t) = \mathbf{f}(\mathbf{x}(t), \mu), \quad \mathbf{x}(t) \in \mathbb{R}^n, \quad \mu \in \mathbb{R},$$

let us consider its phase diagram. Of course it will depend on μ, in the sense that different values of the parameter μ will cause modifications (let's say deformations, distortions etc.) of the phase lines. Such variations of the global phase portrait may be only quantitative (displacements or continuous deformations that are topologically equivalent) or qualitative (an arbitrarily small variation of μ leads to a phase portrait which is not equivalent, due to a local stability change and/or to the creation/destruction of invariant sets, such as equilibrium points or closed orbits). This leads to the following definition

Definition 1.8 The transition between two non-equivalent phase diagrams due to the variation of a parameter is called *bifurcation*.

In other words, a bifurcation is a qualitative modification of the phase diagram of a dynamical system when a parameter crosses a critical (or threshold) value, called *bifurcation value*. It is worth noticing that the kinds of bifurcations can be classified according to a quite limited number of cases. Without entering the details of a more general topological view of this phenomenon, we just mention that the existence of a limited set of possible bifurcations is related to a general theory on structural stability of vector fields depending on parameters, known as singularity theory or theory of catastrophes, see, e.g., [33].

If a real eigenvalue, say $\lambda_1(\mu)$, changes its sign at the bifurcation value μ_0, i.e., it crosses through the origin of the complex plane moving along the real axis as the parameter μ is varied through μ_0, then along the invariant manifold associated to λ_1 we have one of the one-dimensional bifurcations already described for one-dimensional system, namely a fold (or tangent) bifurcation, also denoted as

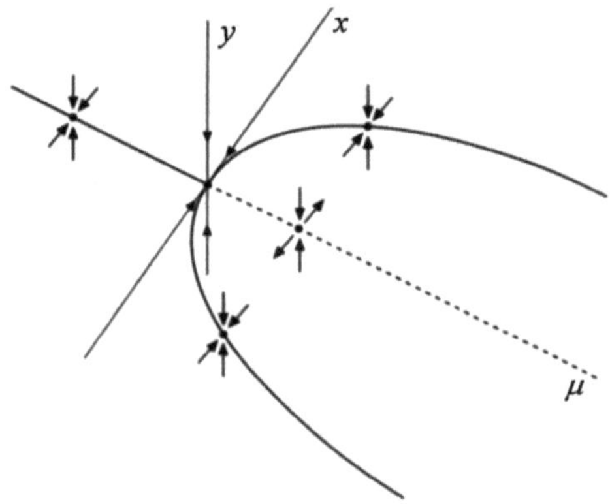

Fig. 1.28 Pitchfork bifurcation in a two-dimensional dynamical system with bifurcation parameter μ

saddle-node bifurcation in dimension greater than 1, or a transcritical bifurcation or a pitchfork bifurcation. This bifurcation only affects the qualitative dynamic behavior along the one-dimensional invariant manifold associated to λ_1. In this case, at each value of the parameter μ a planar phase portrait is associated around the bifurcating equilibrium, hence a three-dimensional bifurcation diagram is required to represent the bifurcation, with a coordinate axis on which the parameter μ is measured and the 2-dimensional phase plane where the corresponding invariant sets are graphically represented, see, e.g., Fig. 1.28, where the case of a supercritical pitchfork bifurcation is qualitatively shown.

In other words, these bifurcations are caused by a single real eigenvalue that changes the sign, associated to an eigenvalue that vanishes, i.e., a change of sign of the determinant of the Jacobian matrix, can be described in terms of the corresponding bifurcations of the one-dimensional restriction of the 2-dimensional dynamical system along the invariant manifold associated with the eigenvalue vanishing at the bifurcation value of the parameter.

In order to give a classification, in the following we generalize and make more precise the classification of such bifurcations. If we denote by

$$\dot{x} = f(x, \mu) \ , \quad x \in \mathbb{R} \ , \quad \mu \in \mathbb{R} \ ,$$

the one-dimensional restriction of the 2-dimensional dynamical system along the invariant manifold along which the bifurcation occurs, such that f is smooth and $x^*(\mu)$ is the equilibrium such that for $\mu = 0$ we have $x^*(0) = 0$ with associated eigenvalue $\lambda(0) = f_x(0, 0) = 0$, we have the following classification:

- If $f_{xx}(0, 0) \neq 0$ and $f_\mu(0, 0) \neq 0$ then the restriction is topologically equivalent to the normal forms:

$$\dot{y} = \mu \pm y^2 \ ,$$

  i.e., the normal forms of the fold bifurcation.

- If $\frac{\partial f(x,\mu)}{\partial x} = 0$ (vanishing eigenvalue), $\frac{\partial^3 f(x,\mu)}{\partial x^3} \neq 0$ and $\frac{\partial^2 f(x,\mu)}{\partial \mu \partial x} \neq 0$, then the restriction is topologically equivalent to the normal form

$$\dot{y} = \mu y - y^3 ,$$

  and according to the sign of $\frac{\partial^3 f(x,\mu)}{\partial x^3}$ and $\frac{\partial^2 f(x,\mu)}{\partial \mu \partial x}$ we have a supercritical or a subcritical pitchfork bifurcation.
- If $\frac{\partial f(x,\mu)}{\partial x} = 0$ (vanishing eigenvalue), $\frac{\partial^2 f(x,\mu)}{\partial x^2} \neq 0$, $\frac{\partial^2 f(x,\mu)}{\partial \mu \partial x} \neq 0$. then the restriction is topologically equivalent to the normal form

$$\dot{y} = \mu y - y^2,$$

  and we have a transcritical (or stability exchange) bifurcation.

It is worth stressing that in all these cases the equilibrium points involved are nodes and saddles that become nonhyperbolic at the bifurcation. Instead, in the case of spiral (or focus) equilibria that change stability due to a couple of complex conjugate eigenvalues that cross the imaginary axis, i.e., $\lambda_{1,2} = \pm i\omega_0$, hence, due to a positive determinant of the Jacobian matrix and a trace that changes its sign. The corresponding bifurcation is known as Hopf (or Andronov-Hopf) bifurcation (see, e.g., [12, 16]).

Theorem 1.6 (Andronov [4]; Hopf [20]) *Let us consider the 2-dimensional dynamical system*

$$\dot{\mathbf{x}} = \mathbf{f}(\mathbf{x}, \mu) , \quad \mathbf{x} \in \mathbb{R}^2 , \quad \mu \in \mathbb{R} ,$$

with **f** *formed by two smooth functions, and let $x^*(\mu)$ be an isolated equilibrium point, i.e.,* $\mathbf{f}(\mathbf{x}^*, \mu) = 0$. *Let us assume that the eigenvalues $\lambda_{1,2}(\mu) = \alpha(\mu) \pm i\omega(\mu)$ are complex for μ in a neighborhood of μ_0 and that for $\mu = \mu_0$ they are purely imaginary, i.e., the real part vanishes: $\alpha(\mu_0) = 0$, $\omega(\mu_0) = \omega_0 > 0$. If $\left.\frac{\partial \operatorname{Re} \lambda_{1,2}}{\partial \mu}\right|_{\mu=\mu_0} > 0$ (transversality condition) holds then x^* is a stable focus for $\mu < \mu_0$ and an unstable focus for $\mu > \mu_0$, and at $\mu = \mu_0$ a closed invariant orbit Γ is created around x^* such that one of the following holds:*

(i) Γ exists for $\mu > \mu_o$ and is a stable limit cycle (supercritical case);
(ii) Γ exists for $\mu < \mu_o$ and is an unstable limit cycle (subcritical case);
(iii) infinitely many closed invariant curves exist for $\mu = \mu_o$ which are neutrally stable (center case).

The period of the trajectories moving around is $T(\mu) = 2\pi/\omega_0 + o(|\mu - \mu_0|)$ and in cases (i) and (ii) the amplitude of Γ increases as the bifurcation parameter moves away from the bifurcation value proportionally to $\sqrt{|\mu - \mu_0|}$.

To sum up, this bifurcation is a device to create limit cycles (see Fig. 1.29).

In the supercritical case, when the equilibrium from stable focus is transformed into an unstable focus, a small stable limit cycle is created around it, which attracts

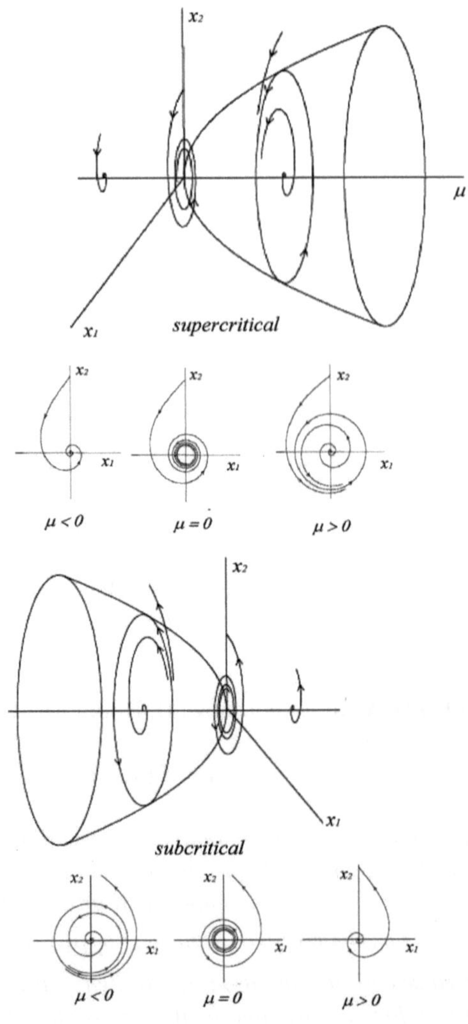

Fig. 1.29 Hopf bifurcation: supercritical (*upper panel*) and subcritical (*lower panel*) case

the trajectories starting inside the cycle, close to (but different from) the equilibrium, as well as those starting outside it. So, the loss of stability is denoted as "*soft*" in the sense that trajectories issuing from a neighborhood of the equilibrium remain close to it even if they oscillate around it without converging. Instead, in the subcritical case an unstable closed orbit surrounds the stable equilibrium and constitutes the boundary that delimits its basin of attraction. As the bifurcation parameter approaches its bifurcation value, the basin shrinks because the unstable orbit collapses to the equilibrium point, and then disappears. Hence after the bifurcation the orbits issuing from the unstable equilibrium are not confined and move towards another attracting set, that may be a different equilibrium or some other closed orbit of large amplitude

Fig. 1.30 Center case

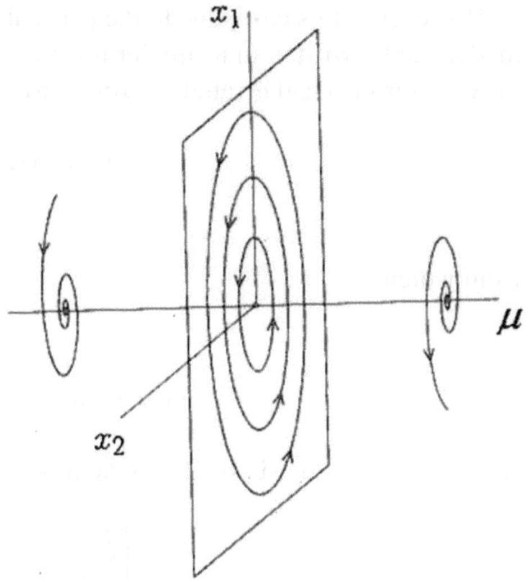

already existing towards infinity (i.e., diverging trajectories). This situation is also denoted as "*hard stability loss*".

It is worth noticing that in the case of a supercritical Hopf bifurcation, at the bifurcation value the nonhyperbolic equilibrium is stable, whereas in the case of subcritical bifurcation at the bifurcation value the nonhyperbolic equilibrium is unstable.

Let us also notice that the case (iii) is similar to what happens in a linear system when the trace of the matrix of coefficients changes its sign while the determinant is positive so that a pair of complex conjugate eigenvalues cross the imaginary axis, see the bifurcation diagram shown in Fig. 1.30.

As an example, let us consider the following "normal form"

$$\begin{cases} \dot{x_1} = \mu x_1 - x_2 - x_1\left(x_1^2 + x_2^2\right) , \\ \dot{x_2} = x_1 + \mu x_2 - x_2\left(x_1^2 + x_2^2\right) . \end{cases} \tag{1.31}$$

The unique equilibrium is $\mathbf{x}^* = (0, 0)$ where the Jacobian matrix is

$$\mathbf{J}\left(x^*\right) = \begin{bmatrix} \mu & -1 \\ 1 & \mu \end{bmatrix} ,$$

whose eigenvalues are $\lambda_{1,2} = \mu \pm i$, hence, it is immediate to see that for $\mu = 0$ an Andronov-Hopf bifurcation occurs as the two complex conjugate eigenvalues cross the imaginary axis at $\mu = 0$ going from left to right for increasing μ. Analytic methods to distinguish super/subcritical cases exist, based on higher order derivatives, however, we can try to see numerically if a stable limit cycle exists for $\mu > 0$ or an unstable one (bounding the basin of $\mathbf{x}^*$) exists for $\mu < 0$.

However in this case, due to the particular structure of the dynamical system, the model can be written in a simpler form by using polar coordinates r (distance from the equilibrium) and θ (angle of rotation):

$$x_1(t) = r(t)\cos\theta(t) \ ,$$
$$x_2(t) = r(t)\sin\theta(t) \ ,$$

from which

$$\dot{x}_1 = \dot{r}\cos(\theta) - r\sin(\theta)\,\dot{\theta} \ ,$$
$$\dot{x}_2 = \dot{r}\sin(\theta) + r\cos(\theta)\,\dot{\theta} \ .$$

Replacing $x_1, x_2, \dot{x}_1, \dot{x}_2$ in (1.31) the model becomes

$$\begin{cases} \dot{r} = r\left(\mu - r^2\right) \ , \\ \dot{\theta} = 1 \ . \end{cases}$$

The second equation indicates a constant rotation speed, the first the presence of an equilibrium $r = 0$, which is stable for $\mu \le 0$ (even if it is not hyperbolic at $\mu = 0$), and a further equilibrium $r = \sqrt{\mu}$ (r can only assume positive values), that coincides with $r = 0$ for $\mu = 0$ and departs from it for $\mu > 0$. This newborn equilibrium is stable and represents a limit cycle of radius $r(\mu) = \sqrt{\mu}$ around the unstable equilibrium $r = 0$ for $\mu > 0$. So, the bifurcation occurring at $\mu = 0$ represents a supercritical Andronov-Hopf bifurcation.

As an exercise it can be proved that the following model exhibits a subcritical Andronov-Hopf bifurcation at $\mu = 0$

$$\begin{cases} \dot{x_1} = \mu x_1 - x_2 + x_1\left(x_1^2 + x_2^2\right) \ , \\ \dot{x_2} = x_1 + \mu x_2 + x_2\left(x_1^2 + x_2^2\right) \ . \end{cases}$$

It can be noticed that at the bifurcation value $\mu = 0$ the equilibrium $(0, 0)$, corresponding to $r = 0$ in polar coordinates, is stable in the supercritical case and unstable in the subcritical case. If we consider only the linear part (identical in both the systems proposed) given by

$$\begin{cases} \dot{r} = \mu r \ , \\ \dot{\theta} = 1 \ , \end{cases}$$

we can notice that the equilibrium $r = 0$ is asymptotically stable for $\mu < 0$ and unstable for $\mu > 0$, but differently from the nonlinear case, at $\mu = 0$ it is a center with infinitely many limit cycles around it. In fact, for $\mu = 0$ we have $\dot{r} = 0$, hence any $r > 0$ is an equilibrium. However, all these closed invariant circles disappear for $\mu > 0$.

An Economic Example: The Kaldor Business Cycle Model (1940)
This is a two-dimensional nonlinear dynamical system to model the endogenous oscillations of an economic system, see [21]. Let $Y(t)$ be the national income (or output) and $K(t)$ the capital stock at time t. The model can be expressed as

$$\begin{cases} \dot{Y} = \alpha\,(I\,(Y,K) - S\,(Y,K))\ , \\ \dot{K} = I\,(Y,K) - \delta K\ , \end{cases} \tag{1.32}$$

where the rate of change $\dot{Y}$ of the output is proportional to the difference between investment $I\,(Y,K)$ and savings $S\,(Y,K)$, the positive proportionality constant α is the measure of the speed of reaction of the national income to such difference.

Kaldor assumes that investments $I\,(Y,K)$ are positively influenced by income Y, i.e., $\frac{\partial I}{\partial Y} := I_Y > 0$, and investments decrease if the capital stock increases, i.e., $\frac{\partial I}{\partial K} := I_K < 0$. The latter assumption is related to the fact that if the capital level is very high entrepreneurs are not motivated to invest to increase production. For sake of simplicity Kaldor assumes that saving S is an increasing function of Y with[4] $0 < S_Y < 1$, and also an increasing function of the capital stock, i.e., $S_K \geq 0$. The fact that the level of economic activities, measured by Y, increases proportionally to the demand excess $I\,(Y,K) - S\,(Y,K)$ is in agreement with the short-period dynamics assumed in Keynesian models.

Also the second dynamic equation is quite standard, as it states that the rate of growth of the capital stock K is given by the level of investments $I\,(Y,K)$ and reduced by a depreciation (or capital decay) rate δ.

Following these general assumptions, let us consider, for sake of simplicity, a linear saving function depending on Y only, and a nonlinear investment function with "saturation effects" for small and large values of Y as well:

$$\begin{aligned} S\,(Y) &= \sigma Y \quad \text{with} \quad 0 \leq \sigma \leq 1 \\ I\,(Y,K) &= \sigma\mu + \gamma\left(\frac{\sigma\mu}{\delta} - K\right) + \arctan\,(Y-\mu)\ , \end{aligned}$$

so, the following "Kaldorian" model is obtained:

$$\begin{cases} \dot{Y} = \alpha\left(\sigma\mu + \gamma\left(\frac{\sigma\mu}{\delta} - K\right) + \arctan\,(Y-\mu) - \sigma Y\right)\ , \\ \dot{K} = \sigma\mu + \gamma\left(\frac{\sigma\mu}{\delta} - K\right) + \arctan\,(Y-\mu) - \delta K\ . \end{cases}$$

From the equilibrium conditions $\dot{Y} = 0$ and $\dot{K} = 0$ we get

$$\begin{aligned} \sigma\mu + \gamma\left(\tfrac{\sigma\mu}{\delta} - K\right) + \arctan\,(Y-\mu) - \sigma Y = 0\ , \\ \sigma\mu + \gamma\left(\tfrac{\sigma\mu}{\delta} - K\right) - \delta K = -\arctan\,(Y-\mu)\ . \end{aligned}$$

[4]The condition $S_Y < 1$ states the principle of Keynesian multiplier, being S_Y the reciprocal of the Keynesian multiplier $1/(1-C_Y)$ where C_Y is the consumption propensity given by $C_Y = 1 - S_Y$.

hence

$$K = \tfrac{\sigma}{\delta} Y$$
$$\sigma \left(1 + \tfrac{\gamma}{\delta}\right)(Y - \mu) = \arctan (Y - \mu)$$

The point $P = (\mu, \sigma\mu/\delta)$ is always an equilibrium of the model (1.32), however, two further equilibrium points may be created as further intersections, symmetric with respect to $Y = \mu$, between the line $z = \sigma(1 + \gamma/\delta)(Y - \mu)$ and the sigmoid curve $z = \arctan (Y - \mu)$, as the slope of the line is varied. The Jacobian matrix

$$\mathbf{J}(Y, K) = \begin{bmatrix} \alpha\left(\frac{1}{1+(Y-\mu)^2} - \sigma\right) & -\alpha\gamma \\ \frac{1}{1+(Y-\mu)^2} & -\gamma - \delta \end{bmatrix}$$

at the equilibrium P becomes

$$\mathbf{J}(P) = \begin{bmatrix} \alpha(1-\sigma) & -\alpha\gamma \\ 1 & -\gamma - \delta \end{bmatrix},$$

hence,

$$\mathrm{Tr}(\mathbf{J}(P)) = \alpha(1-\sigma) - \gamma - \delta\ ,$$
$$\mathrm{Det}(\mathbf{J}(P)) = -\alpha(1-\sigma)(\gamma+\delta) + \alpha\gamma\ ,$$

and from the stability conditions $\mathrm{Tr}(\mathbf{J}(P)) < 0$, $\mathrm{Det}(\mathbf{J}(P)) > 0$ we obtain

$$\mathrm{Tr}(\mathbf{J}(P)) < 0 \Rightarrow \alpha < \frac{\gamma+\delta}{(1-\sigma)} \quad \text{or} \quad \sigma > \frac{\alpha - \gamma - \delta}{\alpha}\ ,$$
$$\mathrm{Det}(\mathbf{J}(P)) > 0 \Rightarrow \sigma > \frac{\delta}{\gamma+\delta}\ .$$

These two stability conditions define a region of stability in the space of the parameters. For example, if we consider the parameters' plane (α, σ) the stability region is bounded by the curve (branch of an equilateral hyperbola) $\sigma = \sigma_h = (\alpha - (\gamma + \delta))/\alpha$ that represents a Hopf bifurcation curve, and the horizontal line $\sigma = \sigma_p = \delta/(\gamma + \delta)$ that represents a pitchfork bifurcation curve. In Fig. 1.31 the stability region in the parameters' plane (α, σ) is represented by the gray-shaded region. If, starting from the stability region with $\alpha > \delta + \gamma$, the propensity to save is decreased below the Hopf bifurcation value $\sigma_h = \delta/(\gamma + \delta)$, then a supercritical Hopf bifurcation occurs after which a stable limit cycle is created, on which periodic oscillations occur. The same occurs for $\sigma > \delta/(\gamma + \delta)$ and speed of adjustment α increasing beyond the bifurcation value $\alpha_h = (\delta + \gamma)/(1 - \sigma)$. In these cases the model is suitable to describe endogenously generated oscillations. However, if starting from a set of parameters inside the stability region the propensity to save σ is decreased below σ_p, then a pitchfork bifurcation occurs at which two stable nodes are created, one below and one above the central equilibrium, which becomes a saddle

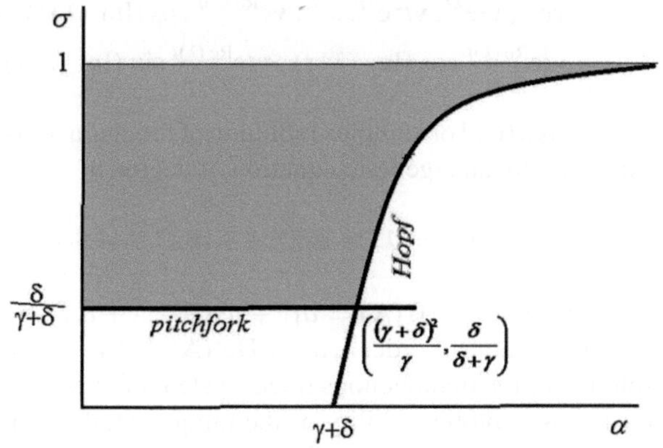

Fig. 1.31 Stability region of the equilibrium point P for the Kaldor model (1.32)

at the bifurcation. After the bifurcation, bistability is observed with two equilibrium points characterized by a smaller and larger value of national income Y, each with its own basin of attraction separated by the stable set of the saddle: a poverty trap and a richness trap. These two different situations may be both present in the lower-right region of the plane, i.e., with sufficiently large values of α and small values of σ, with dynamic scenarios given by three equilibria (two stable spirals with a saddle in the middle whose spiraling stable set separates the basins) surrounded by a large stable limit cycle.

So, this version of the Kaldor model exhibits many different dynamic scenarios, some expected on the basis of the local stability analysis, but other situations can only be revealed through a global numerical explorations with different sets of parameters.

1.2.3 *Multi-dimensional Dynamical Systems in Continuous Time*

Many of the results about linear systems, linearization of nonlinear ones around equilibrium points, their stability and related bifurcations can be extended to n-dimensional dynamical systems with $n > 2$, i.e., with more than two dynamic variables. However, as we will see, some important differences are worth being emphasized, first of all, the possibility of chaotic trajectories and chaotic attractors for $n \geq 3$.

1.2.3.1 Linear Systems

For a linear system $\dot{\mathbf{x}} = \mathbf{A}\mathbf{x}$, $\mathbf{x} \in \mathbb{R}^n$, with $\mathbf{A}$ $n \times n$ matrix of constant coefficients, we again have solutions given by linear combinations of functions like:

$$\mathbf{v}e^{\lambda t}, \mathbf{v}te^{\lambda t}, \mathbf{v}t^2e^{\lambda t}, \ldots, \mathbf{v}e^{\mathrm{Re}\,(\lambda)t}\cos(\mathrm{Im}\ (\lambda)\,t), \mathbf{v}e^{\mathrm{Re}\,(\lambda)t}\sin(\mathrm{Im}\ (\lambda)\,t),$$
$$\mathbf{v}te^{\mathrm{Re}\,(\lambda)t}\cos(\mathrm{Im}\ (\lambda)\,t), \mathbf{v}te^{\mathrm{Re}\,(\lambda)t}\sin(\mathrm{Im}\ (\lambda)\,t), \ldots$$

where λ is a (real or complex) solution of the characteristic equation Det $(\mathbf{A}-\lambda\mathbf{I}) = 0$, expressed by an algebraic equation of degree n

$$P(\lambda) = \lambda^n + a_1\lambda^{n-1} + a_2\lambda^{n-2} + \cdots + a_{n-1}\lambda + a_n = 0\ , \tag{1.33}$$

where again $a_1 = \mathrm{Tr}(\mathbf{A}) = a_{11} + a_{22} + \cdots + a_{nn}$, a_k, $k = 2, \ldots, n-1$, given by a sum of n minors of order k, $a_n = \mathrm{Det}(\mathbf{A})$. Let $\mathbf{v} \in \mathbb{R}^n$ be a corresponding eigenvector, solution of the homogenous linear system of order n: $(\mathbf{A} - \lambda\mathbf{I})\,\mathbf{v} = \mathbf{0}$. The condition for the asymptotic stability of the unique equilibrium $\mathbf{0}$ is that all the eigenvalues[5] have negative real part, i.e., Re $(\lambda) < 0$ for each eigenvalue. This can be equivalently stated as Re $(\lambda_1) < 0$ where λ_1 is the dominant eigenvalue, defined as the one with maximum real part in the set of all eigenvalues (i.e., the rightmost one in the complex plane). Of course, we can have a single real dominant eigenvalue or a couple of complex conjugate dominant eigenvalues.

If the dominant eigenvalue λ_1 is real and negative, then the long run dynamics towards the equilibrium is monotonic as all the possible oscillatory modes associated with complex eigenvalues vanish in the long run faster than the solution associated to λ_1. The return time is estimated as $T_r = -1/\lambda_1$.

If the dominant eigenvalue is a couple of complex conjugate ones, hence with the same real part Re (λ_1), then an oscillatory convergence is observed in the long run with *characteristic return time* $T_r = -1/\,\mathrm{Re}\,(\lambda_1)$ and *rotation period* $T_{rot} = 2\pi\,/\,\mathrm{Im}\,(\lambda_1)$.

A necessary condition for all the eigenvalues (i.e., the solutions of the characteristic equation (1.33)) to have negative real parts is $a_k > 0$ for $k = 1, \ldots, n$.

A necessary and sufficient condition for the same property is formulated through the *Routh-Hurwitz* criterion, which is expressed in terms of the coefficients a_k, $k = 1, \ldots, n$, of (1.33) as follows.

Theorem 1.7 (Routh-Hurwitz criterion) *Let us consider the matrix formed by coefficient of (1.33) arranged in the following matrix*

$$\begin{bmatrix} a_1 & 1 & 0 & 0 & 0 & \ldots & 0 \\ a_3 & a_2 & a_1 & 1 & 0 & \ldots & 0 \\ a_5 & a_4 & a_3 & a_2 & a_1 & \ldots & 0 \\ \ldots & \ldots & & & & & \\ 0 & 0 & 0 & 0 & 0 & \ldots & a_n \end{bmatrix} \tag{1.34}$$

[5]According to the Fundamental Theorem of Algebra a polynomial $P(\lambda)$ of degree n always has n complex solutions (counted with proper multiplicity in the case of coincident ones). Moreover, if the coefficients a_k of the polynomial are real numbers, like in our case, for each complex root with $\mathrm{Im}(\lambda) \neq 0$ the complex conjugate is a root as well.

Then all the solutions of (1.33) have negative real parts if and only if the leading principal minors of the matrix (1.34) are positive.

For example, for $n = 2$ we have

$$a_1 > 0 \quad \text{and} \quad \text{Det}\begin{bmatrix} a_1 & 1 \\ 0 & a_2 \end{bmatrix} > 0\,,$$

i.e., $a_1a_2 > 0$, equivalent to the already stated stability conditions $a_1 > 0$ and $a_2 > 0$, i.e., $\text{Tr}(\mathbf{A}) < 0$ and $\text{Det}(\mathbf{A}) > 0$.

For $n = 3$ the criterion gives

$$a_1 > 0, \quad \text{Det}\begin{bmatrix} a_1 & 1 \\ a_3 & a_2 \end{bmatrix} > 0 \quad \text{and} \quad \text{Det}\begin{bmatrix} a_1 & 1 & 0 \\ a_3 & a_2 & a_1 \\ 0 & 0 & a_3 \end{bmatrix} > 0\,,$$

equivalent to $a_1 > 0$, $a_3 > 0$ and $a_1a_2 > a_3$.

For $n = 4$,

$$a_1 > 0\,, \quad \text{Det}\begin{bmatrix} a_1 & 1 \\ a_3 & a_2 \end{bmatrix} > 0\,, \quad \text{Det}\begin{bmatrix} a_1 & 1 & 0 \\ a_3 & a_2 & a_1 \\ 0 & a_4 & a_3 \end{bmatrix} > 0 \quad \text{and} \quad \text{Det}\begin{bmatrix} a_1 & 1 & 0 & 0 \\ a_3 & a_2 & a_1 & 1 \\ 0 & a_4 & a_3 & a_2 \\ 0 & 0 & 0 & a_4 \end{bmatrix} > 0\,,$$

equivalent to $a_1 > 0$, $a_2 > 0$, $a_3 > 0$, $a_4 > 0$ and $a_1a_2a_3 - a_3^2 - a_4a_1^2 > 0$.

And so on for higher n.

As an example, let us consider the fourth degree equation $z^4 + 5z^3 + 13z^2 + 9z + 10 = 0$. We are not able to solve it, however as the coefficients are all positive and $5 \cdot 13 \cdot 9 - 9^2 - 10 \cdot 5^2 = 254 > 0$ we can deduce that all its 4 roots have negative real part (they may be all real, or two real and two complex conjugate, or two pairs of complex conjugate).

So, again, the problem of stability of the equilibrium of a linear dynamical system is reduced to a set of algebraic conditions, even if these conditions are, of course, more an more complicated as the dimension of the dynamical system increases. If the coefficients of the dynamical system, and consequently of the characteristic equation, depend on one or more parameters, the stability conditions can be used to detect changes of stability as the parameters are varied. In the case of linear systems a transition from stability to instability means a transition from global asymptotic stability to divergent trajectories.

Another interesting result about localization of eigenvalues in the complex plane is expressed by the following.

Theorem 1.8 (Gerschgorin Circle Theorem) *Let* $A = [a_{ij}]$ *be a square matrix with complex entries* $a_{ij} \in \mathbb{C}$. *Let*

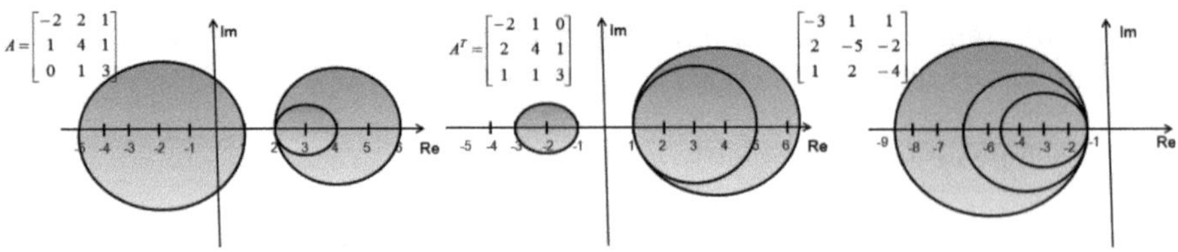

Fig. 1.32 Two applications of Gershgorin Circle Theorem. The matrix A and its transpose on the *left*, a matrix with negative diagonal dominance on the *right*

$$D_k = \left\{ z \in \mathbb{C} \text{ such that } |z - a_{kk}| \leq \sum_{j=1, j \neq k}^{n} |a_{kj}| \right\}, \; k = 1, \ldots, n,$$

be the set of n disks with center in the kth diagonal entry and radius given by the sum of the absolute values of the non-diagonal entries of the same row. Then all the eigenvalues of A must be contained in the union of the n disks.

Corollary 1.1 *As the eigenvalues of a matrix and its transpose are the same, the disks may be defined with reference to columns*

$$D'_k = \left\{ z \in \mathbb{C} \text{ such that } |z - a_{kk}| \leq \sum_{j=1, j \neq k}^{n} |a_{jk}| \right\}, \; k = 1, \ldots, n,$$

hence, the region of the complex plane allowed to eigenvalues is given by the intersection of the two unions, i.e.,

$$\left(\bigcup_{k=1}^{n} D_k \right) \cap \left(\bigcup_{k=1}^{n} D'_k \right).$$

Figure 1.32 gives an example of the application of the Gerschgorin Theorem.

The Gerschgorin Theorem provides useful application to the study of stability in the case of negative diagonal dominant matrices.

Definition 1.9 A matrix is *diagonal dominant* if for each row (or each column) the following inequality holds

$$|a_{kk}| > \sum_{j=1, j \neq k}^{n} |a_{kj}| \quad \left(|a_{kk}| > \sum_{j=1, j \neq k}^{n} |a_{jk}| \right)$$

Moreover, if $a_{kk} < 0$ for each k then the matrix is called *negative diagonal dominant*.

An immediate corollary of the Gerschgorin Theorem is the following.

Corollary 1.2 *If a matrix is negative diagonal dominant then all its eigenvalues have negative real part.*

From the point of view of dynamical systems, this stability statement can be expressed by saying that if self-control (i.e., the inhibitory effect that a dynamic variable exerts on itself) is stronger than joint influence of all other variables, then the system is stable. As an example, let us consider the following linear dynamical system

$$\begin{aligned} \dot{x}_1 &= -3x_1 + x_2 + x_3 \ , \\ \dot{x}_2 &= 2x_1 - 5x_2 + 2x_3 \ , \\ \dot{x}_3 &= x_1 + 2x_2 - 4x_3 \ , \end{aligned}$$

whose Gerschgorin disks are shown in right panel of Fig. 1.32.

1.2.3.2 Nonlinear Systems

Let us consider an n-dimensional dynamical system in the form

$$\dot{\mathbf{x}} = \mathbf{f}(\mathbf{x};\mu), \mathbf{x} \in \mathbb{R}^n, \quad \mu \in \mathbb{R} \ , \tag{1.35}$$

and let $\mathbf{x}^*(\mu)$ be an equilibrium point, implicitly defined as a solution of the nonlinear system $\mathbf{f}(\mathbf{x};\mu) = 0$ of n equations with n unknowns. In order to study the local stability and to have an idea of the kind of local phase portrait in a neighborhood of each equilibrium point, the linear approximation $\dot{\mathbf{X}} = \mathbf{J}(\mathbf{x}^*(\mu))\mathbf{X}$ can be considered, where $\mathbf{J}(\mathbf{x}^*(\mu)) = \left[\frac{\partial f_i}{\partial x_j}|_{\mathbf{x}^*}\right]$ is the $n \times n$ Jacobian matrix computed at the equilibrium point considered, and $\mathbf{X} = \mathbf{x} - \mathbf{x}^*$ is the displacement from the equilibrium. If the equilibrium point is hyperbolic, i.e., all the eigenvalues of $\mathbf{J}(\mathbf{x}^*)$ have nonvanishing real part, then the study of the local stability of the equilibrium is reduced to the study of the stability of the linear approximation, and even the local qualitative behavior of the dynamical system can be deduced from the study of the linear approximation, according to the Hartman-Grobman Theorem. In particular, we have the result that if all the eigenvalues have negative real part, then the equilibrium is locally asymptotically stable. Moreover, the dominant eigenvalue (or dominant couple, in the case of complex conjugate dominant eigenvalues) provides information about the kind of equilibrium we are dealing with and the speed of convergence to the equilibrium.

In analogy with what we have seen for the two-dimensional nonlinear dynamical systems, if the dominant eigenvalue (or couple of dominant eigenvalues) moves from negative to positive real part (i.e., cross the imaginary axis) as some parameter is varied, i.e., if some of the Routh-Hurwitz conditions change sign, then a bifurcation occurs at which the equilibrium considered becomes unstable. This is generally associated with some other change in phase portrait, such as creation/destruction of equilibrium points or closed invariant curves, or merging of equilibria with stability

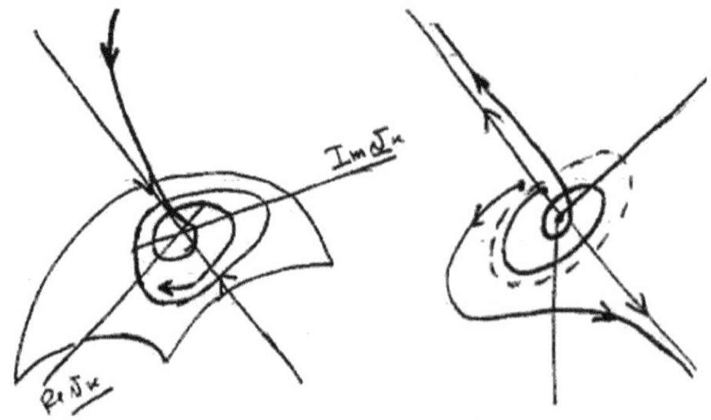

Fig. 1.33 Qualitative representations of saddle-focus equilibrium points in $\mathbb{R}^3$

exchange. In particular, if the eigenvalue crossing the imaginary axis is real, then we have the usual one-dimensional bifurcations along the invariant direction tangent to the corresponding eigenvector (i.e., fold, or saddle-node, or pitchfork or transcritical), whereas if a couple of complex conjugate eigenvalues crosses the imaginary axis with imaginary part different from zero, then an Andronov-Hopf bifurcation occurs leading to the creation of a closed invariant curve in the plane of the two independent real eigenvectors associated (also called center manifold). These bifurcations lead to scenarios similar to the ones already seen for two-dimensional systems, but with a richer variety related to the presence of other dimensions, see, e.g., the qualitative sketches in three dimensions shown in Fig. 1.33.

As it can be noticed, the Jordan curve lemma no longer holds, as trajectories can jump from inside to outside a closed invariant curve in the center manifold by moving outside their plane. This allows the formation of more complicated attractors that cannot exist in two dimensions, which are sometimes called "strange attractors" along which aperiodic motions can be observed with some features that opened a remarkable field of studies under the name of "*deterministic chaos*", an apparent oxymoron. In fact, the two words "deterministic" and "chaos" express two quite counterposed meanings. Deterministic means without uncertainty, predictable, regular, where any cause implies clear effects or consequences. Chaos is generally referred to confusing, unpredictable, irregular systems, where consequences of a given cause are not clear. Indeed, what we are considering is deterministic, because, given an initial condition and the knowledge of the dynamic equations, a unique time evolution (i.e., a trajectory) of the dynamical system is obtained. This allows one to compute the future state of the system for any time without any uncertainty, as it was expresses by the French mathematician Pierre Simon Laplace in 1776in the following famous statement:

> We may regard the present state of the universe as the effect of its past and the cause of its future. An intellect which at a certain moment would know all forces that set nature in motion, and all positions of all items of which nature is composed, if this intellect were also vast enough to submit these data to analysis, it would embrace in a single formula the movements of the greatest bodies of the universe and those of the tiniest atom; for such an intellect nothing would be uncertain and the future just like the past would be present before its eyes (see [23]).

This statement, that was mainly motivated by the usage of dynamical systems to describe the motion of rigid bodies (included astronomical motions) is now known as the *Laplacian determinism*, and the intellect which is assumed to know the equations of motion of the Universe and the its exact state at a given time is sometimes called *Laplace's demon.*

The concept of deterministic chaos[6] in the theory of dynamical systems was first glimpsed by Henri Poincaré during his attempt to find the trajectories of a three-body system in the presence of the gravitational force. In this work Poincaré started the study of dynamical systems by using topological methods or qualitative theory. Even without the possibility to visualize numerical computations of the trajectories, Poincaré described the extreme irregularity of time paths obtained, and the intricacy of highly intermingled trajectories ("I can imagine them in my mind but I cannot describe how complicated they are"). His description of the phenomenon of sensitive dependence on initial conditions is one of the most famous pages of mathematical literature (from [28]):

> If we knew exactly the laws of nature and the situation of the universe at the initial moment, we could predict exactly the situation of that same universe at a succeeding moment. But even if it were the case that the natural laws had no longer any secret for us, we could still only know the initial situation approximately. If that enabled us to predict the succeeding situation with the same approximation, that is all we require, and we should say that the phenomenon had been predicted, that it is governed by laws. But it is not always so; it may happen that small differences in the initial conditions produce very great ones in the final phenomena. A small error in the former will produce an enormous error in the latter. Prediction becomes impossible, and we have an apparently fortuitous phenomenon.

In this sense systems that are deterministic exhibit a behavior so irregular that appear to be similar to chaotic motions, governed by stochastic influences. The discovery of such irregularities, and related difficulties to make reliable predictions in some nonlinear dynamical systems, initially had not a strong impact. The question was stressed, and became quite popular and pervasive in the Sixties of 20th century, after the work of the American mathematician and meteorologist Edward Lorenz, who noticed this difficulties in making predictions in some dynamic models used in weather forecasting. The dynamic equations used by Lorenz are quite simple, even if they are not linear. They can be expressed by the following three-dimensional dynamical system

$$\begin{cases} \dot{x}_1 = \sigma\,(x_2 - x_1)\ , \\ \dot{x}_2 = \rho x_1 - x_2 - x_1 x_3\ , \\ \dot{x}_3 = x_1 x_2 - \beta x_3\ , \end{cases} \tag{1.36}$$

where the dynamic variables $x_i(t)$, $i = 1, 2, 3$, as well as the parameters, represent quantities used to describe weather conditions. For a given set of parameters, namely $\sigma = 10$, $\beta = 2.666$, $\rho = 20$, and initial conditions $x_1(0) = 10$, $x_2(0) = 10$, $x_3(0) = 10$, Fig. 1.34 shows $x_1(t)$ obtained by a numerical simulation of (1.36).

[6]However the term "chaos" in this context was first introduced in the paper "Period three implies chaos" by Li and Yorke, see [24].

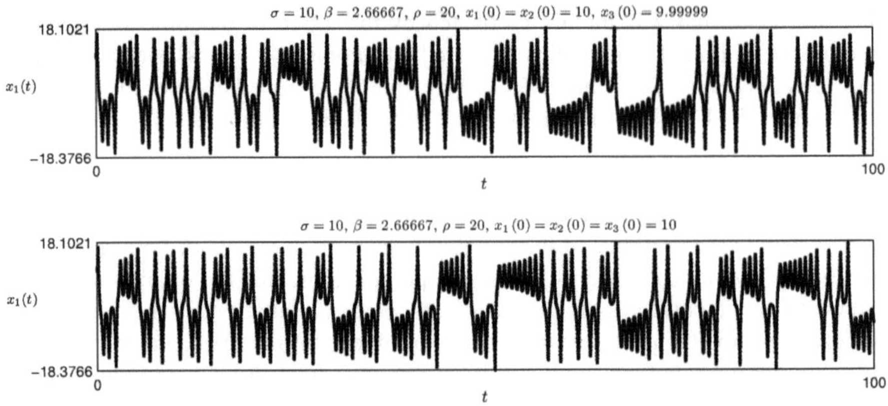

Fig. 1.34 Versus time representation of $x_1(t)$ along a trajectory of the Lorenz model (1.36) obtained with parameters $\sigma = 10, \beta = 2.666, \rho = 28$ and initial conditions $x_1(0) = 10, x_2(0) = 10, x_3(0) = 10$ (*upper panel*); $x_1(0) = 10, x_2(0) = 10, x_3(0) = 9.99999$ (*lower panel*)

It is quite evident how irregular the trajectory is. However, the most striking phenomenon lies in the fact that a modification of the initial condition $x_3(0)$ of a very negligible quantity, e.g., subtracting 10^{-6} so that we start the numerical simulation from $x_3(0) = 9.99999$ instead of $x_3(0) = 10$, a time series for $x_3(0)$ becomes very different from the previous one as time goes on (even if at the early time steps they were quite similar). This phenomenon of sensitivity to initial conditions, already described by Poincaré in 1903, became widely known after the paper by E. Lorenz, see [25], and is now popularly known as the "*butterfly effect*", so called because of the title of a paper given by Edward Lorenz in [26] to the American Association for the Advancement of Science in Washington, D.C., entitled "*Predictability: Does the Flap of a Butterfly's Wings in Brazil set off a Tornado in Texas*?" The flapping wing represents a small change in the initial condition of the system, which causes a chain of events leading to large-scale phenomena. Had the butterfly not flapped its wings, the trajectory of the system might have been vastly different.

The phenomenon of sensitive dependence on initial conditions (or butterfly effect) evidenced how difficult may be to make forecasting even if a dynamic phenomenon is represented by deterministic equations (when they are even slightly nonlinear). Small differences in initial conditions (such as those due to rounding errors in numerical computation) yield widely diverging outcomes for such dynamical systems, rendering long-term prediction impossible in general. This happens even though these systems are deterministic, meaning that their future behavior is fully determined by their initial conditions, with no random elements involved. This was summarized by Edward Lorenz by the sentence: "When the present determines the future, but the approximate present does not approximately determine the future".

So, the statement "if one knows the equations of motion, then one can reliably forecast the future states of a system starting from the knowledge of its state at a given time", is not true in general. This had a strong impact in economics as well. In fact, the paradigm of the rational agent in economics, on which is based the mainstream economic theory after the Sixties of the 20th century, is based on the assumption that economic agents have correct expectations about future states of the economy because they know the equations of motion of the economic systems. The existence of deterministic chaos in economic models based on rational expectations leads to an evident contradiction.

Of course, the same nonlinear model can behave regularly (converging to an equilibrium or to a periodic orbit) for some sets of parameters and exhibit chaotic dynamics for different parameters' values, and a goal of the qualitative study of a continuous time nonlinear dynamical system of dimension greater than two is the detection of the parameters' changes leading to such irregular behavior. The discovery (or, better, the re-discovery after the clear statement of Poincaré in 1903) of this kind of trajectories in deterministic models opened in the Sixties and Seventies of 20th century a huge stream of literature in the field of the theory of dynamical systems, and this caused a sort of revolution in several disciplines, including physics, chemistry, sociology, engineering, economics, biology. The so called "chaos theory" even entered fiction, cinema and philosophical debates, see, e.g., the popularization book [13].

However, even in the presence of chaotic behavior some regularities can be detected. For example, if the trajectories are represented in the phase space one can see that the shape of the attracting set where the chaotic trajectories are confined may be characterized by interesting topological properties. For example, if a chaotic trajectory of the Lorenz model is represented in the phase space (x_1, x_2, x_3) a structure like the one shown in Fig. 1.35 is obtained. If a trajectory starts from an initial condition inside that set, then it remains there and covers any point of it as time goes on, i.e., it is an invariant set. Moreover, if a trajectory starts outside it (not too far), then it moves towards the set, where it exhibits irregular (i.e., non-periodic) time paths and sensitive dependence on initial conditions. For this reason such invariant set is denoted as "chaotic attractor" or "strange attractor". The shape and extension of this attracting compact set may give useful information about the long-run dynamics of the dynamical system, even if it exhibits deterministic chaos.

Fig. 1.35 Lorenz attractor

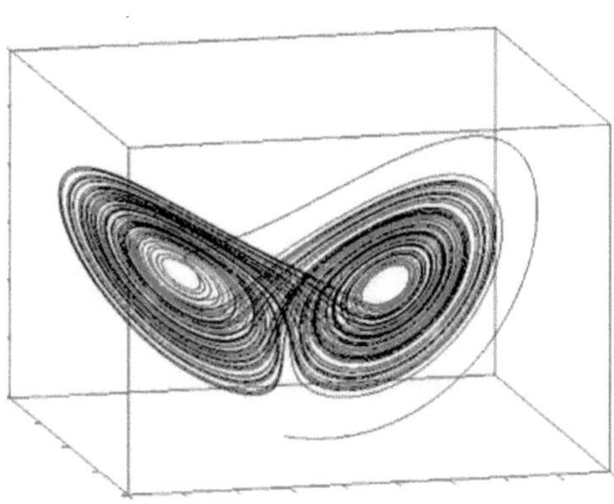

In fact, one can obtain upper and lower bounds (ceiling and floor) for the dynamics of each dynamic variable, even if its time series is quite irregular.

So, even if from one side the discovery of deterministic chaos weakens the predictive capacity of nonlinear dynamical systems, it gives some hope that apparently random phenomena may be generated by a deterministic model (even with a few dynamic variables and with a simple mathematical expression).

We do not enter into more details about deterministic chaos, and in particular we avoid to give here a more rigorous definition of it, because we prefer to postpone such a discussion when dealing with discrete-time dynamical systems, for which deterministic chaos can even be obtained with one-dimensional dynamic models and with very mild nonlinearities.

1.3 Discrete-Time Dynamical Systems

Dynamical systems (1.5) with discrete time $t \in \mathbb{N}$ naturally arise in economic and social modeling, where changes in the state of a system occur as a consequence of decisions that cannot be continuously revised (event-driven time). Given a characteristic time interval Δt, taken as a unit of time advancement $\Delta t = 1$, if $\mathbf{x}(t) \in \mathbb{R}^n$ represents the state of the system at a given time t, then the state at the next time $t + 1$ is obtained by the application of a map, i.e., a transformation or a function $T : M \to M$ defined in the phase space $M \subseteq \mathbb{R}^n$ into itself

$$\mathbf{x}(t+1) = \mathbf{T}(\mathbf{x}(t)). \tag{1.37}$$

So, a single application of the transformation T represents a "unit time advancement" of the state of the dynamical system and the repeated application (or *iteration*) inductively defines a trajectory (Fig. 1.36).

In other terms, a trajectory is obtained by the composition of a map with itself

$$\mathbf{x}(1) = \mathbf{T}(\mathbf{x}(0)); \mathbf{x}(2) = \mathbf{T}(\mathbf{x}(1)) = \mathbf{T}(\mathbf{T}(\mathbf{x}(0)) = \mathbf{T}^2(\mathbf{x}(0); \ \dots \ ; \mathbf{x}(n) = T^n(\mathbf{x}(0))$$

or, more briefly, it can be written as the sequence

$$\tau(\mathbf{x}(0)) = \{\mathbf{x}(t) \in M : \mathbf{x}(t) = \mathbf{T}^n(\mathbf{x}(0)), \ \dots t \in \mathbb{N}\}.$$

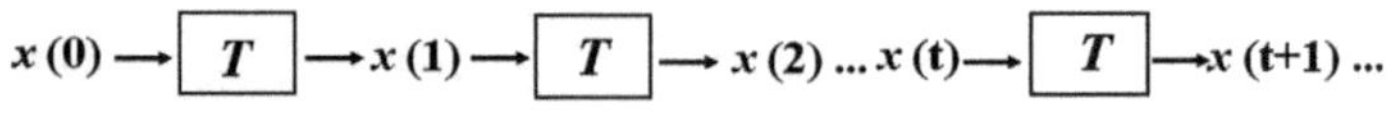

Fig. 1.36 Schematic representation of a discrete trajectory

Discrete-time dynamical systems can be obtained through a discretization of continuous-time dynamical system by replacing time derivative with the corresponding incremental ratio, that is, from $\dot{x}_i = f_i(\mathbf{x})$ with $\dot{x}_i = \frac{dx_i}{dt} \approx \frac{x_i(t+\Delta t)-x_i(t)}{\Delta t}$ and $\Delta t = 1$ we get

$$x_i(t+1) = x_i(t) + f_i(\mathbf{x}(t)) = T_i(\mathbf{x}(t))$$

However many economic dynamic models are directly obtained under a discrete time framework. For example let us consider the well known *Cobweb Model*. A given good is sold in the market at a unit price $p(t)$. The quantity demanded by consumers is a function of the price $Q^d(t) = D(p(t))$ denoted as demand function, usually a continuous and decreasing function (hence invertible). The supply function expresses the output decided by producers as a function of the price $Q^s(t) = S(p^e(t))$, where $p^e(t)$ represents the price expected by producers at time t on the basis of the information they have when deciding about the quantity to be produced. Let $\Delta t = 1$ be the amount of time necessary to realize the production process (i.e., the production lag from production decision to product realization, e.g., maturation period for agricultural products or production time for an industrial process). Then the economic equilibrium condition $Q^d(t) = Q^s(t)$ becomes

$$D(p(t)) = S(p^e(t)). \tag{1.38}$$

Under the assumption of naïve expectations $p^e(t) = p(t-1)$, i.e., without reliable information the producers expect that the price at the end of production will be the same prevailing at the beginning, the model becomes

$$D(p(t)) = S(p(t-1)) \ ,$$

and by applying the inverse of demand function $p = D^{-1}(q)$ and after a simple time translation it assumes the standard explicit form

$$p(t+1) = D^{-1}(S(p(t))) = T(p(t)) \ . \tag{1.39}$$

For example, with linear demand and linear supply functions $D(p) = a - bp$ and $S(p) = -c + dp$ the model becomes

$$p(t+1) = T(p(t)) = -\frac{d}{b}p(t) + \frac{a+c}{b} \ . \tag{1.40}$$

This dynamic model is known as the "cobweb model".

Another example is the Cournot duopoly model (see, e.g., [6]), where two firms produce at time t the quantities $q_1(t)$ and $q_2(t)$ of the same good (or homogeneous goods) and sell it in the same market characterized by an inverse demand function

$p = D^{-1}(Q)$, where $Q = q_1 + q_2$ is the total quantity produced. If $C_i(q_i)$, $i = 1, 2$, are the respective cost functions, then the profits of the two firms are given by

$$\Pi_1(q_1, q_2) = pq_1 - C_1(q_1) = D^{-1}(q_1 + q_2)q_1 - C_1(q_1) \, ,$$
$$\Pi_2(q_1, q_2) = pq_2 - C_2(q_2) = D^{-1}(q_1 + q_2)q_2 - C_2(q_1) \, ,$$

hence, the profit function of each firm also depends on the production of the other one, the source of interdependence being the demand function. At each time t each firm decides its next period production $q_i(t+1)$ (to be realized after the production lag $\Delta t = 1$) in order to maximize its own profit. However, at time t each firm does not know the production decision of the other firm, so an expected value must be considered in the maximization problems

$$q_i(t+1) = \arg \max_{q_i(t+1)} \Pi_i\,(t+1) = \arg \max_{q_i} [D^{-1}\left(q_i + q^e_{-i}\,(t+1)\right) q_i - C_i\,(q_i)] \, . \tag{1.41}$$

For example, if we consider linear demand and linear cost functions, $p = a - b(q_1 + q_2)$ and $C_i(q_i) = c_i q_i$, then producer 1 faces the optimization problem

$$\max_{q_i(t+1)} \Pi_i\,(t+1) = \max_{q_1} [(a - c_1)q_1 - bq_1 q^e_2(t+1) - bq_1^2] \, .$$

From the first order condition (necessary condition for a maximum) $\frac{\partial \Pi_1}{\partial q_1} = 0$, we get $(a - c_1) - bq^e_2(t+1) - 2bq_1 = 0$ from which it is $q_1(t+1) = -1/2q^e_2(t+1) + (a - c_1)/(2b)$. The second order condition $\frac{\partial^2 \Pi_1}{\partial q_1^2} = -2b < 0$ ensures that it is indeed a maximum. If we solve the same problem for the second firm and we assume naïve expectations, i.e., that $q^e_j(t+1) = q_j(t)$, we obtain the following two-dimensional linear discrete time dynamical system

$$\begin{cases} q_1(t+1) = B_1(q_2(t)) = -\frac{1}{2}q_2(t) + \frac{a-c_1}{2b} \\ \\ q_2(t+1) = B_2(q_1(t)) = -\frac{1}{2}q_1(t) + \frac{a-c_2}{2b} \end{cases} \tag{1.42}$$

Instead, if the (inverse) demand function is assumed to be isoelastic (in particular with unitary elasticity) with the form $p = 1/Q$, then the same arguments lead to the following nonlinear discrete dynamical system

$$\begin{cases} q_1(t+1) = R_1(q_2(t)) = \sqrt{q_2(t)/c_1} - q_2(t) \\ \\ q_2(t+1) = R_2(q_1(t)) = \sqrt{q_1(t)/c_2} - q_1(t) \end{cases} \tag{1.43}$$

1.3.1 One-Dimensional Discrete Dynamical Systems

1.3.1.1 The Simplest One: Linear Homogeneous

The simplest discrete time recurrence is the linear homogeneous iterated map

$$x(t+1) = ax(t) \tag{1.44}$$

with initial condition $x(0) = x_0$. The general solution of (1.44) can be obtained inductively, being $x(1) = ax_0$, $x(2) = ax(1) = a^2x_0$, $x(3) = ax(2) = a^3x_0$, ...,

$$x(t) = x_0a^t, \quad t \in \mathbb{N}\,. \tag{1.45}$$

The sequence (1.45) converges to the unique asymptotic equilibrium $x^* = 0$ if $|a| < 1$, i.e., $-1 < a < 1$. In this case we say that (1.44) is a contraction mapping, as at each iteration the distance of $x(t)$ from $x^* = 0$ is reduced of the factor $|a|$. For example, if $a = 1/2$ then $x(1) = x_0/2$, $x(2) = x_0/4$ etc. The same holds for $a = -1/2$, even if this occurs through oscillations of decreasing amplitude: $x(1) = -x_0/2$, $x(2) = x_0/4$, $x(3) = -x_0/8$ etc. So, for negative values of a the sequence (1.45) oscillates around $x^* = 0$, as it assumes the same sign of x_0 at even iterations and opposite sign at odd iterations. It is worth stressing that oscillations can be obtained with a one-dimensional discrete-time dynamical system, whereas this was impossible in the case of one-dimensional smooth systems in continuous time. Of course this is due to the fact that the points generated by (1.44) can jump between different points without touching the intermediate points. Diverging sequences are obtained for $|a| > 1$, monotonically diverging if $a > 1$, diverging through oscillations if $a < -1$. For example $a = 2$ gives $x(1) = 2x_0$, $x(2) = 4x_0$ etc., whereas $a = -2$ gives $x(1) = -2x_0$, $x(2) = 4x_0$, $x(3) = -8x_0$ etc. Finally, for $a = 1$ the identity map is obtained, whose iteration gives a constant sequence $x(t) = x_0$ for each $t \in \mathbb{N}$, whereas $a = -1$ gives the oscillating sequence $x(t) = (-1)^t x_0$.

We can see the map $x' = ax$ as a transformation of the real line into itself, i.e., a function that transforms each point $x \in \mathbb{R}$ into its unique image $x' \in \mathbb{R}$. If we consider a segment AB, i.e., the closed interval $AB = \{x \in \mathbb{R}, A \leq x \leq B\}$, and we apply the transformation $x' = ax$ to all the points of the segment, then the length of a new segment $A'B'$ is $\overline{A'B'} = |a|\,\overline{AB}$, i.e., it is contracted of the factor a if $|a| < 1$, expanded if $|a| > 1$, and with the same length if $|a| = 1$. Moreover, its orientation remains the same, i.e., $A < B$ implies $A' < B'$, if $a > 0$, whereas its orientation is reversed, i.e., $A < B$ implies $A' > B'$, whenever $a < 0$.

From the general solution (1.45) of the linear recurrence (1.44), the solution of nonhomogeneous linear map can be easily obtained. In fact, given

$$x(t+1) = ax(t) + b \tag{1.46}$$

it can be noticed that if it converges, then it converges to the unique steady state (or fixed point) x^* characterized by the condition $x(t+1) = x(t)$. Thus, x^* is the solution of the equation $x = ax + b$, i.e., it is $x^* = b/(1-a)$, provided that $a \neq 1$. The change of coordinates $X(t) = x(t) - x^* = x(t) - b/(1-a)$ that translates the fixed point into the origin, transforms the affine (or linear nonhomogeneous) recurrence into a linear homogeneous one. In fact, by replacing $x(t) = X(t) + b/(1-a)$ into (1.46) we get $X(t+1) = aX(t)$, i.e., in the form (1.44), and consequently the general solution is $X(t) = X(0)a^t$, from which going back to the originary variable

$$x(t) = \left(x_0 - \frac{b}{1-a}\right) a^t + \frac{b}{1-a.} \tag{1.47}$$

Such solution converges to $x^* = b/(1-a)$ for $|a| < 1$, oscillates between $-x_0$ and x_0 for $a = -1$; finally, in the particular case $a = 1$, (1.46) becomes the arithmetic sequence $x(t+1) = x(t) + b$, whose solution is $x(t) = x_0 + bt$, which is increasing or decreasing according to the sign of b.

This completely solves, for example, the linear cobweb model with naïve expectations (1.40) whose equilibrium is $p^* = (a+c)/(b+d)$, located at the intersection of the demand and supply curves, and the solution starting from the initial price $p(0) = p_0$ is

$$p(t) = \left(p_0 - \frac{a+c}{b+d}\right)\left(-\frac{d}{b}\right)^t + \frac{a+c}{b+d}. \tag{1.48}$$

The corresponding time series exhibit oscillatory behavior, being $-d/b < 0$: they are convergent to p^* when $b > d$, i.e., the decreasing demand function is steeper than the increasing supply, diverging otherwise.

1.3.1.2 Qualitative Analysis of One-Dimensional Nonlinear Models in Discrete Time

Let us consider now a general discrete-time dynamical system with one dynamic variable

$$x(t+1) = f(x(t)) \tag{1.49}$$

with initial condition $x(0) = x_0$. The equilibrium points (or fixed points) are defined by the equilibrium condition $x(t+1)) = x(t)$, i.e., are the solutions of the equation

$$f(x) = x\ . \tag{1.50}$$

Let x^* be a solution of (1.50). Then a linear approximation of (1.49) in a neighborhood of x^* can be obtained as $f(x) = f(x^*) + f'(x^*)(x - x^*) + o(x - x^*) = x^* + f'(x^*)(x - x^*) + o(x - x^*)$, leading to the linear approximation

$$x(t+1) = x^* + f'(x^*)\left(x - x^*\right)$$

that reduces to the linear homogeneous case $X(t+1) = aX(t)$ after the translation $X(t) = x(t) - x^*$ that measures the displacement from the equilibrium point. From the discussion about the linear case of the previous section the following result immediately follows:

Proposition 1.2 (1D Local Asymptotic Stability in Discrete Time) *Let x^* be an equilibrium point of (1.49), i.e., $f(x^*) = x^*$. If $\left|f'(x^*)\right| < 1$, then x^* is a locally asymptotically stable equilibrium; if $\left|f'(x^*)\right| > 1$, then x^* is unstable.*

Indeed, if x^* is hyperbolic, which in the case of discrete dynamical systems (DDS) means that $|f'(x^*)| \neq 1$, the Hartman-Grobman theorem (1959–1960) can be stated as follows:

Theorem 1.9 (Hartman-Grobman for DDS) *Let x^* be a hyperbolic fixed point of (1.49), with f being differentiable. Then there exists a neighborhood of x^* where map (1.49) is topologically conjugate to its linear approximation.*

Notice that in the case of discrete time the stability condition $-1 < f'(x^*) < 1$ includes both an upper and a lower threshold for the slope of the function f at the equilibrium point, and the two limiting values -1 and $+1$ constitute two different conditions of nonhyperbolicity of the equilibrium point. The condition of nonhyperbolicity $f'(x^*) = 1$ corresponds to the analogous condition $f'(x^*) = 0$ for continuous time one-dimensional models. We will see in the following that if such condition is crossed as a parameter varies then the bifurcations that occur are similar to those detected in continuous-time models. Instead, the other nonhyperbolicity condition $f'(x^*) = -1$ has no analogue in continuous time models, as it is characterized by oscillatory behavior. Indeed, the presence of negative derivative is often related to phenomena of overshooting (or over-reaction). This means that even if we have a decreasing map around x^*, i.e., $f(x) > x^*$ in a left neighborhood of x^* and $f(x) < x^*$ in a right neighborhood of x^*, so that $x(t+1) > x(t)$ if $x(t)$ is on the left of x^* and $x(t+1) < x(t)$ if $x(t)$ is on the right of x^*, a trajectory starting from a neighborhood of x^* will jump from one side to another of x^*; thus, either convergence does not occur or, if convergence is obtained, being $-1 < f'(x^*) < 0$, then it takes place through oscillations.

Before discussing the possible bifurcations of 1-dimensional nonlinear models, we describe a useful graphical method that is widely used to obtain the trajectories of a one-dimensional discrete dynamical system (1.49), even in a nonlinear case, without any analytic computation. This method is based on the knowledge of the graph of the function $y = f(x)$. It consists in drawing such graph on a cartesian plane together with the diagonal $y = x$. Starting from the initial condition x_0 on the horizontal axis, the successive value of the recurrence $x(1)$ is obtained by moving upward up to the graph and then to the left, on the vertical axis (the codomain) where the values (images) are represented. Then, in order to continue the iteration of the function, this value must be brought back to the horizontal axis, i.e., from the codomain to the domain, in order to apply the function again. This can be done by using the diagonal, locus of point such that $y = x$: the point $x(1)$ is moved horizontally towards the diagonal

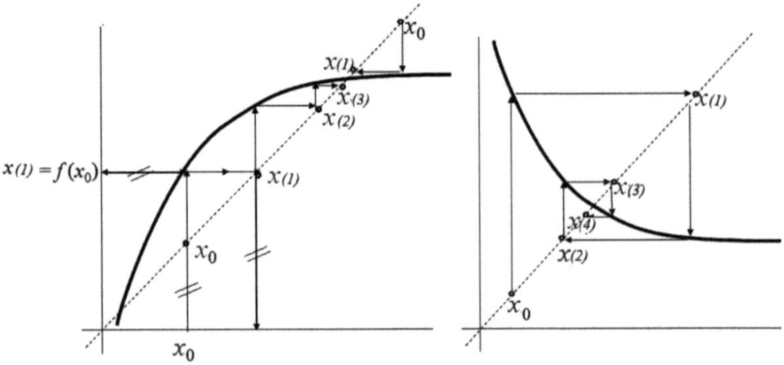

Fig. 1.37 Staircase diagram

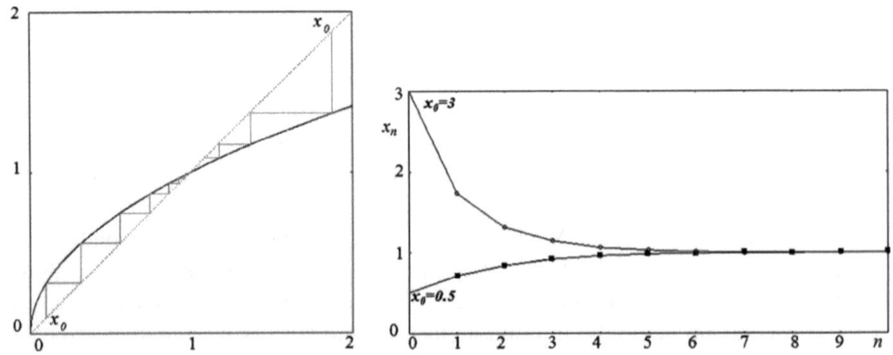

Fig. 1.38 Staircase diagram and versus time representation for recurrence $x_{t+1} = \sqrt{x_t}$

and then vertically towards the horizontal axis (see Fig. 1.37). Then the process is repeated again to get $x(2)$ etc.

Notice that some portions of the horizontal and vertical movements have been traveled back and forth, so that they can be deleted and the movements reduced to the following: starting from x_0 on the diagonal, *vertical to the graph, horizontal to the diagonal where* $x(1) = f(x_0)$ *is placed*, then *vertical to the graph, horizontal to the diagonal where* $x(2) = f(x(1))$ *is placed* and so on. This graphical construction, called *staircase diagram* allows us to get the whole trajectory as a set of points along the diagonal. See the other examples in Fig. 1.37, in particular the oscillatory trajectory shown in the right panel, obtained with a decreasing map.

In Fig. 1.38 this method is applied to the map $f(x) = \sqrt{x}$, whose iteration can be simply obtained by a pocket calculator, starting from any initial condition $x(0) > 0$ and repeatedly pressing the square-root key. It is easily realized that it always converges to the globally stable fixed point $x^* = 1$.

From these preliminary arguments it is evident that in the case of decreasing one-dimensional discrete time dynamical systems, whose trajectories are obtained by

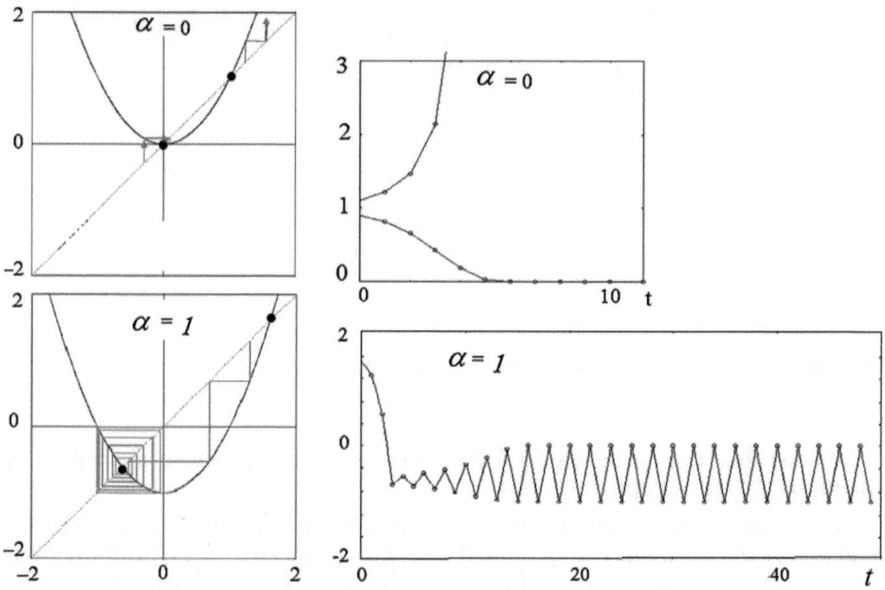

Fig. 1.39 Staircase diagram and versus time representation for recurrence $x_{t+1} = x_t^2 - a$

iterated maps, oscillations and even periodic cycles are obtained. Consider, for example, the map $f(x) = 1/x$, i.e., the recurrence $x(t+1) = 1/x(t)$ that leads to a cycle of period two $(x_0, 1/x_0)$ for each initial condition $x_0 \neq 0$. Or the map $f(x) = x^2 - 1$ that, starting from $x_0 = 0$ gives the cycle $x(1) = -1$, $x(2) = 0$, $x(3) = -1$. Moreover, starting from another initial condition such as $x_0 = 3/2$, it generates $x(1) = 5/4 = 1.25$, $x(2) = 0.5625$, $x(3) = -0.6836$, $x(4) = -0.5327$, $x(5) = -0.7162$, $x(6) = -0.4870$, $x(7) = -0.7628$, $x(8) = -0.4181$, and then slowly approaches the 2-cycle $(-1, 0)$.

In other words, not only periodic sequences exist where after a given number of iterations the same value is reached and then the same set of numbers repeats indefinitely, but also there are sequences that approach asymptotically such periodic sets. Several situations of this kind are obtained with the iterated quadratic[7] map

$$f(x) = x^2 - \alpha \tag{1.51}$$

with parameter's values in the range $\alpha \in (0, 2)$. This map can be iterated very easily even with a pocket calculator and many different situations, from convergence to a fixed point to convergence to a periodic sequence or even aperiodic and very irregular sequences are obtained. Try, for example with $\alpha = 1.3$ and with $\alpha = 2$, starting, e.g., from $x(0) = 0.5$. Some staircase diagrams of this quadratic map are shown in Figs. 1.39 and 1.40.

[7]Quadratic means polynomial of degree two.

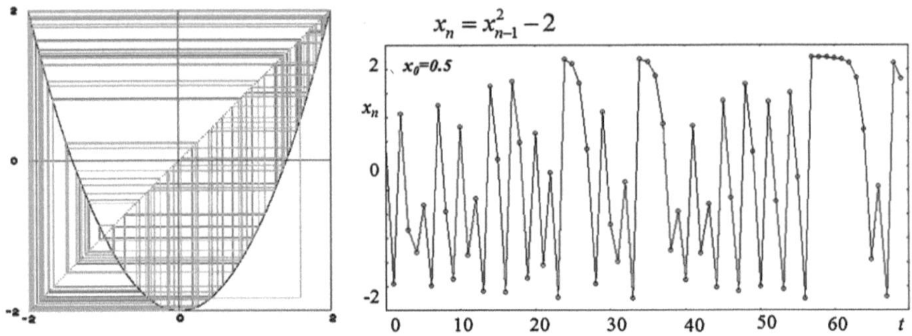

Fig. 1.40 A chaotic trajectory for the quadratic map (1.51)

We end this section by giving a definition of a periodic cycle as well as the conditions for its stability.

A periodic cycle of period k is a set of points $C_k = \{c_1, c_2, \ldots, c_k\}$ such that $c_i \neq c_1, i = 2, \ldots, k, f(c_i) = c_{i+1}, i = 1, \ldots, k-1$, and $f(c_k) = c_1$. So, the periodic points can be obtained as $C_k = \{c_1, f(c_1), f^2(c_1) \ldots, f^{k-1}(c_1)\}$ with $f^k(c_1) = c_1$. This last equality states that c_1 is a fixed point of the composite function $f^k(x)$. Indeed, as the initial periodic point of the cycle is arbitrary, any periodic point of a k cycle is a fixed point of f^k, i.e., $f^k(c_i) = c_i$ for each $i = 1, \ldots, k$. This is quite intuitive, because, after k iterations of f, all the points of the k-cycle are obtained and the initial point is reached again. In other words, if the map f is applied iteratively starting from a k-periodic point and we look at the result at intervals of k iterations, then we see always the same point.

Notice that every periodic point c_i of a k-cycle C_k is a fixed point of $f^k(x)$ but it is not a fixed point of any $f^j(x)$ with $j < k$. Indeed, a fixed point x^* of $f(x)$ is also a fixed point of any composite function $f^j(x)$ for any $j > 1$, as $f(x^*) = x^*$ implies $f^2(x^*) = f(f(x^*)) = f(x^*) = x^*$ and so on. So, the k-periodic points are all and the only fixed points of $f^k(x)$ which are not fixed points of $f^j(x)$ for any $j < k$.

The stability of a k-cycle C_k can be determined by the study of the stability of one of its periodic points c_i as a fixed point of $f^k(x)$, i.e., by the condition $\left|\frac{df^k}{dx}(c_i)\right| < 1$. By using the chain rule for the derivation of composite functions, the derivative of the composite function $f^k(x)$ can be reduced to the product of the derivatives of the simpler function $f(x)$ along the k-periodic points

$$\frac{df^k}{dx}(c_i) = f'(c_1) \cdot f'(c_2) \cdot \cdots \cdot f'(c_k) = \prod_{i=1}^{k} f'(c_i)$$

This can be easily proved inductively. In fact, for $k = 2$ from $f(c_1) = c_2$ and $f(c_2) = c_1$ we obtain $\frac{df^2}{dx}(c_1) = f'(f(c_1))f'(c_1) = f'(c_2)f'(c_1)$ and analogously for $\frac{df^2}{dx}(c_2)$.

So, if $\frac{df^{k-1}}{dx}(c_1) = \prod_{i=1}^{k-1} f'(c_i)$ then for the derivative in c_1 of $f^k(x) = f(f^{k-1}(x))$ we get

$$f'(f^{k-1}(c_1))\frac{df^{k-1}}{dx}(c_1) = f'(c_k)\prod_{i=1}^{k-1} f'(c_i) = \prod_{i=1}^{k} f'(c_i).$$

1.3.1.3 Local Bifurcations of One-Dimensional Discrete Dynamical Systems

Let us consider a one-dimensional discrete dynamical system whose structure depends on a parameter $\alpha \in \mathbb{R}$

$$x(t+1) = f(x(t); \alpha)$$

and let $x^*(\alpha)$ be a fixed point defined implicitly by the equilibrium equation $f(x; \alpha) = x$. The stability condition $|f'(x^*(\alpha))| < 1$ indicates that as the parameter α varies the fixed point can lose stability through two bifurcation conditions, at which the fixed point is nonhyperbolic, $f'(x^*(\alpha)) = +1$ and $f'(x^*(\alpha)) = -1$. As one of these two bifurcation conditions is crossed a local bifurcation occurs at which the fixed point changes its stability property and something else happens, as shown in the pictures here below where some canonical maps are given, as well as their graphs and bifurcation diagrams. Notice that the three local bifurcations occurring with multiplier $f'(x^*(\alpha)) = +1$ are essentially the same as those occurring in dynamical system in continuous time, the only difference being that in this case the tangency occurs along the diagonal, where the fixed points are located, and consequently involve slope 1. However, differently from the linear case where the oscillatory expansion has no limits, leading to oscillatory divergence, in the nonlinear case a creation of a 2-periodic cycle occurs at the bifurcation value. This 2-periodic cycle may be stable if it is created around the unstable fixed point (supercritical case) thus attracting the trajectories escaping from a neighborhood of x^*, or unstable if it exists around the stable fixed point (subcritical case) thus bounding its basin of attraction (Figs. 1.41, 1.42 and 1.43).

What is new is the bifurcation occurring with multiplier $f'(x^*(\alpha)) = -1$, denoted as *flip bifurcation*, at which the fixed point changes its oscillatory stability (i.e., convergence through damped oscillations) into oscillatory instability (i.e., trajectories starting very close to x^* exhibit oscillatory expansion) (Fig. 1.44).

It is worth noticing that this bifurcation, leading to the creation of two periodic points of a 2-cycle, causes the creation of two new fixed points of the map $f^2(x) = f(f(x))$, besides the previously existing x^* as any fixed point of $f(x)$ is also a fixed point of $f^2(x)$. So, the flip bifurcation of f is associated with a pitchfork bifurcation of $f^2(x)$. Indeed, if $f'(x^*) = -1$ then $\frac{df^2}{dx}(x^*) = f'(f(x^*))f'(x^*) = f'(x^*)f'(x^*) = (-1)(-1) = +1$, a bifurcation condition of f^2 corresponding to a pitchfork bifurcation.

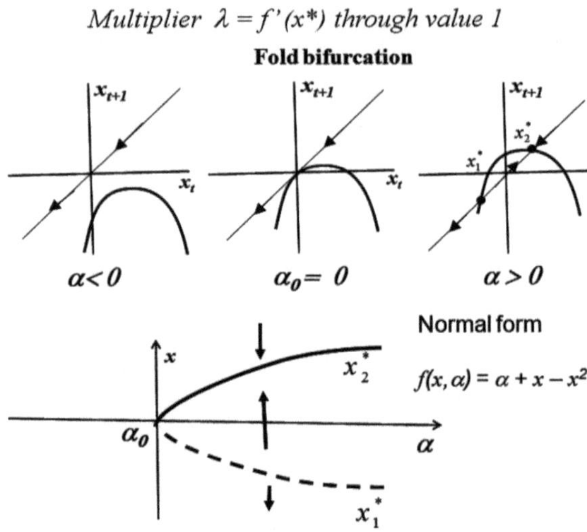

Fig. 1.41 Fold bifurcation for discrete time mappings

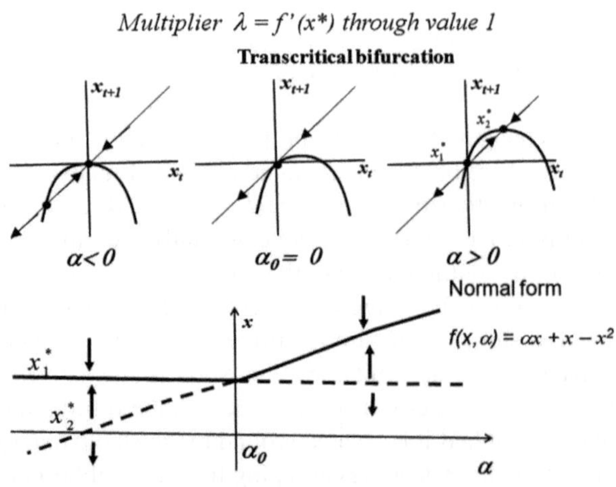

Fig. 1.42 Transcritical bifurcation for discrete time mappings

All the local bifurcations described above can be observed in periodic cycles as well. In fact, when a periodic cycle $C_k = \{c_1, c_2, \ldots, c_k\}$ with associated multiplier $\lambda(C_k) = \prod_{i=1}^{k} f'(c_i)$ changes its stability properties due to a variation of a parameter, if the multiplier exits the stability range $-1 < \lambda(C_k) < 1$ through the value $+1$ then a bifurcation of the cycle is observed, that may be of fold type (a couple of k-cycle, one stable and one unstable, are created or destroyed through their merging) of transcritical type (two k-periodic cycles of opposite stability merge and exchange their stability) or of pitchfork type (two further k-periodic cycles are created at the bifurcation). Instead if the multiplier exits the stability range $-1 < \lambda(C_k) < 1$ through the value -1 then a flip bifurcation of the k-cycle is observed at which a cycle of double period 2k is created.

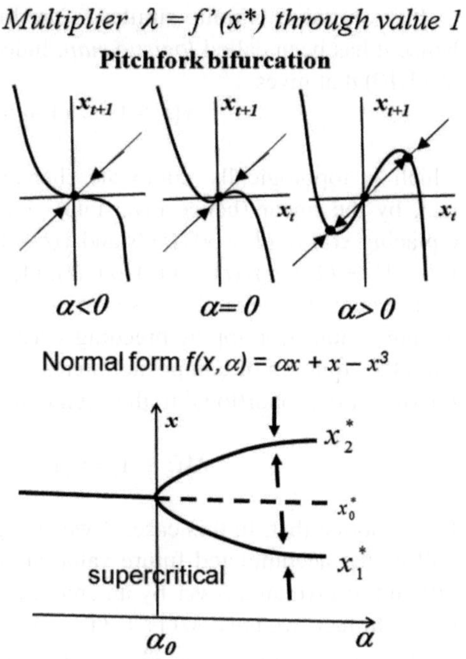

Fig. 1.43 Pitchfork bifurcation for discrete time mappings

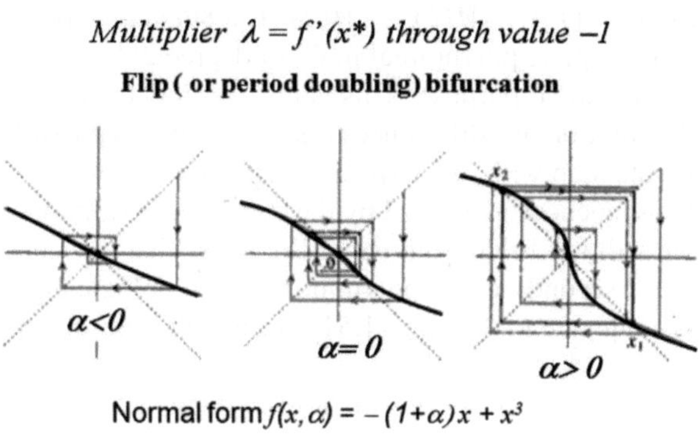

Fig. 1.44 Flip (or period doubling) bifurcation, only exists for discrete time mappings

1.3.1.4 The Logistic Map

In this section we consider the quadratic map (see, e.g., [11])

$$x(t+1) = \mu x(t)(1 - x(t)), \quad \mu > 0\,, \tag{1.52}$$

whose graph is represented by a concave parabola that intersects the diagonal in the two fixed points

$$x_0^* = 0 \text{ and } x_1^* = 1 - \frac{1}{\mu}\,. \tag{1.53}$$

Its expression is quite similar to the logistic model in continuous time (1.10), hence it has been called *logistic map*. Indeed, it can be obtained by a discretization of (1.10) that gives

$$x(t+1) = (1+\alpha)\,x(t) - sx(t)^2 \,, \tag{1.54}$$

which is topologically conjugate (hence dynamically equivalent) to (1.52). In fact, by the linear (hence invertible) change of variable $y = sx/(1+\alpha)$, i.e., by replacing $x(t) = (1+\alpha)y(t)/s$ and $x(t+1) = (1+\alpha)\,y(t+1)/s$ in (1.54), we get $y(t+1) = (1+\alpha)y(t) - (1+\alpha)y(t)$, identical to (1.52) with $\mu = 1+\alpha$.

The map (1.54) is indeed used to model the time evolution of a population reproducing at non-overlapping breeding seasons. The same map (1.54) can be obtained from the equation of composite interests $M(t+1) = (1+r)M(t)$ if we imagine to impose a tax proportional to the square of the money

$$M(t+1) = (1+r)\,M(t) - sM(t)^2 \,. \tag{1.55}$$

Let us notice that, in this case, if we ask, given an initial capital $M(0) = C_0$, what will be the accumulated future value after n years according to (1.55), it is quite difficult to give an answer by an analytic expression that gives $M(n)$ as a function of C_0. In fact, we have $M(1) = (1+r)\,C_0 - sC_0^2$, $M(2) = (1+r)\,M(1) - sM(1)^2$ $= (1+r)\left[(1+r)\,C_0 - sC_0^2\right] - s\left[(1+r)\,C_0 - sC_0^2\right]^2$, i.e., a 4th degree polynomial, $M(3) = (1+r)\,M(2) - sM(2)^2$ is a 8th degree polynomial in C_0 and so on. $M(10)$ is a complete polynomial in C_0 of degree $2^{10} = 1024$, a computation impossible for any practical purpose. This is just to show that, even if the analytic computation of the solution of a difference equation is always possible in principle by composing the iterate map with itself, this is practically impossible when it is nonlinear. Moreover, the sequences generated by the recurrence (1.52) may become quite complicated, as we will see in the following. We stress that the same holds for any quadratic map, as all second degree polynomials are topologically conjugate. For example, (1.52) is conjugate to the map (1.51) through the coordinate change $y = -x/\mu + 1/2$ with $\alpha = \mu^2/4 - \mu/2$.

What makes famous the logistic map (1.52) is the article "Simple mathematical models with very complicated dynamics", published in 1976 by Robert M. May in *Nature*, from which many other papers followed where the same model was used in several fields, included economics and finance. The paper of May ends with:

> evangelical plea for the introduction of these difference equations into elementary mathematics courses, so that students' intuition may be enriched by seeing the wild things that simple nonlinear equations can do. [...] The elegant body of mathematical theory pertaining to linear systems, and its successful application to many fundamentally linear problems in the physical sciences, tends to dominate even moderately advanced University courses in mathematics and theoretical physics. The mathematical intuition so developed ill equips the student to confront the bizarre behavior exhibited by the simplest of discrete nonlinear systems, such as equation (1.52). Yet such nonlinear systems are surely the rule, not the exception, outside the physical sciences. Simple mathematical models with very complicated dynamics. I would therefore urge that people be introduced to, say, equation (1.52) early in their mathematical education. This equation can be studied phenomenologically by

iterating it on a calculator, or even by hand. Its study does not involve as much conceptual sophistication as does elementary calculus. Such study would greatly enrich the student's intuition about nonlinear systems. Not only in research, but also in the everyday world of politics and economics, we would all be better off if more people realized that simple nonlinear systems do not necessarily possess simple dynamical properties.

We now follow May's invitation and propose a qualitative study of some properties of the map (1.52), and this study will lead us to encounter the phenomenon of deterministic chaos in a much simpler model than the one we have seen when dealing with systems of at least three ordinary differential equations.

The stability of the two fixed points (1.53) is readily determined through the computation of the derivative $f'(x) = \mu(1-2x)$ at the fixed points, $f'(x_0^*) = \mu$ and $f'(x_1^*) = 2 - \mu$. From the stability conditions $|f'(x_i^*)| < 1$, $i = 0, 1$, we have that $x_0^* = 0$ is locally asymptotically stable for $\mu < 1$ and $x_1^*(\mu) = 1 - 1/\mu$ is locally asymptotically stable for $1 < \mu < 3$. At $\mu = 1$ a transcritical bifurcation occurs at which the two fixed points merge and exchange their stability properties: in fact $x_1^*(\mu) < 0$ and is unstable for $0 < \mu < 1$, whereas $x_1^*(\mu) > 0$ and is stable as μ increases across the bifurcation value $\mu = 1$ at which $x_0^* = x_1^* = 0$. Notice that at $\mu = 2$ $f'(x_1^*) = 0$ (x_1^* is said to be superstable) and then the slope $f'(x_1^*)$ of the tangent at x_1^* becomes negative for $\mu > 2$ (hence we have oscillatory convergence, see Fig. 1.45, left panel). At $\mu = 3$ a flip bifurcation of x_1^* occurs at which a stable cycle of period two, say $C_2 = \{\alpha, \beta\}$, is created around the unstable fixed point (see Fig. 1.45, right panel).

The periodic points α and β can be computed as fixed points of $F(x) = f^2(x) = f(\mu x(1-x))$ given by the fourth degree map $F(x) = \mu(\mu x(1-x)(1 - \mu x(1-x))$. Its fixed points, solutions of $F(x) = x$, are solutions of the equation $x\left[\mu^2(1-x)(1-\mu x(1-x)) - 1\right] = 0$. We already know that $x_0^* = 0$ and $x_1^* = (\mu - 1)/\mu$ are fixed points of $F(x)$. Hence the equation can be factorized as $x\,(x - (\mu-1)/\mu)\left[x^2 + ((\mu-1)/\mu)\,x + \left((\mu+1)/\mu^2\right)\right] = 0$ from which the two fixed points of $F(x)$ that are not fixed points of $f(x)$, i.e., the 2-periodic points α and

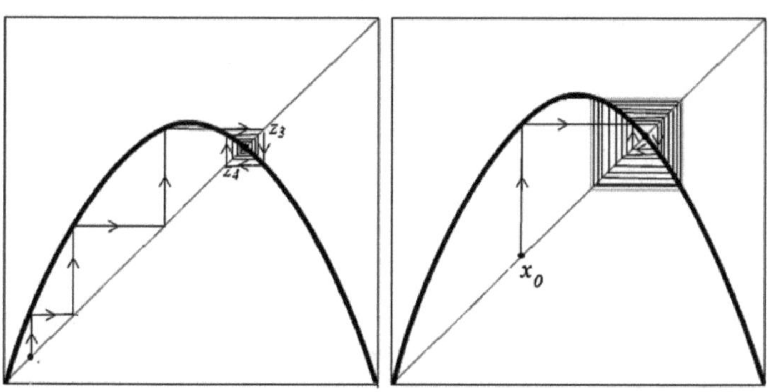

Fig. 1.45 *Left panel* Staircase diagram of a trajectory obtained with $\mu = 2.7$. *Right panel* Trajectory obtained with $\mu = 3.2$

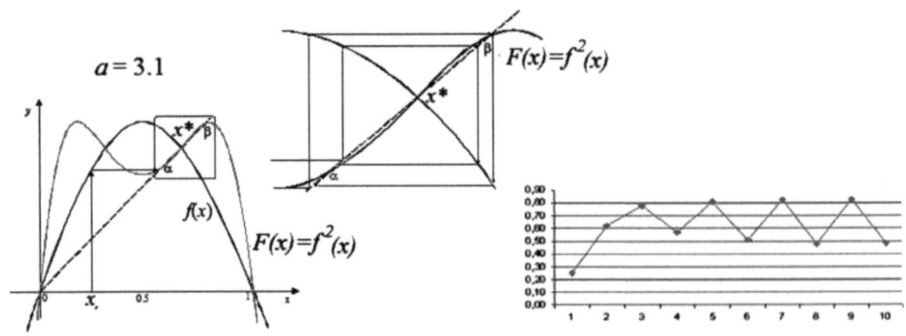

Fig. 1.46 Flip bifurcation for the logistic map

β, are $\left(\mu + 1 \pm \sqrt{(\mu - 3)(\mu + 1)}\right)/2\mu$, existing for $\mu \geq 3$ (at $\mu = 3$ they coincide with the bifurcating equilibrium x_1^*). As shown in Fig. 1.46, at $\mu = 3$ a pitchfork bifurcation for $F(x)$ occurs, leading to the creation of two new stable fixed points of $F(x)$ corresponding to the periodic points of a stable cycle of period 2.

The stability of this cycle can be checked by the computation of the derivative of $F(x)$ in one of them, given by $F'(\alpha) = F'(\beta) = f'(\alpha)f'(\beta)$. Just after the bifurcation this derivative is slightly less than 1, then it decreases as μ increases beyond the bifurcation value $\mu = 3$ until it becomes -1 at $\mu = 1 + \sqrt{6} \simeq 3.449$. This corresponds to a second flip bifurcation, this time of $F(x)$, at which the cycle C_2 loses stability and a stable cycle of period 4 of $f(x)$ is created. If μ is further increased then also the cycle of period 4 becomes unstable and a stable cycle of period 8 is created, and so on. Indeed, infinitely many stable cycles of period 2^n are created, which become unstable as μ is increased. All this sequence of period doubling bifurcations (also called period doubling cascade) occurs in a finite range of the parameter μ. In fact, if we denote by $\mu_1 = 3$ the first bifurcation value, $\mu_2 = 1 + \sqrt{6}$ the second one and so on, the distance between two successive bifurcation points $\Delta_n = \mu_{n+1} - \mu_n$ decreases and tends to 0, i.e., as μ increases the bifurcations become more and more frequent and accumulate at the limit point $\mu_\infty = 3.56994571869\ldots$.

After this limit point all cycles of period 2^n, $n \in \mathbb{N}$ have been created and have become unstable, periodic trajectories of any period can appear as well as aperiodic trajectories, i.e., bounded trajectories generated by the infinite iteration of (1.52) and that never hit an already visited point. Such trajectories are called chaotic, their points fill an invariant interval (or set of intervals) in which the following properties hold (used sometimes as a definition of existence of deterministic chaos).

(1) *infinitely many unstable periodic points exist, which are dense in the invariant set;*
(2) *an aperiodic trajectory exists that is dense in the set.*

An invariant set for which these two properties hold is said to be chaotic (Fig. 1.47).

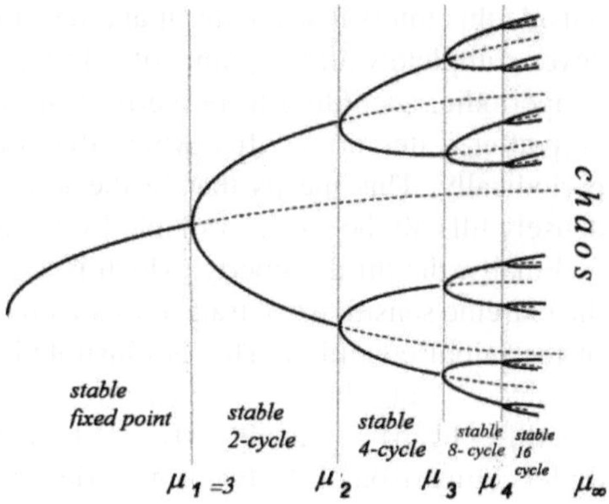

Fig. 1.47 Schematic picture of early steps of period doubling sequence

As a consequence of the these two conditions we have that the sensitivity with respect to the initial conditions (or butterfly effect) also exists, that is often added as the third (and most famous) property:

(3) Sensitivity to initial conditions. *Two trajectories starting from different, although arbitrarily close, initial conditions remain bounded but their reciprocal distance exponentially increases and, in a finite time, becomes as large as the state variables.*

The first property, about the existence of dense and repelling periodic points inside the invariant set where chaotic dynamics occur, is the key to understand the "microscopic reason" for the occurrence of chaotic dynamics. In fact, it is quite intuitive that the motion inside a trapping bounded set where infinitely many and dense repellors are nested, will be quite irregular. The second property, also called "mixing" property, states that a trajectory exists that moves erratically inside the invariant set filling it completely, see Fig. 1.48 where the initial portion of a trajectory is shown by a staircase diagram that, if continued, will cover completely the interval $[f(c), c]$ where c is the maximum value (vertex of the parabola). Trajectories starting

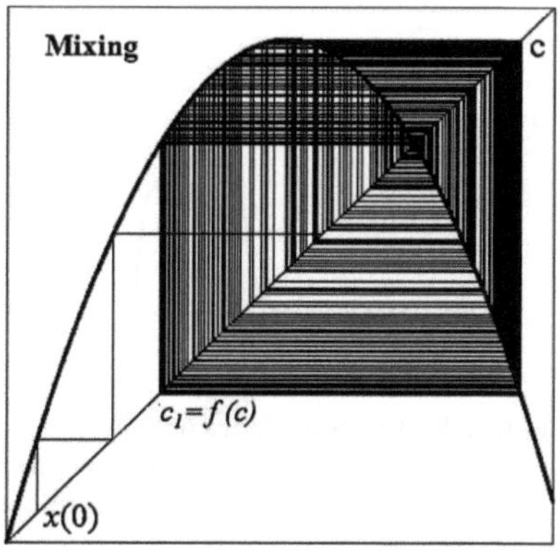

Fig. 1.48 A chaotic trajectory for the logistic map

outside this interval will enter it and never escape (hence it is an attractor) and will cover completely all the points of it in the long run. In fact, as the dense trajectory is aperiodic, according to property 2, it will never reach an already visited point (say after k iterations) after which the countable set of k points will be repeated periodically. This means that in the long run (after infinitely many iterations) it densely fills all the space available for the motion inside the invariant set.

Finally, the third property, which is a consequence of the other two, is given by the extreme sensitivity of trajectories with respect to small, even negligible, changes of the initial condition. This is illustrated in Fig. 1.49, where two trajectories, say $x(t)$ and $y(t)$, are shown both generated for $\mu = 4$ but starting from initial conditions that differ by 10^{-6}, namely $x(0) = 0.1$ and $y(0) = 0.100001$. As it can be seen (by a direct comparison of the two time series or by looking at versus-time representation of the distance between their points $|x(t) - y(t)|$), the difference between the two time series remains negligible during the early iterations, then this difference grows up until the distance between the two trajectories becomes of the same order of magnitude as the single values, i.e., an error of 100 % is obtained by this negligible difference in the initial conditions. Of course, the property of sensitive dependence on initial conditions makes any long-term prediction quite meaningless, even if based on the knowledge of the deterministic law of motion that governs the time evolution

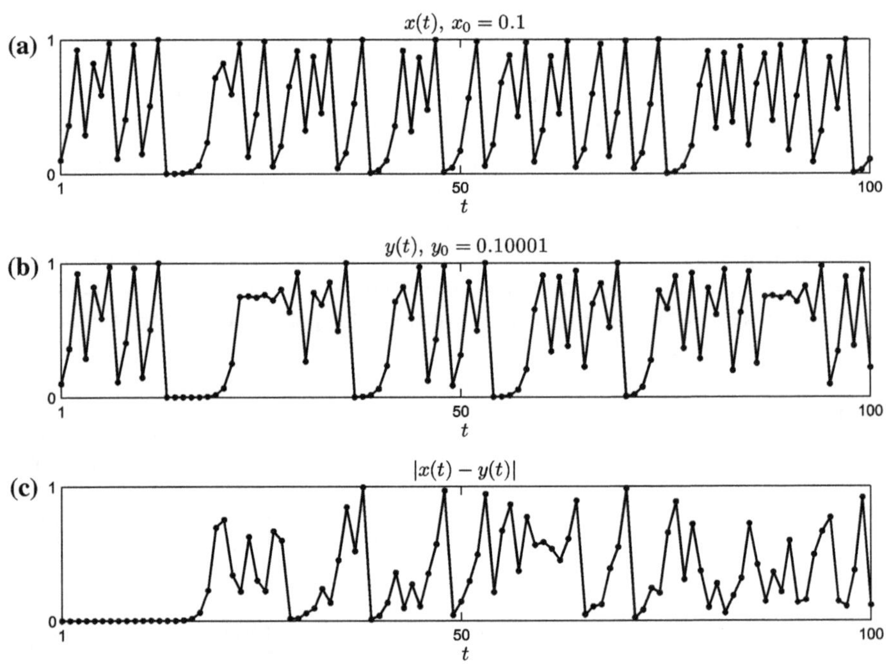

Fig. 1.49 Two trajectories obtained form $\mu = 4$ are represented versus time: $x(t)$, obtained starting from initial condition $x_0 = 0.1$ (panel **a**) and $y(t)$ starting from $y_0 = 0.100001$ (panel **b**). In panel (**c**) the difference $|x(t) - y(t)|$ is represented

of the system. Remember the clear description of this phenomenon given by Poincaré at the beginning of 20th century.

One the other side, the discovery of the phenomenon of deterministic chaos may be used to give the hope that at the basis of time evolutions that appear to be quite irregular (erratic, random) a deterministic law of motion exists, of course nonlinear and in a condition of deterministic chaos. In other words, even at the basis of very irregular and disordered phenomena it may be worth looking for (even simple) deterministic law of motion.

A sign of regularity in the realm of chaos is worth noticing. In fact, let us remark that, as stressed while looking at Fig. 1.48, the trapping interval inside which periodic or aperiodic dynamics occur has an obvious upper bound, given by the maximum value c, and a lower bound given by its image $c_1 = f(c)$. So, even if the motion inside this trapping interval may be chaotic, in any case upper and lower bounds can be given. This may give useful information, for example, when a model that shows deterministic chaos is used to simulate the irregular paths of prices in a stock market. Natural upper and lower bounds may be a useful information. The same holds in the case of model for weather forecastings, as these models cannot be used to obtain daily weather forecastings in the long run, however the boundaries of the invariant attracting set inside which asymptotic dynamics are bounded can give information on the long-run evolution of climate.

Moreover, the knowledge of maximum and minimum values (i.e., the foldings of the graph of the iterated function) as well as their images, may show more complex structures of allowed and forbidden regions for asymptotic dynamics, as shown in Fig. 1.50, where the vertex of the parabola c and its images $c_i = f^i(c)$, $i = 1, \ldots, 3$ bound a trapping region with a hole inside (i.e., the union of two disjoint intervals). And even if the dynamics is chaotic, no iterated points are allowed to enter the hole between the two intervals. This important property, that will be stressed even in the case of higher dimensional discrete dynamical systems, is related to the shape of chaotic attractors (see, e.g., the Lorenz attractor in continuous time) and can put some order in the topological properties of chaotic systems.

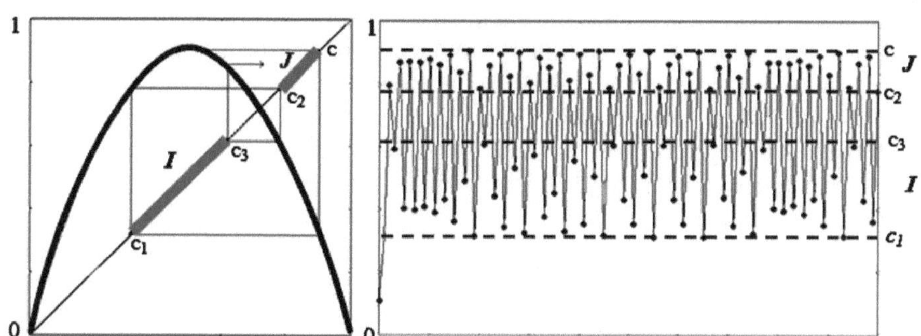

Fig. 1.50 Attracting invariant intervals bounded by critical points (maximum and its images) for $\mu = 3.61$

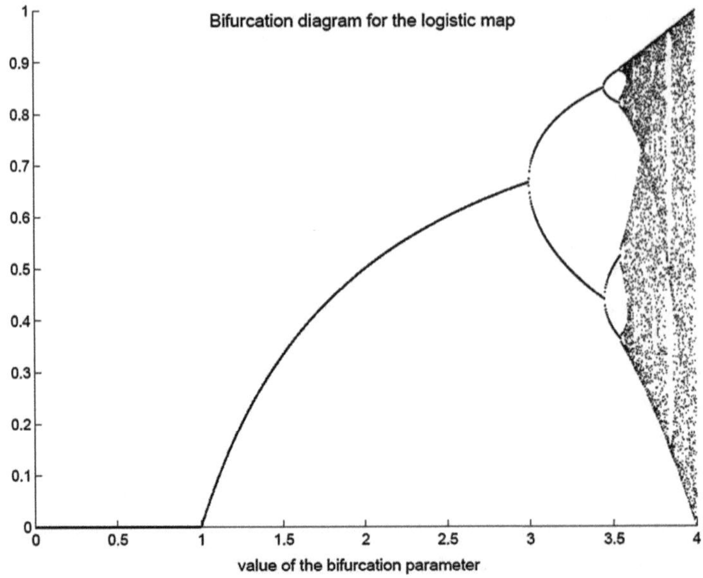

Fig. 1.51 Bifurcation diagram for the logistic map

A kind of "summary" of the different dynamic scenarios, obtained as the bifurcation parameter μ is increased, is given by the bifurcation diagram (see Fig. 1.51) obtained measuring the different values of μ in the horizontal axis while along the vertical axis the points of a trajectory are reported (after a given transient portion has been discarded). This means that, starting from a given initial condition, the attractor reached by the trajectory is represented for each value of μ. The complete bifurcation diagram for $\mu \in [0, 4]$ is given in Fig. 1.51, where the period doubling sequence, the transition to chaos, as well as the cyclic chaotic intervals (or chaotic bands) bounded by the critical point and its images are visible.

Another evident feature that can be seen in the bifurcation diagram is the presence, for certain ranges of the bifurcation parameter, of white strips where chaos seems to disappear for a while and the overall dynamics are captured by an attracting periodic cycle. These strips are called "periodic windows". Quite evident is the periodic window of a 3-cycle obtained for values of μ around 3.85. Indeed, enlargements of the bifurcation diagram show that such periodic windows are infinitely many, for example a stable cycle of period 5 is visible in a narrow white strip around $\mu = 3.74$ etc. A periodic window of period k is created through a fold bifurcation of $f^k(x)$, see for example in Fig. 1.52 the graph of $f^3(x)$ leading to the sudden creation of three couples of fixed points (each couple formed by one stable and one unstable fixed point) due to a tangency between the graph of $f^3(x)$ and the diagonal. Notice that the number of relative maximum and minimum points of $f^k(x)$ increases as k increases, and the simultaneous tangencies are k. Each couple of fixed points, created at the fold bifurcation, corresponds to a couple of periodic points of f, one stable

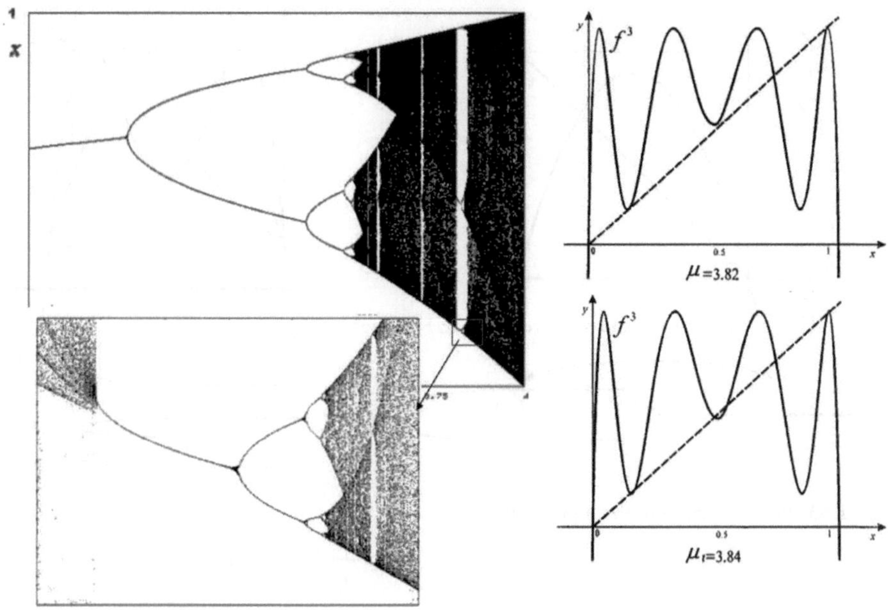

Fig. 1.52 Periodic windows in the bifurcation diagram of the logistic map and their explanation in terms of fold bifurcation

and one unstable, belonging to a stable and unstable k-cycle respectively. As μ is further increased then each fold behaves as a small quadratic map characterized by one maximum or one minimum (called a unimodal map) hence the stable cycle will loose stability via a flip bifurcation followed by the period doubling cascade. So, from each periodic point inside a periodic window a small bifurcation diagram with the same structure of the whole bifurcation diagram can be observed (see the enlargement in Fig. 1.52), thus giving rise to an inner self-similarity structure typical of fractal structures.

We end this section by giving a geometric interpretation of the observed phenomena. First of all, let us notice that the logistic map is a noninvertible map. In fact, the map $x' = f(x) = \mu x(1 - x)$ is such that a unique image x' is associated with each x in the function domain, whereas given a value x' in the codomain we obtain two preimages, computed as

$$x_1 = f_1^{-1}(x') = \frac{1}{2} - \frac{\sqrt{\mu\,(\mu - 4x')}}{2\mu}; \quad x_2 = f_2^{-1}(x') = \frac{1}{2} + \frac{\sqrt{\mu\,(\mu - 4x')}}{2\mu}. \tag{1.56}$$

Of course, if $x' > \mu/4$, i.e., taking x' above the maximum value, no real preimages are obtained. We say that the logistic map is a $Z_0 - Z_2$ noninvertible map, and the critical point $c = \mu/4$ separates the real line into the two subsets: $Z_0 = (c, +\infty)$, where no inverses are defined, and $Z_2 = (-\infty, c)$, whose points have two rank-1

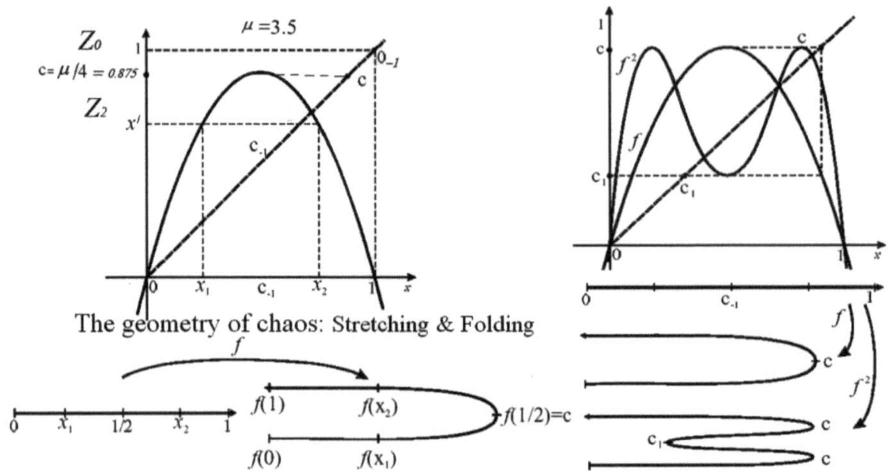

Fig. 1.53 Folding action of the logistic map

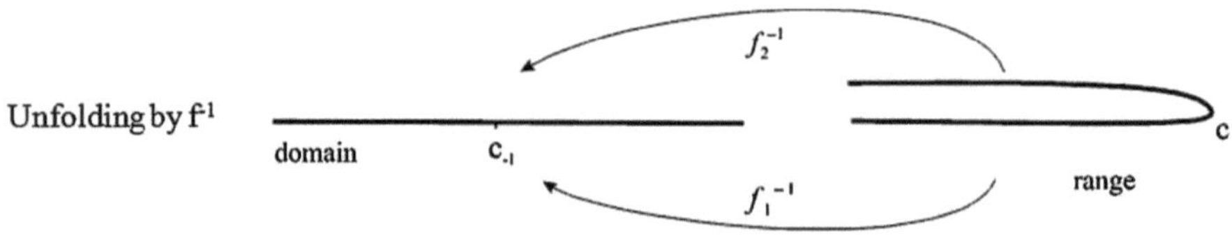

Fig. 1.54 Unfolding action of the two inverses of the logistic map

preimages (see Fig. 1.53). If $x' \in Z_2$, its two rank-1 preimages (1.56) are located symmetrically with respect to the point $c_{-1} = 1/2 = f_1^{-1}(\mu/4) = f_2^{-1}(\mu/4)$. Hence, c_{-1} is the point where the two merging preimages of c are located. As the logistic map is differentiable, at c_{-1} the first derivative vanishes. Geometrically, the action of a noninvertible map can be expressed by saying that it "folds and pleats" its domain, so that distinct points are mapped into the same point. This is equivalently stated by saying that several inverses are defined, and these inverses "unfold" S (see Fig. 1.54).

It can be noticed that, as the map is partially increasing (for $x < c_{-1}$ where $f'(x) > 0$) and partially decreasing (for $x > c_{-1}$ where $f'(x) < 0$), it is orientation preserving for $x < c_{-1}$ and orientation reversing for $x > c_{-1}$. So, a nonlinear map with a relative maximum or minimum, will "fold" any segment that includes c_{-1}. In fact, as it can be seen in Fig. 1.53, as the point x in the domain varies from 0 to 1 the corresponding image moves up and down and the sum of the two segments is greater than 1. This can be expressed by saying that the map folds and stretches. So, the repeated application of the map consists in the repeated geometric application of stretching and folding actions (see, e.g., the action of $f^2(x) = f(f(x))$ in Fig. 1.53). This implies that a small initial segments (i.e., a set of points initially very close) after many applications of stretching and folding actions will be quite dispersed. This is another way to state sensitivity dependence on initial conditions.

1.3.1.5 Basins of Attraction in One-Dimensional Discrete Dynamical Systems

Given the discrete dynamical system $x(t+1) = T(x(t))$, $x \in \mathbb{R}$, let us consider an invariant attracting set $A \subset \mathbb{R}$. The *Basin of attraction of* A is the set of all the points that generate trajectories converging to A

$$\mathscr{B}(A) = \{x | T^n(x) \to A \quad \text{as } n \to +\infty\}. \tag{1.57}$$

Starting from the definition of attracting set, let $U(A)$ be a neighborhood of A whose points converge to A. Of course $U(A) \subseteq \mathscr{B}(A)$, but note that also the points of the phase space which are mapped inside U after a finite number of iterations belong to $\mathscr{B}(A)$. Hence, the *total basin of* A (or briefly the basin of A) is given by

$$\mathscr{B}(A) = \bigcup_{n=0}^{\infty} T^{-n}(U(A)), \tag{1.58}$$

where $T^{-1}(x)$ represents the preimages of x (remember that the preimages of x may not exist or may be more than one if the map T is noninvertible, i.e., if it has several distinct inverses) and $T^{-n}(x)$ represents the set of points that are mapped into x after n iterations of the map T. Let us first consider one-dimensional, continuous and invertible maps. If $f : I \to I$ is a continuous and increasing function, then the only possible invariant sets are the fixed points. When many fixed points exist, say $x_1^* < x_2^* < \cdots < x_k^*$, they are alternatingly stable and unstable: the unstable fixed points are the boundaries that separate the basins of the stable ones. Starting from an initial condition where the graph of f is above the diagonal, i.e., $f(x_0) > x_0$, the generated trajectory is an increasing sequence converging to the stable fixed point on the right, or it is diverging to $+\infty$. On the other hand, starting from an initial condition such that $f(x_0) < x_0$, the trajectory is a decreasing sequence converging to the fixed point on the left, or it is diverging to $-\infty$ (see Fig. 1.55, where p^* is a stable fixed point, and its basin is bounded by two unstable fixed points q^* and r^*).

An example is shown in Fig. 1.56 where the increasing function $f(x) = \mu \cdot \arctan(x-1)$ is considered for increasing values of μ. For $\mu < 1$ a unique fixed point exists which is globally asymptotically stable. At $\mu = 1$ a fold bifurcation occurs at which a pair of fixed points is created, one stable and one unstable, leading to a situation of bistability where the unstable equilibrium is the boundary that separates the two basins of attraction.

If $f : I \to I$ is a continuous and decreasing map, the only possible invariant sets are one fixed point (unique) and cycles of period 2. In fact, if $f(x)$ is a decreasing map then $f^2(x)$ is increasing, hence it can only have fixed points one of which, say x^*, is the (unique) fixed point of $f(x)$ and the other ones (if any) always appear in pairs at opposite sides with respect to x^* and represent couples of periodic points of cycles of period 2. Such periodic points of the cycles of period 2 are alternatingly stable and unstable, the unstable ones being boundaries of the basins of the stable ones (see

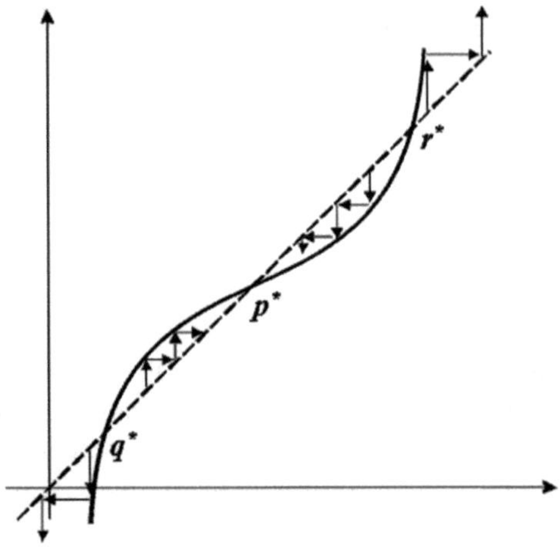

Fig. 1.55 Basin of attraction for an increasing map

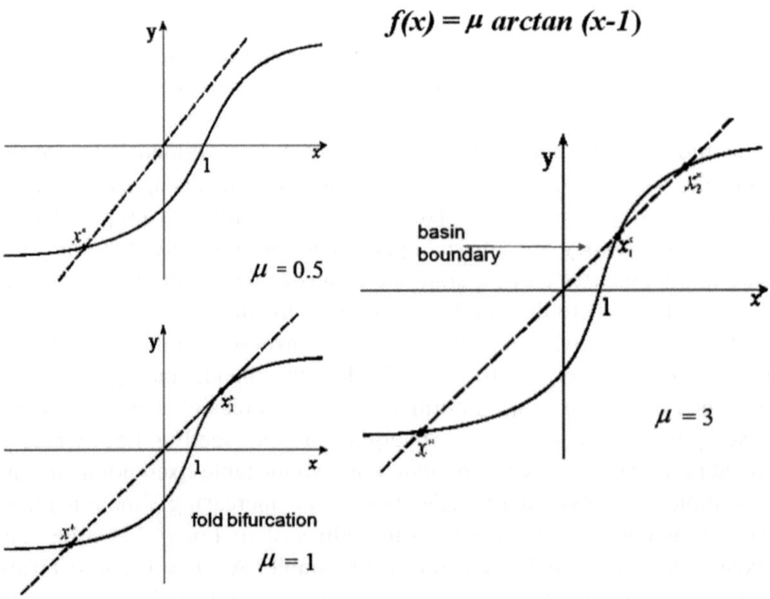

Fig. 1.56 An example of basins of attraction for an increasing map which undergoes a fold bifurcation

Fig. 1.57, where the basin of the unique fixed point x^* of the map $f(x) = 1 - ax^3$ is bounded by the periodic points α_1, α_2 of an unstable cycle of period 2). When x^* becomes unstable through a flip bifurcation as the parameter a increases, a stable 2-cycle $\{\beta_1, \beta_2\}$ is created around it, whose basin is still bounded by the unstable cycle

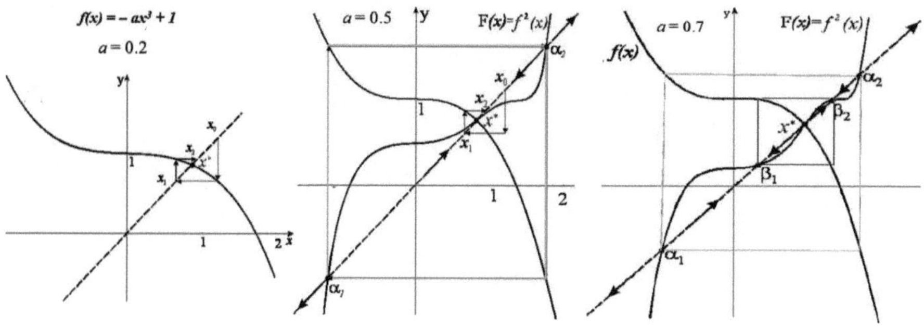

Fig. 1.57 Example of a decreasing map with basin's delimitation by a repelling cycle of period 2

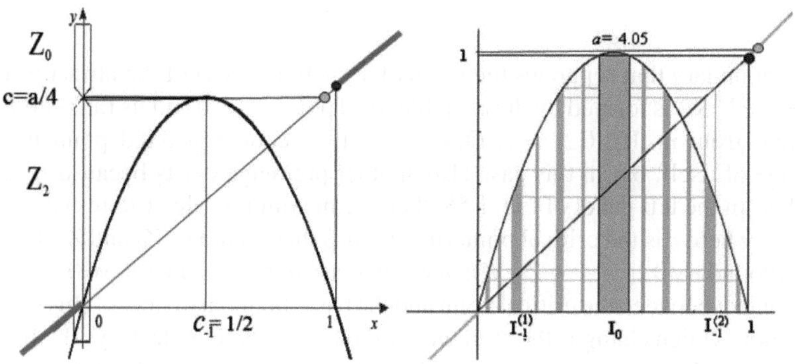

Fig. 1.58 Final bifurcation for the attractor at finite distance of the logistic map

$\{\alpha_1, \alpha_2\}$. Initial conditions outside the interval (α_1, α_2) diverge, i.e., belong to the "basin of infinity".

In general, in the case of one-dimensional invertible maps the only kinds of attractors are fixed points and cycles of period two. In the first case, the basin is an open interval which includes the fixed point, and in the second case, the basin is the union of two open intervals, each one including an attracting periodic point.

If the map is invertible, then the basins of the attracting sets are always intervals that include the attractors. This may be no longer true if the map is noninvertible, as in this case nonconnected portions of the basins may exist that are far from the attractor to which their points converge. This is due to the "unfolding action" of the inverses that may create preimages of a neighborhood of an attractor far from the related attractor. As a first example, let us consider the logistic map (1.52) whose graph is represented again in Fig. 1.58. As far as $\mu < 4$, every initial condition $x_0 \in (0, 1)$ generates bounded sequences, converging to a unique attractor A (which may be the fixed point $x_1^* = (\mu - 1)/\mu$ or a more complex attractor, periodic or chaotic). Initial conditions out of the interval $[0, 1]$ generate sequences diverging to $-\infty$.

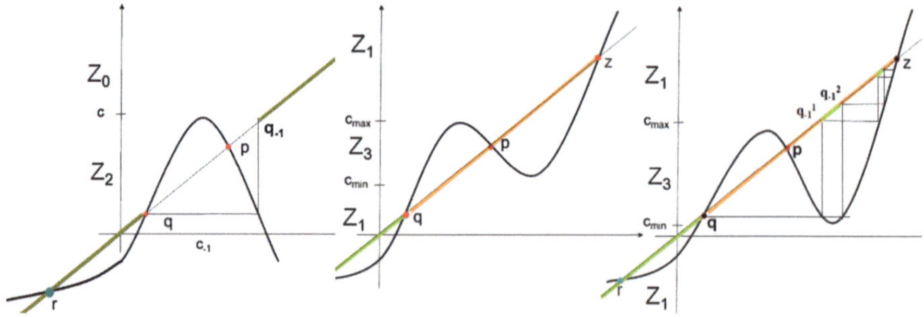

Fig. 1.59 Qualitative picture to represent a global (or contact) bifurcation leading to a nonconnected basin of attraction

The boundary that separates the basin of attraction $\mathscr{B}(A)$ of the attractor A, from the basin $\mathscr{B}(\infty)$ is formed by the unstable fixed point $x_0^* = 0$ and its rank-1 preimage (different from itself), $0_{-1} = 1$. Observe that, of course, a fixed point is always preimage of itself, but in this case also another preimage exists because $x_0^* \in Z_2$. If $\mu < 4$, as in the left panel of Fig. 1.58, then the maximum value (vertex) $c = \mu/4 < 0_{-1} = 1$, where c is the critical point (maximum) that separates Z_0 and Z_2. Hence the basin's boundary $0_{-1} = 1 \in Z_0$. When we increase μ, at $\mu = 4$ we have $0_{-1} = c = 1$, i.e., a contact between the critical point and the basin boundary occurs. This is a global bifurcation, which changes the structure of the basin (really it destroys the basin). In fact, for $\mu > 4$ (right panel of Fig. 1.58) we have $0_{-1} < c$, and the portion $(0_{-1}, c)$ of $\mathscr{B}(\infty)$ enters Z_2. This implies that new preimages of that portion are created, which belong to $\mathscr{B}(\infty)$ according to (1.58). Now almost every point belongs to the basin of divergent trajectories, the only points which are left on the interval I are the points belonging to a chaotic invariant set Λ, a subset of zero measure on which the restriction of the map is still chaotic, a chaotic repellor.

A similar situation occurs for a unimodal $Z_0 - Z_2$ map where the attractor at infinity is replaced by an attracting fixed point, as the one shown in the left panel of Fig. 1.59. As in the previous example, we have an attractor A, which may be the fixed point p (or some other invariant set around it) with a simply connected basin bounded by the unstable fixed point q and its rank-1 preimage q_{-1}. This example differs with respect to the previous one because in this case initial conditions taken in the complementary set generate trajectories converging to the stable fixed point r. This means that the basin $\mathscr{B}(r)$ is formed by the union of two nonconnected portions: $B_0 = (-\infty, q) \subset Z_2$, which contains r (called *immediate basin*, the largest connected component of the basin which contains the attractor) and $B_1 = (q_{-1}, +\infty) = f^{-1}(B_0) \subset Z_0$. In the figure the two nonconnected portions of the basin $\mathscr{B}(r)$ are marked by green bold lines. A global basin bifurcation occurs, if a parameter variation causes an increase of the critical point c (maximum value) until it crosses the basin boundary q^*_{-1}. If this happens, the interval (q_{-1}, c), which is part of B_1, enters Z_2, and infinitely many nonconnected portions of $\mathscr{B}(r)$ emerge, nested inside the interval (q, q_{-1}). After this bifurcation the total basin can still be expressed as the union of all the preimages of any rank of the immediate basin B_0.

Another interesting situation is obtained if we change the right branch of the map by folding it upwards such that another critical point, a minimum, is created. Such a situation is shown in the central panel of Fig. 1.59. This is a noninvertible $Z_1 - Z_3 - Z_1$ map, where Z_3 is the portion of the codomain bounded by the relative minimum value c_{min} and relative maximum value c_{max}. In the situation shown in the central panel we have three attractors: the fixed point r, with $\mathcal{B}(r) = (-\infty, q)$ represented by green color along the diagonal, the attractor A, around p, with basin $\mathcal{B}(A) = (q, z)$, represented by orange color, bounded by two unstable fixed points, and $+\infty$ (i.e., positively diverging trajectories) with basin $\mathcal{B}(+\infty) = (z, +\infty)$. In this case all the basins are immediate basins, each being given by an open interval that includes the attractor. In the situation shown in the central panel, both basin boundaries q and z are in Z_1, so they have only themselves as unique preimages (like for an invertible map). However, the situation drastically changes if, for example, some parameter variation causes the minimum value c_{min} to move downward, until it goes below q (as in the right panel). After the contact $c_{min} = q$ that marks the occurrence of a global bifurcation, the portion (c_{min}, q) enters Z_3, so new preimages $f^{-k}(c_{min}, q)$ appear. These preimages constitute nonconnected portions of $\mathcal{B}(r)$ nested inside $\mathcal{B}(A)$ and are represented by the thick green portions of the diagonal intermingled with orange portions that belong to $\mathcal{B}(A)$.

1.3.1.6 An Economic Example: Nonlinear Cobweb with Adaptive Expectations

Let us consider again the cobweb model (1.38), $D(p(t)) = S(p^{ae}(t))$, that with naïve expectations and linear demand and supply functions gives the linear discrete-time model (1.40) showing oscillatory convergence to $p^* = (a + c)/(b + d)$ when $b > d$ and divergence otherwise (see Fig. 1.60, where the shape of the staircase diagram justifies the name of the model).

Following [19], we now introduce a nonlinear supply function that represents a production saturation effect

$$S(p) = \arctan(\lambda(p - 1))$$

where λ represents the slope of the supply at the reference price $p = 1$.

With the same linear demand function, the cobweb model with naïve expectations $D(p(t)) = S(p(t - 1))$ gives rise to the following nonlinear discrete dynamic model

$$p(t) = f(p(t - 1)) = \frac{1}{b}\left[a - \arctan(\lambda(p(t - 1) - 1)\right].$$

The map $f(x)$ is decreasing, and by using the supply slope λ as a bifurcation parameter the equilibrium price p^* (located at the intersection between demand and supply, see

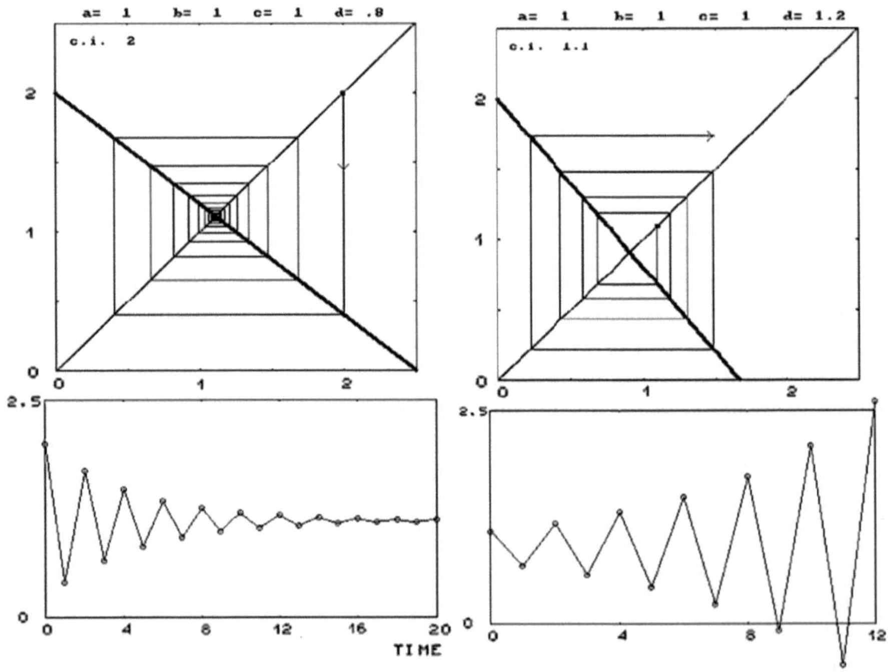

Fig. 1.60 The two possible dynamic scenarios for a linear cobweb model: oscillatory convergence (on the *left*) and oscillatory divergence (on the *right*)

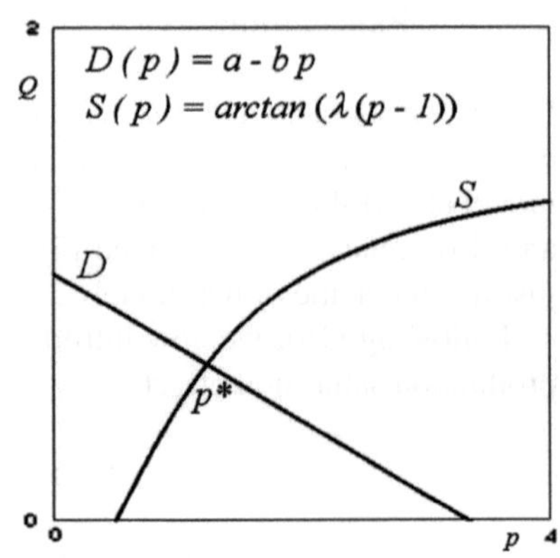

Fig. 1.61 Nonlinear supply function with saturation

Fig. 1.61) undergoes a flip bifurcation for increasing values of λ as shown in the bifurcation diagram of Fig. 1.62, where two staircase diagrams, before and after the bifurcation, are shown. So, differently from the linear model, after the stability loss of the equilibrium price a bounded oscillatory dynamics is obtained, which converges to a cycle of period 2.

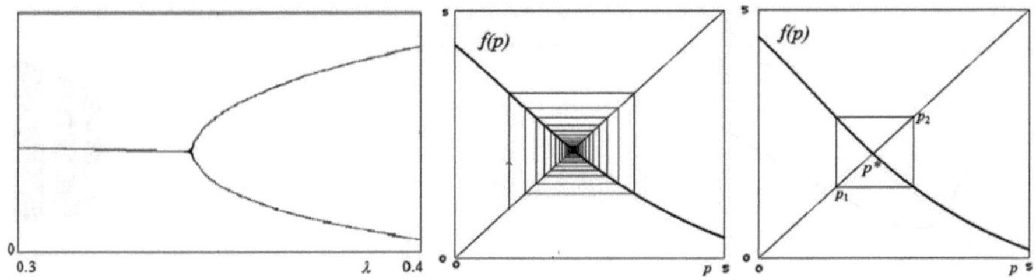

Fig. 1.62 Possible dynamic scenarios for the nonlinear cobweb model: instability of the equilibrium price without divergence

A further modification of the model consists in the introduction of adaptive expectations (see again [19])

$$p^e(t+1) = p^e(t) + \alpha(p(t) - p^e(t)) \quad 0 \le \alpha \le 1 \tag{1.59}$$

in the model

$$p(t) = f(p^e(t)) = \frac{1}{b}\left[a - \arctan\left(\lambda(p^e(t) - 1)\right)\right] \tag{1.60}$$

The equation of price expectations dynamics (1.59) can be described as follows. At any time t producers observe the discrepancy between the realized price $p(t)$ and the expected price for the same period $(p(t) - p^e(t))$ and according to such observed "estimation error" correct the previous price estimate $p^e(t)$ in order to obtain the next one: if the expected price was underestimated, i.e., $p^e(t)$ reveals to be less than the observed one $p(t)$, then they increase the current estimation in order to form the next expected price $p^e(t+1)$; if the expected price $p^e(t)$ was overestimated, i.e., it reveals to be greater than the one observed by producers, then they decrease it to form the next expected price. The value of the parameter α modulates the entity of the correction: notice that for $\alpha = 1$ adaptive expectations (1.59) reduce to naïve expectations $p^e(t+1) = p(t)$. In this sense (1.59) is a generalization of naïve expectations as these are included as a particular case. Instead in the other limiting case $\alpha = 0$ we obtain a complete inertia $p^e(t+1) = p^e(t)$, as producers never change their initial guess $p^e(0)$ on the basis of observed prices.

By inserting $p(t) = f(p^e)$ into (1.59) we get a law of motion in the space of expected prices

$$p^e(t+1) = F(p^e(t)) = p^e(t) + \alpha(f(p^e) - p^e(t)) = (1-\alpha)p^e(t) + \alpha f(p^e) \tag{1.61}$$

From the dynamics of expected prices (1.61) the corresponding dynamics of realized prices (i.e., prices really observed in the market) is obtained by the transformation $p(t) = f(p^e(t))$ in (1.60), a transformation from beliefs to realizations.

In order to analyze the dynamic behaviour of (1.61) let us notice that the function $F(p)$ is a convex combination (i.e., a weighted average) of the identity function

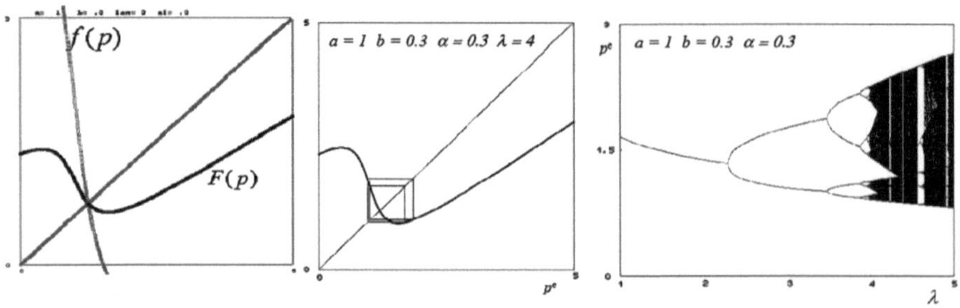

Fig. 1.63 Cobweb model with adaptive expectations

(whose graph is the diagonal) and the decreasing function f, so its graph is placed between the two graphs (see the left panel of Fig. 1.63), being closer to the diagonal as $\alpha \to 0$ and closer to the graph of f as $\alpha \to 1$.

From the derivative $F'(p) = 1 - \alpha + \alpha f'(p) = 1 - \alpha - \alpha\lambda/(b\left(1+\lambda^2 (p-1)^2\right))$ we can see that for $\alpha > b/(b+\lambda)$ it vanishes at two points, relative minimum and maximum (see Fig. 1.63). Moreover, it is always stable for sufficiently small values of α, whereas for a certain value of α the equilibrium becomes unstable through a flip bifurcation for increasing values of λ. Differently from the model with naïve expectations, where the decreasing map (even if nonlinear) could not have attractors more complex than a cycle of period 2, in this case, being the map noninvertible (i.e., characterized by the presence of turning points, relative maximum and minimum in this case) the first period doubling bifurcation is followed by a sequence of successive period doublings (the period doubling route to chaos) as shown in the bifurcation diagram of Fig. 1.63.

1.3.2 Two-Dimensional Discrete Dynamical Systems

A discrete dynamical system (1.37) with two dynamic variables, say $x_1(t)$ and $x_2(t)$ with $t \in \mathbb{N}$, assumes the form

$$\begin{aligned} x_1(t+1) &= T_1(x_1(t), x_2(t)) \\ x_2(t+1) &= T_2(x_1(t), x_2(t)) \end{aligned} \tag{1.62}$$

and needs an initial condition $(x_1(0), x_2(0))$ in order to generate a trajectory in the two-dimensional phase space. The equilibrium points of the dynamical system (1.62) are the fixed points of the map $T : \mathbb{R}^2 \to \mathbb{R}^2$, defined by the system of two equations with two unknowns

$$\begin{cases} T_1(x_1, x_2) = x_1 \\ T_2(x_1, x_2) = x_2 \end{cases} \tag{1.63}$$

Periodic cycles can be defined as in the case of one-dimensional iterated maps, just replacing $x' = f(x)$ with $\mathbf{x}' = \mathbf{T}(\mathbf{x})$ with $\mathbf{x} \in \mathbb{R}^2$ and $\mathbf{T}(\mathbf{x}) = (T_1(\mathbf{x}), T_2(\mathbf{x}))$. The stability of fixed points as well as the stability of k-periodic cycles (each periodic point being a fixed point of $\mathbf{T}^k$), as well as the kind of motion in a neighborhood of the fixed point or the periodic cycle, can be determined through the linearization of the map T in a neighborhood of the fixed point (or of any periodic point of the cycle). So, let us first analyze the dynamic properties of iterated linear maps.

1.3.2.1 Linear Systems

Let us consider the following linear (homogeneous) system of two difference equations in the (normal) form:

$$\begin{cases} x_1(t+1) = a_{11}x_1(t) + a_{12}x_2(t) \\ x_2(t+1) = a_{21}x_1(t) + a_{22}x_2(t) \end{cases} \tag{1.64}$$

that can be written in the matrix form

$$\mathbf{x(t+1) = Ax(t)} \tag{1.65}$$

where $\mathbf{A} = \begin{pmatrix} a_{11} & a_{12} \\ a_{21} & a_{22} \end{pmatrix}$; $\mathbf{x}(t) = \begin{pmatrix} x_1(t) \\ x_2(t) \end{pmatrix}$.

Like in the case of linear dynamical systems in continuous time, the general solution, i.e., set of all the solutions of (1.64), is obtained from the linear combination of two independent solutions. Moreover, also in this case, these two solutions are searched by proposing a "trial solution" in the same form as the one obtained for the one-dimensional linear difference equation, i.e.,

$$x_i(t) = v_i\lambda^t, i = 1, 2 \tag{1.66}$$

After replacing this trial solution into (1.64) we get

$$\begin{cases} \lambda^{t+1}v_1 = a_{11}\lambda^t v_1 + a_{12}\lambda^t v_2 \\ \lambda^{t+1}v_2 = a_{21}\lambda^t v_1 + a_{22}\lambda^t v_2 \end{cases}$$

and dividing for λ^t we get the usual eigenvalue problem

$$\begin{cases} (a_{11} - \lambda)v_1 + a_{12}v_2 = 0 \\ a_{21}v_1 + (a_{22} - \lambda)v_2 = 0 \end{cases}$$

that has nontrivial solutions if λ is a solution of the "characteristic equation"

$$P(\lambda) = \lambda^2 - \mathrm{Tr}(A)\lambda + \mathrm{Det}(A) = 0 \,,$$

where $\mathrm{Tr}(A) = a_{11} + a_{22}$ and $\mathrm{Det}(A) = a_{11}a_{22} - a_{12}a_{21}$.

So again, like in the case of linear dynamical systems in continuous time, the problem of finding the solutions is reduced to a problem of linear algebra, the only difference being that the solutions are now in the form (1.66) instead of (1.20). In particular, if we denote by $\Delta = \mathrm{Tr}\,(A)^2 - 4\mathrm{Det}\,(A)$ we have that

- If $\Delta > 0$ then we have two real and distinct eigenvalues and the general solution has the form
$$\mathbf{x}(t) = c_1\mathbf{v}_1\lambda_1^t + c_2\mathbf{v}_2\lambda_2^t,$$
where $\mathbf{v}_1$ and $\mathbf{v}_2$ are the corresponding eigenvectors and c_1, c_2 are real constants that are uniquely determined by imposing the initial conditions $x_i(0) = x_{i0}$, $i = 1, 2$.
- If $\Delta = 0$ then we have real and coincident eigenvalues $\lambda_1 = \lambda_2 = \lambda$ and the general solution has the form
$$\mathbf{x}(t) = c_1\mathbf{v}\lambda^t + c_2\mathbf{v}t\lambda^t.$$
- If $\Delta < 0$ then we have two complex conjugate eigenvalues $\lambda_{1,2} = -\mathrm{Tr}(A)/2 \pm i\sqrt{-\Delta}/2 = |\lambda|\,(\cos\theta \pm i\sin\theta)$ where $|\lambda| = \sqrt{\mathrm{Re}\,(\lambda)^2 + \mathrm{Im}\,(\lambda)^2} = \sqrt{\mathrm{Det}(A)}$ and $\theta = \arctan\,(\mathrm{Im}\,(\lambda)/\mathrm{Re}\,(\lambda))$ or equivalently $\cos\theta = -\mathrm{Tr}(A)/(2\sqrt{\mathrm{Det}(A)})$. The general real solution is obtained as
$$\mathbf{x}(t) = |\lambda|^t\,[(c_1\mathbf{v}_1 - c_2\mathbf{v}_2)\sin(\theta t) + (c_1\mathbf{v}_1 + c_2\mathbf{v}_2)\cos(\theta t)]$$
where $\mathbf{v} = \mathbf{v}_1 + i\mathbf{v}_2$ is a complex eigenvector associated with $\lambda_1 \in \mathbb{C}$.

In any case, we can see that the general solution converges asymptotically to the equilibrium $\mathbf{x} = \mathbf{0}$ if and only if $|\lambda_i| < 1$, $i = 1, 2$, i.e., both the eigenvalues are inside the unit circle of the complex plane defined by $\mathrm{Re}\,(\lambda)^2 + \mathrm{Im}\,(\lambda)^2 < 1$. The phase portraits associated with the different positions of the eigenvalues in the complex plane with respect to the unit circle are shown in the Fig. 1.64. The phase line represented in this qualitative picture looks quite similar to those shown for the phase portraits of linear dynamical systems in continuous time. Of course, the phase point along trajectories moves at discrete time pulses, i.e., it jumps from one point to another. However such discrete motion occurs along phase curves quite similar to those of continuous time dynamical systems, see Fig. 1.65.

The stability criterion, i.e., the necessary and sufficient conditions to have all the eigenvalues less than 1 in modulus, are given by

$$\begin{aligned} P(1) &= 1 - \mathrm{Tr}(A) + \mathrm{Det}(A) > 0\,, \\ P(-1) &= 1 + \mathrm{Tr}(A) + \mathrm{Det}(A) > 0\,, \\ \mathrm{Det}(A) &< 1\,. \end{aligned} \tag{1.67}$$

In the plane $(\mathrm{Tr}(A), \mathrm{Det}(A))$ these three conditions define the interior of a triangle (known as *stability triangle,* see Fig. 1.66) bounded by the three straight lines whose equations are given by $P(1) = 0$, $P(-1) = 0$ and $\mathrm{Det}(A) = 1$. When the

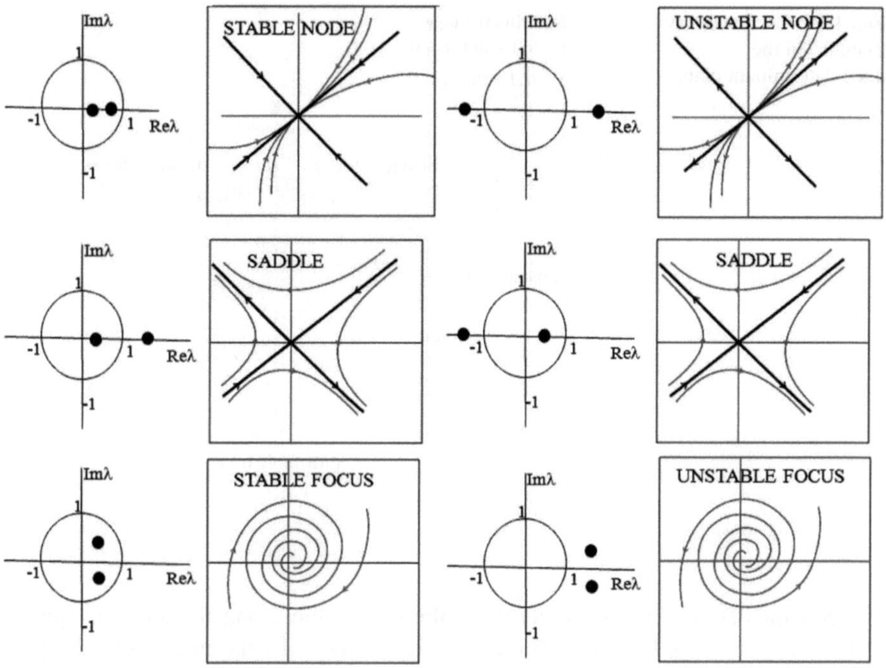

Fig. 1.64 Linear discrete dynamical systems of the plane: classification of the phase portraits for the unique fixed point

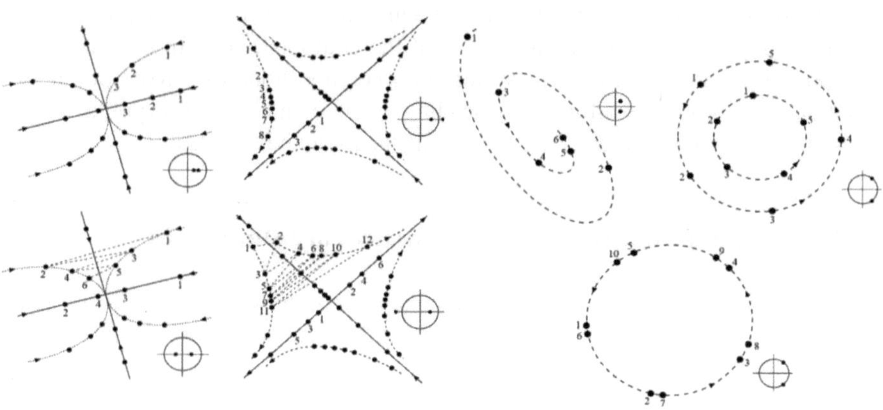

Fig. 1.65 Examples of relations between iterated points and phase curves

point $(\mathrm{Tr}(A), \mathrm{Det}(A))$ is inside the triangle then the fixed point $\mathbf{x} = \mathbf{0}$ is globally asymptotically stable, whereas when the point $(\mathrm{Tr}(A), \mathrm{Det}(A))$ is outside the triangle, the equilibrium $\mathbf{x} = \mathbf{0}$ is unstable and the trajectories diverge. Along the triangle we have non-generic (structurally unstable) situations of marginal stability.

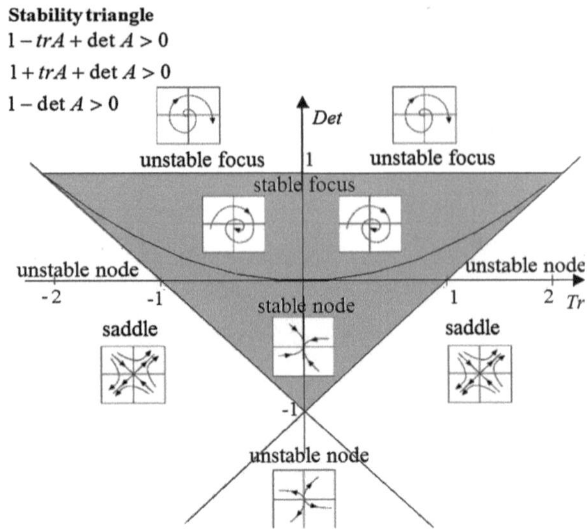

Fig. 1.66 The "stability triangle" in the trace-determinant plane

If the point $(\text{Tr}(A), \text{Det}(A))$ exits the stability triangle along the side of equation $P(1) = 1 - \text{Tr}(A) + \text{Det}(A) = 0$ then an eigenvalue exits the unit circle along the real axis in the point $\lambda = 1$; if the point $(\text{Tr}(A), \text{Det}(A))$ exits the stability triangle along the side of equation $P(-1) = 1 + \text{Tr}(A) + \text{Det}(A) = 0$ then an eigenvalue exits the unit circle along the real axis in the point $\lambda = -1$; if the point $(\text{Tr}(A), \text{Det}(A))$ exits the stability triangle along the side of equation $\text{Det} = 1$ then a pair of complex conjugate eigenvalues exit the unit circle of the complex plane. In the case of linear approximation of a nonlinear system in a neighborhood of a fixed point, such situations will represent bifurcations leading to the contact of fixed points and the creation of new kind of attractors, as we will see in the next section.

1.3.2.2 Nonlinear Discrete Dynamical Systems in 2 Dimensions

Let us consider a nonlinear discrete dynamical system in two dimensions (1.62) and let $\mathbf{x}^* = (x_1^*, x_2^*)$ be a fixed point, solution of (1.63). The linear approximation around the fixed point is given by

$$\mathbf{x}(t+1) - \mathbf{x}^* = \mathbf{J}_T(\mathbf{x}^*)(\mathbf{x}(t) - \mathbf{x}^*)$$

where $\mathbf{J}_T$ is the jacobian matrix

$$\mathbf{J}_T(\mathbf{x}) = [J_{ij}] = \begin{bmatrix} \partial T_1/\partial x_1 & \partial T_1/\partial x_2 \\ \partial T_2/\partial x_1 & \partial T_2/\partial x_2 \end{bmatrix}$$

The necessary and sufficient conditions for asymptotic stability of 2-dimensional linear systems in (1.67) can be used as sufficient conditions for *local* asymptotic stability of an equilibrium point $\mathbf{x}^*$ of a 2-dimensional *nonlinear* discrete dynamical system by using the Jacobian matrix evaluated at the fixed point $\mathbf{J}_T(\mathbf{x}^*)$ as coefficient matrix.

When all the eigenvalues are less than one in absolute values (i.e., inside the unit circle of the complex plane) then the fixed point is locally attracting. When at least one eigenvalue is greater than one in absolute value then the fixed point is unstable.

If the structure of the discrete dynamical system (1.62) depends on a parameter, say $\mu \in \mathbb{R}$, and consequently any fixed point $\mathbf{x}^* = \mathbf{x}^*(\mu)$ as well as $\mathbf{J}_T(\mathbf{x}^*)$ depend on μ, then as μ varies a real eigenvalue, say $\lambda_1(\mu)$ may exit the unit circle, or a pair of complex conjugate eigenvalues, say $\lambda_{1,2}(\mu)$, with $\lambda_2(\mu) = \bar{\lambda}_1(\mu)$, may exit the unit circle. In the former case, i.e., for real eigenvalues, we have properties similar to those already described in the one-dimensional case. That is, when one eigenvalue crosses through $\lambda = -1$ (hence the "stability triangle" of Fig. 1.66 is crossed through the left side $1 + \text{Tr} + \text{Det} = 1$) then a *flip bifurcation* may occur, while when one eigenvalue λ crosses through $\lambda = +1$ (hence the "stability triangle" of Fig. 1.66 is crossed through the right side $1 - \text{Tr} + \text{Det} = 1$) then we may have a saddle-node or a transcritical or a *pitchfork bifurcation*. However, as in the analogue situation for continuous-time dynamical systems of dimension greater than one, in the case of complex conjugate eigenvalues that exit the unit circle of the complex plane (i.e., the "stability triangle" of Fig. 1.66 is crossed through the upper side $\text{Det} = 1$) a bifurcation occurs which is the discrete-time analogue of the Hopf bifurcation for dynamical systems in continuous time. In the case of discrete time it is called *Neimark-Sacker bifurcation*. Also in this case the presence of complex eigenvalues implies oscillatory dynamics along spiralling phase curves, hence oscillations around the equilibrium, and at the Neimark-Sacker bifurcation a closed invariant curve around the fixed point is created (or around the periodic point of a cycle, because as usual any k-periodic point of a map T corresponds to a fixed point of the map T^k). Here we give a simplified description of the *Neimark-Sacker bifurcation theorem*, see more specialized books for a more rigorous statement.

Theorem 1.10 (Neimark-Sacker Bifurcation) *Let $T(x,\mu) : \mathbb{R}^2 \to \mathbb{R}^2$ be a one-parameter family of 2-dimensional maps which has a family of fixed points $x^*(\mu)$ at which the eigenvalues are complex conjugate, say $\lambda(\mu)$, $\overline{\lambda}(\mu)$. Assume that for $\mu = \mu_0$:*

1. $|\lambda(\mu_0)| = 1$, *but* $\lambda^j(\mu_0) \neq 1$ *for* $j = 1, 2, 3, 4$;
2. $\frac{d|\lambda(\mu)|}{d\mu}(\mu_0) = d \neq 0$ *(transversality condition).*

Then in a neighborhood of $x^*(\mu_0)$ the map T is topologically conjugate to the map given by (in polar coordinates) $T_e(r,\theta) = (r(1 + d(\mu - \mu_0) + ar^2), \theta + c + br^2) +$ higher-order terms. If, in addition,

3. $a \neq 0$,

then there is a simple closed invariant curve in a neighborhood of $x^*(\mu_0)$.

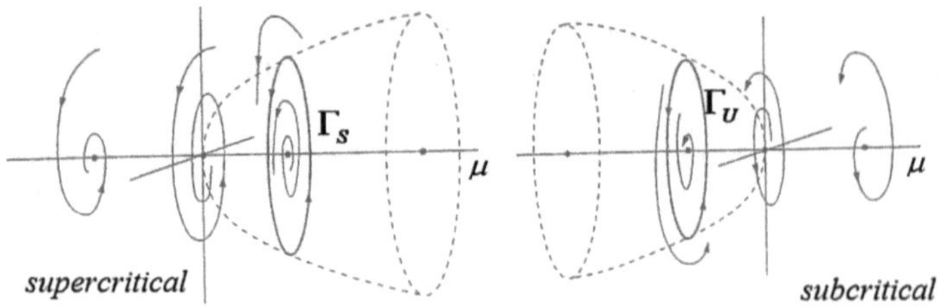

Fig. 1.67 Neimark-Sacker bifurcation

The signs of the coefficients d and a determine the direction and stability of the bifurcating orbits. The Neimark-Sacker bifurcation is called supercritical (when $a < 0$) or subcritical (when $a > 0$) (Fig. 1.67). We remark that numerically one can deduce the type of the bifurcation just from the stability of the fixed point at the bifurcation value: If the fixed point is locally attracting (resp., repelling), then the Neimark-Sacker bifurcation is supercritical (resp. subcritical). Let us notice that for *linear maps* the condition $a \neq 0$ is never satisfied, not only at the fixed point, but in the whole region of definition of the map. And, indeed, considering a linear map, say with complex conjugate eigenvalues $\lambda(\mu), \overline{\lambda}(\mu)$, if $|\lambda(\mu_0)| = 1$ then the fixed point is a center, so that the trajectory of any point different from the fixed point belongs to a different invariant ellipse and the motion is either periodic or quasi-periodic, depending on the parameters. For $\mu \neq \mu_0$ the fixed point is either a globally attracting focus or a repelling focus (in which case the trajectories go to infinity). Thus the bifurcation which occurs in a linear map, when its complex-conjugate eigenvalues cross the unit circle, is also called *center bifurcation*. A similar bifurcation can be observed in nonlinear maps with $a = 0$ as well.

The coefficients c and b give information on the rotation of the jumping phase point along the bifurcating closed invariant curve. In fact, the discrete time motion of the phase point along the closed invariant curve may be such that the jumping point completely fills the closed curve by a nonperiodic trajectory (which is denoted as quasi-periodic trajectory because it oscillates with a given period and amplitude but never hits an already visited point) or after n iterations (and m revolutions along the closed curve) it may hit an already visited point and consequently it enters a n-cycle (a phenomenon called *frequency locking*).

As an example, let us consider the map

$$\begin{aligned} x(t+1) &= y(t) \ , \\ y(t+1) &= y(t) - \alpha x(t) + x(t)^2 \ . \end{aligned} \tag{1.68}$$

It has two fixed points: $O = (0, 0)$ and $P = (\alpha, \alpha)$. The Jacobian matrix

$$J(x, y) = \begin{bmatrix} 0 & 1 \\ 2x - \alpha & 1 \end{bmatrix}$$

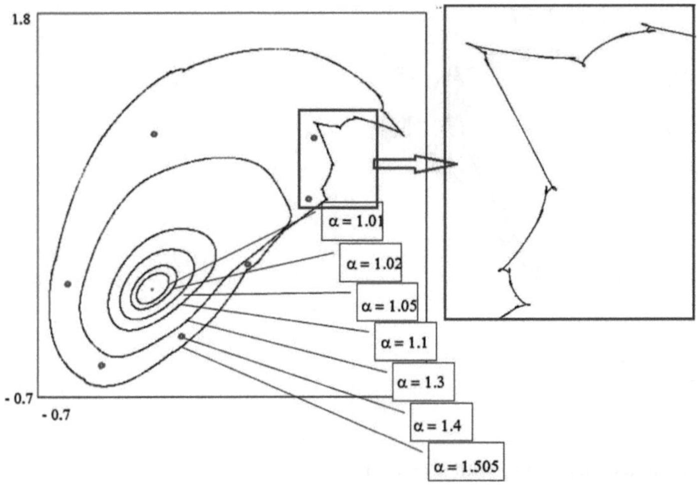

Fig. 1.68 Closed invariant curves of increasing amplitude generated by a Neimark-Sacker bifurcation of the map (1.68), until the invariant curve becomes a chaotic attractor

computed at the fixed point O is such that $\mathrm{Tr}(J(O)) = 1$ and $\mathrm{Det}(J(O)) = \alpha$. According to the stability conditions in (1.67), O is a stable fixed point for $0 < \alpha < 1$: at $\alpha = 0$ a transcritical bifurcation occurs at which O takes the stability of P; at $\alpha = 1/4$ the eigenvalues become complex conjugate, so that O is transformed from a stable node to a stable focus (but this is not a bifurcation as the phase portrait of a stable node is topologically conjugate to that of a stable focus). Then at $\alpha = 1$, O loses stability and for $\alpha > 1$ it becomes an unstable focus, with a stable invariant curve around it (supercritical Neimark-Sacker bifurcation). As α is further increased the stable closed orbit enlarges and the motion is quasi-periodic on it (see Fig. 1.68, where the development of the asymptotic trajectories is shown for increasing values of α). Just after the bifurcation the stable invariant curve is completely filled by the trajectories, and the amplitude increases as α is increased. For $\alpha = 1.4$ a frequency locking occurs and the trajectories converge to a periodic cycle of period 7 with periodic points located along the invariant curve. Then for higher values of α the closed invariant curve is broken and a more complex attracting set can be observed (see in Fig. 1.68 the enlarged portion of the attractor obtained for $\alpha = 1.505$) whose shape depends on the nonlinearities of the map prevailing far from the fixed point.

This is even more evident in Fig. 1.69, obtained for $\alpha = 1.54$, where the points of the trajectories starting from initial conditions in the white region around the fixed point O asymptotically form the chaotic attractor clearly visible in the left panel. The trajectories starting from the grey region diverge, and the boundary that separates the two basins of attraction is the stable set of the saddle point P. The right panel of the picture shows the points $x(t)$ and $y(t)$ versus time, joined by segments (just a visual trick to give more emphasis to the chaotic oscillatory pattern).

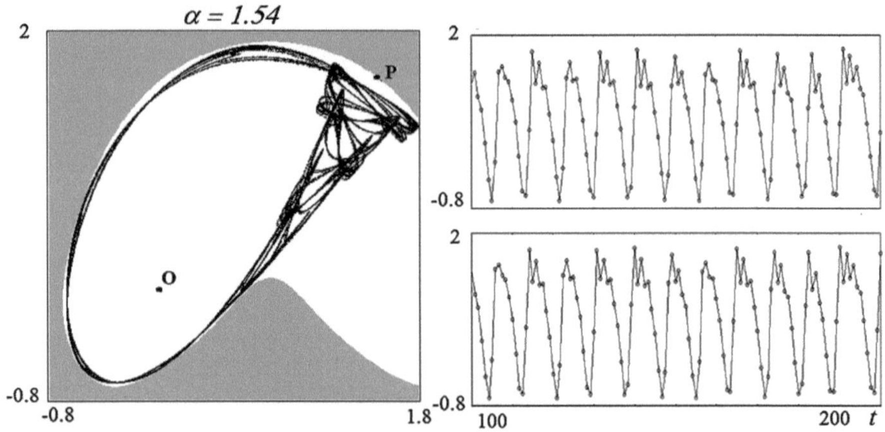

Fig. 1.69 Again the map (1.68) with $\alpha = 1.54$

1.3.2.3 Some Economic Examples

A Duopoly Game with Linear Demand and Gradient Dynamics Let us consider a Cournot duopoly, i.e., an industry consisting of two quantity-setting firms, labelled by $i = 1, 2$, producing the same good for sale on the market. Production decisions of both firms occur at discrete time periods $t = 0, 1, 2, \ldots$ Let $q_i(t)$ represent the output of the ith firm during period t, at a production cost $C_i(q_i)$. The price prevailing in period t is determined by the total supply $Q(t) = q_1(t) + q_2(t)$ through a demand function

$$p = f(Q) \tag{1.69}$$

from which the single-period profit of the ith firm is given by

$$\Pi_i(q_1, q_2) = q_i f(Q) - C_i(q_i) \ . \tag{1.70}$$

Following [10], we assume that each duopolist does not have a complete knowledge of the demand function, and tries to infer how the market will respond to its production changes by an empirical estimate of the marginal profit. This estimate may be obtained by a market research or by brief experiments of small (or local) production or price variations performed at the beginning of period t, and we assume that even if the firms are quite ignorant about the market demand, they are able to obtain a correct empirical estimate of the marginal profits $\left(\frac{\partial \Pi_i}{\partial q_i}\right)^{(e)} = \frac{\partial \Pi_i}{\partial q_i}(q_1, q_2) \;\; i = 1, 2$. This local estimate of expected marginal profits is much easier to obtain than a global knowledge of the demand function (involving values of Q that may be very different from the current ones). With this kind of information the producers behave as local profit maximizers, the local adjustment process being one where a firm increases its output if it perceives a positive marginal profit and decreases its production if it is negative:

$$q_i(t+1) = q_i(t) + \alpha_i(q_i)\,\frac{\partial \Pi_i}{\partial q_i}(q_1, q_2)\ ; \qquad i = 1, 2 \tag{1.71}$$

where $\alpha_i(q_i)$ is a positive function which gives the extent of production variation of firm i following a given profit signal. An adjustment mechanism similar to (1.71) has been proposed by some authors with constant α_i. Instead we assume α_i proportional to q_i, $\alpha_i(q_i) = v_i q_i$; $i = 1, 2$, where v_i is a positive constant which will be called *speed of adjustment*, equivalent to the assumption that the "relative production change" is proportional to the estimated marginal profit:

$$\frac{q_i(t+1) - q_i(t)}{q_i(t)} = v_i \frac{\partial \Pi_i}{\partial q_i}(q_1, q_2).$$

Let us consider a linear demand function $f(Q) = a - bQ$, with a, b positive constants, and linear cost functions $C_i(q_i) = c_i q_i$; $i = 1, 2$, where the positive constants c_i are the marginal costs. With these assumptions

$$\Pi_i(q_1, q_2) = q_i\left[a - b(q_1 + q_2) - c_i\right],\ i = 1, 2\ , \tag{1.72}$$

and the marginal profit for firm i is

$$\frac{\partial \Pi_i}{\partial q_i} = a - c_i - 2bq_i - bq_j,\ \ i, j = 1, 2,\ \ j \neq i. \tag{1.73}$$

With the above assumptions, the dynamic model is expressed by the iteration of the following two-dimensional nonlinear map $T\ (q_1, q_2) \to \left(q_1', q_2'\right)$ defined as

$$T: \begin{cases} q_1' = (1 + v_1(a - c_1))q_1 - 2bv_1 q_1^2 - bv_1 q_1 q_2 \\ \\ q_2' = (1 + v_2(a - c_2))q_2 - 2bv_2 q_2^2 - bv_2 q_1 q_2 \end{cases} \tag{1.74}$$

where $'$ denotes the unit-time advancement operator, that is, if the right-hand side variables are productions of period t then the left-hand ones represent productions of period $(t + 1)$.

The fixed points of the map (1.74) are the solutions of the algebraic system

$$\begin{cases} q_1(a - c_1 - 2bq_1 - bq_2) = 0\ , \\ q_2(a - c_2 - bq_1 - 2bq_2) = 0\ , \end{cases}$$

obtained by setting $q_i' = q_i$, $i = 1, 2$, in (1.74). We can have at most four fixed points: $E_0 = (0, 0)$, $E_1 = ((a - c_1)/(2b), 0)$ if $c_1 < a$, $E_2 = (0, (a - c_2)/(2b))$ if $c_2 < a$, which will be called *boundary equilibria*, and the fixed point $E_* = (q_1^*, q_2^*)$, with

$$q_1^* = \frac{a + c_2 - 2c_1}{3b}, \; q_2^* = \frac{a + c_1 - 2c_2}{3b} \;, \tag{1.75}$$

which is positive (i.e., it belongs to the strategy space of the duopoly model) provided that

$$\begin{cases} 2c_1 - c_2 < a \;, \\ 2c_2 - c_1 < a \;. \end{cases} \tag{1.76}$$

The equilibrium point E_*, when it exists, is the unique Nash equilibrium, located at the intersection of the two reaction curves given by the two straight lines which represent the locus of points of vanishing marginal profits (1.73). The study of the local stability of the fixed points is based on the localization, on the complex plane, of the eigenvalues of the Jacobian matrix of (1.74)

$$\mathbf{J}(q_1, q_2) = \begin{bmatrix} 1 + v_1(a - c_1 - 4bq_1 - bq_2) & -v_1 bq_1 \\ -v_2 bq_2 & 1 + v_2(a - c_2 - bq_1 - 4bq_2) \end{bmatrix} \tag{1.77}$$

It is easy to prove that whenever the equilibrium E_* exists (i.e., (1.76) are satisfied), the boundary fixed points E_i , $i = 0, 1, 2$, are unstable. In fact at E_0 the Jacobian matrix becomes a diagonal matrix

$$\mathbf{J}(0, 0) = \begin{bmatrix} 1 + v_1(a - c_1) & 0 \\ 0 & 1 + v_2(a - c_2) \end{bmatrix} \tag{1.78}$$

whose eigenvalues, given by the diagonal entries, are greater than 1 if $c_1 < a$ and $c_2 < a$. Thus E_0 is a repelling node with eigendirections along the coordinate axes. At E_1 the Jacobian matrix becomes a triangular matrix

$$\mathbf{J}\left(\frac{a - c_1}{2b}, 0\right) = \begin{bmatrix} 1 - v_1(a - c_1) & -\frac{v_1}{2}(a - c_1) \\ 0 & 1 + \frac{v_2}{2}(a - 2c_2 + c_1) \end{bmatrix} \tag{1.79}$$

whose eigenvalues, given by the diagonal entries, are $\lambda_1 = 1 - v_1(a - c_1)$, with eigenvector $\mathbf{r}_1^{(1)} = (1, 0)$ along the q_1 axis, and $\lambda_2 = 1 + v_2/2(a - 2c_2 + c_1)$, with eigenvector $\mathbf{r}_1^{(2)} = (1, 2\frac{1-v_1(a-c_1)}{v_1(a-c_1)})$. When (1.76) are satisfied E_1 is a saddle point, with local stable manifold along q_1 axis and the unstable one tangent to $\mathbf{r}_1^{(2)}$, if

$$v_1 < \frac{2}{a - c_1} \;, \tag{1.80}$$

otherwise E_1 is an unstable node. The bifurcation occurring at $v_1 = 2/(a - c_1)$ is a flip bifurcation at which E_1 from attracting becomes repelling along the q_1 axis, on which a saddle cycle of period 2 appears. The same arguments hold for the other boundary fixed point E_2. It is a saddle, with local stable manifold along the q_2 axis and the unstable one tangent to $\mathbf{r}_2^{(2)} = (1, 2\frac{1-v_2(a-c_2)}{v_2(a-c_2)})$, if

$$v_2 < \frac{2}{a - c_2}, \tag{1.81}$$

otherwise it is an unstable node. Also in this case the bifurcation that transforms the saddle into the repelling node is a flip bifurcation creating a 2-cycle saddle on the q_2 axis.

To study the local stability of the Nash equilibrium we consider the Jacobian matrix at E_*

$$\mathbf{J}(q_1^*, q_2^*) = \begin{bmatrix} 1 - 2v_1 b q_1^* & -v_1 b q_1^* \\ -v_2 b q_2^* & 1 - 2v_2 b q_2^* \end{bmatrix}. \tag{1.82}$$

Its eigenvalues are real because the characteristic equation $\lambda^2 - \text{Tr}\lambda + \text{Det} = 0$, where Tr represents the trace and Det the determinant of (1.82), has positive discriminant

$$\text{Tr}^2 - 4\text{Det} = 4b^2 \left[\left(v_1 q_1^* - v_2 q_2^* \right)^2 + v_1 v_2 q_1^* q_2^* \right] > 0.$$

It is easy to realize that $\lambda_i < 1,\ i = 1, 2$, since $1 - \text{Tr} + \text{Det} > 0$ when (1.76) hold, thus a sufficient condition for the local asymptotic stability of E_* is $1 + \text{Tr} + \text{Det} > 0$, which ensures $\lambda_i > -1,\ i = 1, 2$. This condition, which becomes

$$3b^2 q_1^* q_2^* v_1 v_2 - 4bq_1^* v_1 - 4bq_2^* v_2 + 4 < 0, \tag{1.83}$$

defines a region of stability in the plane of the speeds of adjustment (v_1, v_2) whose shape is like the shaded area of Fig. 1.70. This stability region is bounded by the portion of hyperbola, with positive v_1 and v_2, whose equation is given by the vanishing of the left hand side of (1.83). For values of (v_1, v_2) inside the stability region the Nash equilibrium E_* is a stable node, and the hyperbola represents a bifurcation curve at which E_* loses its stability through a period doubling (or *flip*) bifurcation. This

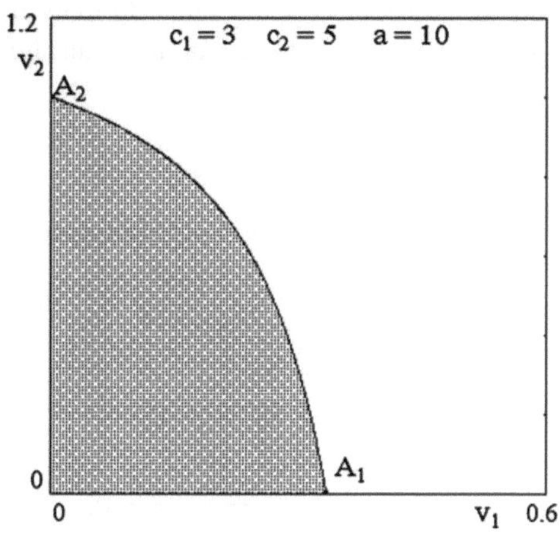

Fig. 1.70 The *shaded area* represents, in the plane of speeds of adjustment (v_1, v_2), the region of local asymptotic stability of the Nash equilibrium. The values of the other parameters are $c_1 = 3,\ c_2 = 5,\ a = 10$

bifurcation curve intersects the axes v_1 and v_2 in the points A_1 and A_2 respectively, whose coordinates are given by

$$A_1 = \left(\frac{3}{a + c_2 - 2c_1}, 0\right) \quad and \quad A_2 = \left(0, \frac{3}{a + c_1 - 2c_2}\right). \tag{1.84}$$

From these results we can obtain information on the effects of the model's parameters on the local stability of E_*. For example, an increase of the speeds of adjustment, with the other parameters held fixed, has a destabilizing effect. In fact, an increase of v_1 and/or v_2, starting from a set of parameters which ensures the local stability of the Nash equilibrium, can bring the point (v_1, v_2) out of the stability region, crossing the flip bifurcation curve. Similar arguments apply if the parameters v_1, v_2, c_1, c_2 are fixed and the parameter a, which represents the maximum price of the good produced, is increased. In this case the stability region becomes smaller, as can be easily deduced from (1.84), and this can cause a loss of stability of E_* when the moving boundary is crossed by the point (v_1, v_2). An increase of the marginal cost c_1, with c_2 held fixed, causes a displacement of the point A_1 to the right and of A_2 downwards. Instead, an increase of c_2, with c_1 held fixed, causes a displacement of A_1 to the left and of A_2 upwards. In both cases the effect on the local stability of E_* depends on the position of the point (v_1, v_2). In fact, if $v_1 < v_2$, i.e., the point (v_1, v_2) is above the diagonal $v_1 = v_2$, an increase of c_1 can destabilize E_*, whereas an increase of c_2 reinforces its stability. The situation is reversed if $v_1 > v_2$. From these arguments the combined effects due to simultaneous changes of several parameters can be deduced. For example if E_* becomes unstable because of a price increase (due to a shift of the demand curve), its stability can be regained by a reduction of the speeds of reaction, whereas an increase of a marginal cost c_i can be compensated by a decrease of the corresponding v_i: in the presence of a high marginal cost, stability is favored by a more prudent behavior (i.e., lower reactivity to profit signals).

A Duopoly Game with Isoelastic Demand and Gradient Dynamics Following [8], we consider now the same duopoly model (1.71) but with a different demand function (often used in economics as an alternative of linear demand) called isoelastic

$$p = \frac{1}{Q} \tag{1.85}$$

In this case the one-period profit of firm i is given by

$$\Pi_i(q_1, q_2) = \frac{q_i}{q_1 + q_2} - c_i q_i; \quad i = 1, 2. \tag{1.86}$$

Hence the estimated marginal profits are

$$\frac{\partial \Pi_1}{\partial q_1} = \frac{q_2}{(q_1 + q_2)^2} - c_1 \quad \text{and} \quad \frac{\partial \Pi_2}{\partial q_2} = \frac{q_1}{(q_1 + q_2)^2} - c_2.$$

With these assumptions, model (1.71) becomes

$$T: \begin{cases} q_1' = q_1 \left(1 - c_1 v_1 + v_1 \frac{q_2}{(q_1+q_2)^2}\right) \\ q_2' = q_2 \left(1 - c_2 v_2 + v_2 \frac{q_1}{(q_1+q_2)^2}\right) \end{cases} \tag{1.87}$$

The fixed points of (1.87) are defined as the non-negative solutions of the algebraic system

$$\begin{cases} q_1 \left(\frac{q_2}{(q_1+q_2)^2} - c_1\right) = 0 \\ q_2 \left(\frac{q_1}{(q_1+q_2)^2} - c_2\right) = 0 \end{cases}$$

obtained by setting $q_i' = q_i$, $i = 1, 2$, in (1.87). As the map (1.87) is not defined in $(0, 0)$, the unique equilibrium point is

$$E^* = (q_1^*, q_2^*) = \left(\frac{c_2}{(c_1 + c_2)^2}, \frac{c_1}{(c_1 + c_2)^2}\right) \tag{1.88}$$

which is also the unique Nash equilibrium of the classical Cournot duopoly game, as E^* is located at the intersection of the two reaction curves $\frac{\partial \Pi_i}{\partial q_i} = 0,\ i = 1, 2,$ (first order conditions) and also the second order sufficient conditions are satisfied at E^*, since $\frac{\partial^2 \Pi_i}{\partial q_i^2}(E^*) = -2(c_1 + c_2)c_i < 0,\ i = 1, 2$. At E^* the optimal profits of the two firms are

$$\Pi_1^* = c_2 \quad \text{and} \quad \Pi_2^* = c_1. \tag{1.89}$$

The study of the local stability of the Nash equilibrium is based on the localization, on the complex plane, of the eigenvalues of the Jacobian matrix of (1.87)

$$J(q_1, q_2) = \begin{bmatrix} 1 - v_1 c_1 + v_1 \frac{q_2(q_2-q_1)}{(q_1+q_2)^3} & v_1 \frac{q_1(q_1-q_2)}{(q_1+q_2)^3} \\ v_2 \frac{q_2(q_2-q_1)}{(q_1+q_2)^3} & 1 - v_2 c_2 + v_2 \frac{q_1(q_1-q_2)}{(q_1+q_2)^3} \end{bmatrix}$$

computed at E^*

$$J^* = \begin{bmatrix} 1 + v_1 c_1 \left(\frac{c_1-c_2}{c_1+c_2} - 1\right) & v_1 c_2 \frac{c_2-c_1}{c_1+c_2} \\ v_2 c_1 \frac{c_1-c_2}{c_1+c_2} & 1 + v_2 c_2 \left(\frac{c_2-c_1}{c_1+c_2} - 1\right) \end{bmatrix}$$

The characteristic equation $\lambda^2 - \mathrm{Tr}(J^*)\, \lambda + \mathrm{Det}(J^*) = 0$, where

$$\mathrm{Tr}(J^*) = 2\left(1 - (v_1 + v_2)\frac{c_1 c_2}{c_1 + c_2}\right) \quad \text{and}$$

$$\mathrm{Det}(J^*) = 1 + v_1 v_2 c_1 c_2 - 2(v_1 + v_2)\frac{c_1 c_2}{c_1 + c_2}$$

has complex conjugate roots if

$$(c_2v_2 - c_1v_1)(c_1v_2 - c_2v_1) < 0\ . \tag{1.90}$$

This condition can be easily visualized in the space $\mathscr{V} = \{v_1, v_2 \mid v_1 \geq 0, v_2 \geq 0\}$ of the speeds of adjustment shown in Fig. 1.71, where (1.90) is satisfied in the region, which we call region $\mathscr{F}$, between the two lines of equation

$$v_2 = \frac{c_1}{c_2}v_1 \quad \text{and} \quad v_2 = \frac{c_2}{c_1}v_1\ . \tag{1.91}$$

The Nash equilibrium E^* is locally asymptotically stable if the usual stability conditions hold

$$\begin{cases} 1 - \mathrm{Tr}(J^*) + \mathrm{Det}(J^*) = v_1v_2c_1c_2 > 0 \\ 1 + \mathrm{Tr}(J^*) + \mathrm{Det}(J^*) = c_1c_2v_1v_2 - 4\frac{c_1c_2}{c_1+c_2}(v_1+v_2) + 4 > 0 \\ \mathrm{Det}(J^*) - 1 \quad = \quad c_1c_2v_1v_2 - 2\frac{c_1c_2}{c_1+c_2}(v_1+v_2) < 0 \end{cases} \tag{1.92}$$

The first of (1.92) is always satisfied, which means that loss of stability through the critical value $\lambda = 1$ cannot occur (in fact a unique equilibrium always exists, and neither fold nor transcritical nor pitchfork bifurcation can be obtained with a unique equilibrium). The other two conditions define a bounded region of stability in the parameters' space, that can be represented in the plane $\mathscr{V}$ of the speeds of adjustment by the region $\mathscr{S} = OB_1A_1A_2B_2$, shaded in Fig. 1.71. This region, which is symmetric with respect to the diagonal $v_1 = v_2$, is bounded by the positive branches of two equilateral hyperbolae whose equations are obtained from the second and the third of (1.92) taken as equalities. From these equations the coordinates of the points A_i and B_i, $i = 1, 2$, can be easily obtained

$$A_1 = \left(\frac{2}{c_1}, \frac{2}{c_2}\right) \quad A_2 = \left(\frac{2}{c_2}, \frac{2}{c_1}\right) \quad B_1 = \left(\frac{c_1+c_2}{c_1c_2}, 0\right) \quad B_2 = \left(0, \frac{c_1+c_2}{c_1c_2}\right)\ . \tag{1.93}$$

If the marginal costs c_1 and c_2 are fixed the shape of the stability region $\mathscr{S}$ remains the same, and by increasing v_1 and/or v_2 the point $P = (v_1, v_2)$ can move out of it. If P crosses the boundary of $\mathscr{S}$ along the arc A_1A_2 (belonging to the hyperbola of equation $\mathrm{Det}(J^*) = 1$) then the fixed point E^* changes from a stable focus to an unstable focus via a Neimark-Sacker bifurcation. If P exits the region $\mathscr{S}$ by crossing one of the arcs B_1A_1 or B_2A_2 (both belonging to the other hyperbola, of equation $1 + \mathrm{Tr}(J^*) + \mathrm{Det}(J^*) = 0$) the fixed point E^* changes from an attracting node to a saddle point through a period doubling (or flip) bifurcation.

Similar arguments apply if the marginal costs (c_1, c_2) are varied. For example, if c_1 and c_2 are increased the stability region $\mathscr{S}$ becomes smaller, as can be easily

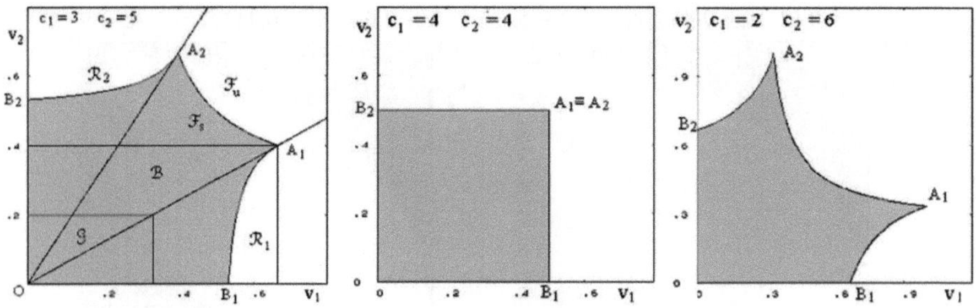

Fig. 1.71 Stability region for the model (1.87) in the (v_1, v_2)-parameter plane

deduced from (1.93), and this can cause the exit of P from $\mathscr{S}$ even if the speeds of adjustment v_1 and v_2 are held constant. Also in this case the loss of stability can occur via a Neimark-Sacker or a flip bifurcation depending on the boundary arc which is crossed by the point P.

We observe that if $c_1 > c_2$ the positions of the vertices A_1 and A_2 are swapped, and if $c_1 = c_2$ these vertices merge, and the region $\mathscr{S}$ becomes a square, like in the central panel of Fig. 1.71, bounded by the branches of a degenerate hyperbola. In this particular case the region $\mathscr{F}$ disappears, and the possibility of Neimark-Sacker bifurcations is lost. On the contrary, if the difference between the marginal costs of the two firms is increased, the region $\mathscr{F}$ enlarges and the arc A_1A_2, representing the curve where Neimark-Sacker bifurcations occur, becomes larger (see right panel of Fig. 1.71).

The fact that an increase of the speeds of adjustment has a destabilizing role in an oligopoly dynamic model is a typical result, well known in the literature. However this stability analysis reveals a new phenomenon: starting from a set of parameters for which the Nash equilibrium E^* is unstable, stability of E^* can be obtained by increasing one (or both) v_i. This happens when the point $P = (v_1, v_2)$ belongs to one of the regions denoted by $\mathscr{R}_1$ or $\mathscr{R}_2$ in the left panel of Fig. 1.71. Furthermore, if the parameters of the model are varied in such a way that the point P moves from region $\mathscr{R}_1$(or $\mathscr{R}_2$) to the region $\mathscr{F}_u$ by increasing v_1 (or v_2) we obtain two bifurcations, which cause a transition from two instability situations separated by a "window" of stability.

This particular bifurcation sequence is characterized by two different local bifurcations: a period halving (or backward flip) bifurcation followed by a supercritical Neimark-Sacker bifurcation.

A rich variety of other dynamic scenarios can be numerically shown, see, e.g., the sequences of pictures in the Fig. 1.72. In the first picture (upper left panel) $v_1 = 0.61$, $v_2 = 0.443$, $c_1 = 3$, $c_2 = 5$. At this stage the motion along the stable closed invariant curve is locked at the periodic cycle of period 7 whose periodic points are shown in the upper-left panel. Then, starting from this situation, the speed of adjustment v_1 is increased and the 7-cycle undergoes a period-doubling bifurcation leading to an

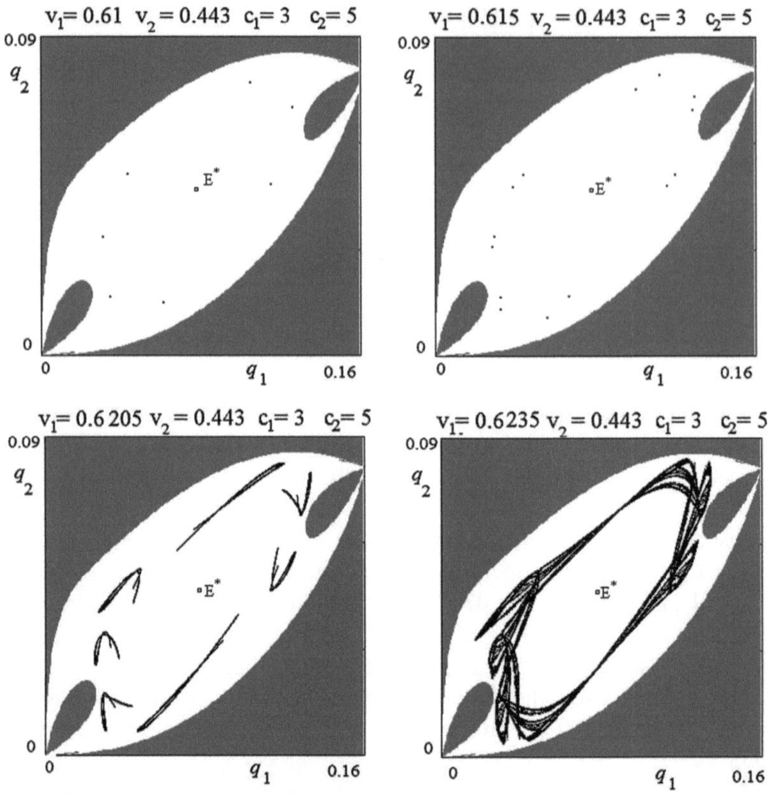

Fig. 1.72 Numerical representation of attractors and basins of the map (1.87)

attracting 14-cycle, then a 7-pieces (or 7-cyclic) chaotic attractor and finally a unique large annular chaotic attractor.

It is worth noticing that in the sequences of dynamic scenarios shown above, leading to the creation of chaotic attractors starting from sequences of local bifurcations, an attractor is eventually obtained whose points are very close to the boundary of its own basin, like in the last picture. Indeed, when a chaotic attractor has a contact with its basin's boundary it is destroyed, at a global (or contact) bifurcation denoted as "final bifurcation" or "boundary crisis". After this contact the generic initial condition in the basin of the"died attractor" belong to the basin that was "on the other side" of the basin boundary where the contact occurred. However, the skeleton of the former attractor, formed by the dense set on infinitely many repelling periodic points that where nested inside it, still exists. It is called the "ghost" of the "just died" chaotic attractor, and it implies that many trajectories spend a long number of steps (i.e., a long transients) in the region occupied by the former attractor before converging to the other attractor (that may be an attractor at finite distance or at infinity, i.e., with trajectories that diverge).

1.3.3 Discrete Dynamical System Represented by Noninvertible Maps

As we have seen through the examples of nonlinear dynamical systems discussed in the previous sections, an analysis of their global properties is necessary to understand the structure of the attractors and their basins of attraction, as well as their qualitative changes. Global properties and bifurcations are such that they cannot be deduced from the linearization procedure, based on the study of eigenvalues and eigenvectors of the Jacobian matrix.

In the case of discrete-time dynamical systems, the two kinds of complexities observed in the previous sections, given by the complex structures of the attracting sets and the complex structures of the basins of attraction, can be often characterized by the global folding properties of maps whose iteration inductively define the trajectories. In particular, as we have already discussed in the Sects. 1.3.1.4 and 1.3.1.5, a delimitation of the trapping sets (including chaotic sets) as well as a study of the complex topological structure of basins of attraction (including the case of non-connected basins), can be characterized through the analysis of noninvertible maps and their properties. The definition of critical sets, that are generalizations of local maximum and minimum values of one-dimensional maps to maps defined in higher dimensional spaces, will give us a very useful tool to detect the global (or contact) bifurcations giving rise to qualitative changes of the invariant sets and their basins. We refer the reader to [1, 3, 29].

1.3.3.1 Critical Sets: Definitions and Simple Examples

A map $T : S \to S$, $S \subseteq \mathbb{R}^n$, defined by $\mathbf{x}' = T(\mathbf{x})$, transforms a point $\mathbf{x} \in S$ into a unique point $\mathbf{x}' \in S$. The point $\mathbf{x}'$ is called the *rank-1 image* of $\mathbf{x}$, and a point $\mathbf{x}$ such that $T(\mathbf{x}) = \mathbf{x}'$ is a *rank-1 preimage* of $\mathbf{x}'$.

If $\mathbf{x} \neq \mathbf{y}$ implies $T(\mathbf{x}) \neq T(\mathbf{y})$ for each $\mathbf{x}, \mathbf{y}$ in S, then T is an *invertible map* in S, because the inverse mapping $\mathbf{x} = T^{-1}(\mathbf{x}')$ is uniquely defined; otherwise T is a said to be a *noninvertible map*, because points $\mathbf{x}$ exist that have several rank-1 preimages, i.e., the inverse relation $\mathbf{x} = T^{-1}(\mathbf{x}')$ is multivalued. So, noninvertible means "many-to-one", that is, distinct points $\mathbf{x} \neq \mathbf{y}$ may have the same image, $T(\mathbf{x}) = T(\mathbf{y}) = \mathbf{x}'$.

A one-dimensional example has been given by the logistic map (1.52) where points symmetric with respect to its symmetry axis $x = 1/2$ have the same image (see Fig. 1.53). The corresponding two inverses have been computed in (1.56).

To give an example in two dimensions, let us again consider a quadratic map $T : (x, y) \to (x', y')$ defined by

$$T : \begin{cases} x' = ax + y \\ y' = x^2 + b \end{cases} \tag{1.94}$$

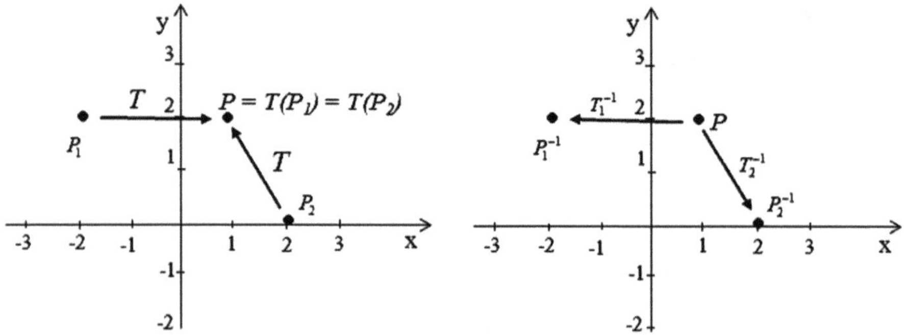

Fig. 1.73 *Left panel* Action of the map T given in (1.94). *Right panel* Action of the two inverses T_1^{-1} and T_2^{-1} defined in (1.95)

It can map distinct points into the same point. For example if we consider the map with parameters $a = 1/2$ and $b = -2$, then the two points $P_1 = (-2, 2)$ and $P_2 = (2, 0)$ are mapped into the same point $P = (1, 2)$. This means that at least two inverses must be defined in P, one mapping it into the rank-q preimage P_1 and the other one into the other preimage P_2 (see Fig. 1.73). Indeed, like in the case of the logistic map, also for this two-dimensional map we can explicitly compute the two inverses: given x' and y', if we solve the algebraic system (1.94) with respect to the unknowns x and y we get two solutions, given by

$$T_1^{-1}: \begin{cases} x = -\sqrt{y' - b} \\ y = x' + a\sqrt{y' - b} \end{cases} ; \quad T_2^{-1}: \begin{cases} x = \sqrt{y' - b} \\ y = x' - a\sqrt{y' - b} \end{cases} \tag{1.95}$$

Geometrically, the action of a noninvertible map can be expressed by saying that it "folds and pleats" the space S, so that distinct points are mapped into the same point. This is equivalently stated by saying that several inverses are defined in some points of S, and these inverses "unfold" S.

For a noninvertible map, S can be subdivided into regions Z_k, $k \geq 0$, whose points have k distinct rank-1 preimages. Generally, for a continuous map, as the point $\mathbf{x}'$ varies in $\mathbb{R}^n$, pairs of preimages appear or disappear as it crosses the boundaries separating different regions. Hence, such boundaries are characterized by the presence of at least two coincident (merging) preimages. This leads us to the definition of the *critical sets*, one of the distinguishing features of noninvertible maps.

Definition 1.10 (*Gumowski and Mira* [17]) The *critical set CS* of a continuous map T is defined as the locus of points having at least two coincident *rank* $-$ 1 preimages, located on a set CS_{-1}, called *set of merging preimages*.

The critical set CS is generally formed by $(n-1)$-dimensional hypersurfaces of $\mathbb{R}^n$, and portions of CS separate regions Z_k of the phase space characterized by a different number of *rank* $-$ 1 preimages, for example Z_k and Z_{k+2} (this is the standard occurrence for continuous maps). The critical set CS is the n-dimensional generalization of the notion of local minimum or local maximum value of a one-dimensional map, and of the notion of *critical curve* LC of a noninvertible two-dimensional map.[8] The set CS_{-1} is the generalization of local extremum point of a one-dimensional map, and of the *fold curve* LC_{-1} of a two-dimensional noninvertible map. In a differentiable one-dimensional map the critical points c_{-1} can be searched among the points where the derivative vanishes, as we have seen for the logistic map.

However, we remark that in general the condition of vanishing derivative is not sufficient to define the critical points of rank-0 since such condition may be also satisfied by points which are not local extrema (e.g., the inflection points with horizontal tangent). Moreover, for continuous and piecewise differentiable maps, as well as for discontinuous maps, the condition of vanishing derivative is not necessary as well, because such maps may have the property that the images of points where the map is not differentiable are critical points, according to the definition given above. This occurs whenever such kink points are local maxima or minima. Even in the case of piecewise continuous maps, a point of discontinuity may behave as a critical point of T, even if the definition in terms of merging preimages cannot be applied. The difference with respect to the case of a continuous map is that now the number of distinct rank-1 preimages through a critical point differs generally by one (instead of two), that is, a critical value c (in general the critical set CS) separates regions Z_k and Z_{k+1}.

In order to explain the geometric action of a critical point in a continuous map, let us consider, again, the logistic map, and as already stressed in Sect. 1.3.1.4 let us notice that as x moves from 0 to 1 the corresponding image $f(x)$ spans the interval $[0, c]$ twice, the critical point c being the turning point (see again Figs. 1.53 and 1.54). In other words, if we consider how the segment $\gamma = [0, 1]$ is transformed by the map f, we can say that it is *folded and pleated* to obtain the image $\gamma' = [0, c]$. Such folding gives a geometric reason why two distinct points of γ, say x_1 and x_2, located symmetrically with respect to the point $c_{-1} = 1/2$, are mapped into the same point $x' \in \gamma'$ due to the folding action of f. The same arguments can be explained by looking at the two inverse mappings f_1^{-1} and f_2^{-1} defined in $(-\infty, \mu/4]$ according to (1.56). We can consider the range of the map f formed by the superposition of two half-lines $(-\infty, \mu/4]$, joined at the critical point $c = \mu/4$, and on each of these half-lines a different inverse is defined. In other words, instead of saying that two distinct maps are defined on the same half-line we say that the range is formed by two distinct half lines on each of which a unique inverse map is defined. This point of view gives a geometric visualization of the critical point c as the point in which two distinct inverses merge. The action of the inverses, say $f^{-1} = f_1^{-1} \cup f_2^{-1}$, causes an unfolding of the range by mapping c into c_{-1} and by opening the two half-lines one on the right and one on the left of c_{-1}, so that the whole real line $\mathbb{R}$ is covered. So, the map f folds the real line, the two inverses unfold it.

[8] The terminology and notation originate from the notion of critical point as it is used in the classical works of Julia and Fatou.

Another interpretation of the folding action of a critical point is the following. Since $f(x)$ is increasing for $x \in [0, 1/2)$ and decreasing for $x \in (1/2, 1]$, its application to a segment $\gamma_1 \subset [0, 1/2)$ is orientation preserving, whereas its application to a segment $\gamma_2 \subset (1/2, 1]$ is orientation reversing. This suggests that an application of f to a segment $\gamma_3 = [a, b]$ including the point $c_{-1} = 1/2$ preserves the orientation of the portion $[a, c_{-1}]$, i.e., $f([a, c_{-1}]) = [f(a), c]$, whereas it reverses the portion $[c_{-1}, b]$, i.e., $f([c_{-1}, b]) = [f(b), c]$, so that $\gamma_3' = f(\gamma_3)$ is folded, the folding point being the critical point c.

Let us now consider the case of a continuous two-dimensional map $T : S \to S$, $S \subseteq \mathbb{R}^2$, defined by

$$T : \begin{cases} x_1' = T_1(x_1, x_2) \\ x_2' = T_2(x_1, x_2) \quad , \end{cases} \tag{1.96}$$

If we solve the system of the two equations (1.96) with respect to the unknowns x_1 and x_2, then, for a given $\left(x_1', x_2'\right)$, we may have several solutions, representing rank-1 preimages (or backward iterates) of $\left(x_1', x_2'\right)$, say $(x_1, x_2) = T^{-1}\left(x_1', x_2'\right)$, where T^{-1} is in general a multivalued relation. In this case we say that T is noninvertible, and the critical set (formed by critical curves, denoted by LC from the French "Ligne Critique") constitutes the set of boundaries that separate regions of the plane characterized by a different number of rank-1 preimages. According to the definition, along LC at least two inverses give merging preimages, located on LC_{-1}.

For a continuous and (at least piecewise) differentiable noninvertible map of the plane, the set LC_{-1} is included in the set where Det $\mathbf{J}(x_1, x_2)$ changes sign, since T is locally an orientation preserving map near points (x_1, x_2) such that Det $\mathbf{J}(x_1, x_2) > 0$ and orientation reversing if Det $\mathbf{J}(x_1, x_2) < 0$. In order to explain this point, let us recall that when an affine transformation $\mathbf{x}' = A\mathbf{x} + \mathbf{b}$, where $A = \left\{a_{ij}\right\}$ is a 2×2 matrix and $\mathbf{b} \in \mathbb{R}^2$, is applied to a plane figure, then the area of the transformed figure grows, or shrinks, by a factor $\rho = |\text{Det}A|$, and if Det$A > 0$ then the orientation of the figure is preserved, whereas if Det$A < 0$ then the orientation is reversed. This property also holds for the linear approximation of (1.96) in a neighborhood of a point $\mathbf{p} = (x_1, x_2)$, given by an affine map with $A = \mathbf{J}$, $\mathbf{J}$ being the Jacobian matrix evaluated at the point $\mathbf{p}$

$$\mathbf{J}(\mathbf{p}) = \begin{bmatrix} \partial T_1/\partial x_1 & \partial T_1/\partial x_2 \\ \partial T_2/\partial x_1 & \partial T_2/\partial x_2 \end{bmatrix} \tag{1.97}$$

A qualitative visualization is given in Fig. 1.74. Of course, if the map is continuously differentiable then the change of the sign of $\mathbf{J}$ occurs along points where $\mathbf{J}$ vanishes, thus giving the characterization of the fold line LC_{-1} as the locus where the jacobian vanishes.

In order to give a geometrical interpretation of the action of a multi-valued inverse relation T^{-1}, it is useful to consider a region Z_k as the superposition of k sheets, each associated with a different inverse. Such a representation is known as *Riemann foliation* of the plane. Different sheets are connected by folds joining two sheets, and the projections of such folds on the phase plane are arcs of LC. This is shown in

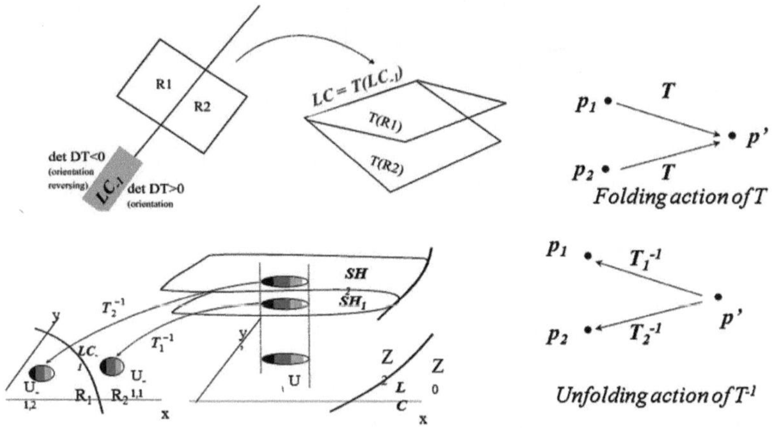

Fig. 1.74 Folding and unfolding action of a two-dimensional noninvertible map

the qualitative sketch of Fig. 1.74, where the case of a $Z_0 - Z_2$ noninvertible map is considered. This graphical representation of the unfolding action of the inverses also gives an intuitive idea of the mechanism which causes the creation of nonconnected basins for noninvertible maps of the plane.

Let us consider again the map (1.94) as a canonical example of a two-dimensional noninvertible map. Given a point (x', y'), according to (1.95) it has two rank-one preimages if $y' \geq b$, and no preimages if $y' < b$. So, (1.94) is a $Z_0 - Z_2$ noninvertible map, where Z_0 (region whose points have no preimages) is the half plane $Z_0 = \{(x, y) \,|y < b\}$ and Z_2 (region whose points have two distinct rank-1 preimages) is the half plane $Z_2 = \{(x, y) \,|y > b\}$. The line $y = b$, which separates these two regions, is LC, i.e., the locus of points having two merging rank-1 preimages, located on the line $x = 0$, that represents LC_{-1}. Being (1.94) a continuously differentiable map, the points of LC_{-1} necessarily belong to the set of points at which the Jacobian determinant vanishes, i.e., $LC_{-1} \subseteq J_0$, where $J_0 = \{(x, y) \,|\text{Det}J(x, y) = -2x = 0\}$. In this case LC_{-1} coincides with J_0 (the vertical axis $x = 0$) and the critical curve LC is the image by T of LC_{-1}, i.e., $LC = T(LC_{-1}) = T(\{x = 0\}) = \{(x, y) \,|y = b\}$.

In order to show the folding action related to the presence of the critical lines fact, we consider a plane figure (a circle) U divided by LC_{-1} into two portions, say $U_1 \in R_1$ and $U_2 \in R_2$ (Fig. 1.75, *left panel*) and we apply the map (1.94) to the points of U. The image $T(U_1) \cap T(U_2)$ is a nonempty set included in the region Z_2, which is the region whose points p' have rank-1 preimages $p_1 = T_1^{-1}(p') \in U_1$ and $p_2 = T_2^{-1}(p') \in U_2$. This means that two points $p_1 \in U_1$ and $p_2 \in U_2$, located at opposite sides with respect to LC_{-1}, are mapped in the same side with respect to LC, in the region Z_2. This is also expressed by saying that the ball U is "folded" by T along LC on the side with more preimages (see Fig. 1.75, left panel). The same concept can be equivalently expressed by stressing the "unfolding" action of T^{-1}, obtained by the application of the two distinct inverses in Z_2 which merge along LC.

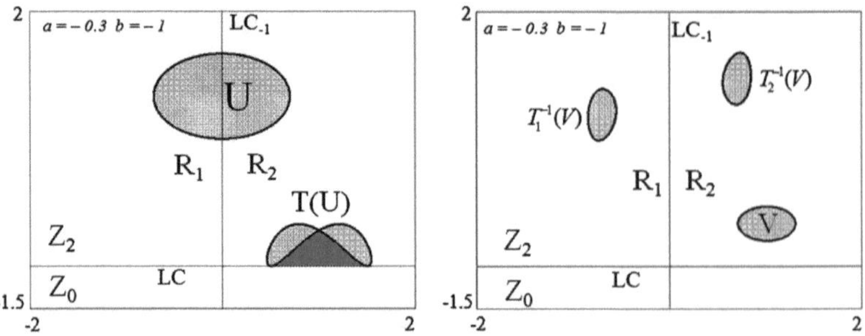

Fig. 1.75 Folding and unfolding action of the map (1.94)

Indeed, if we consider a ball $V \subset Z_0$, then the set of its *rank* $-$ 1 preimages $T_1^{-1}(V)$ and $T_2^{-1}(V)$ is made up of two balls $T_1^{-1}(V) \in R_1$ and $T_2^{-1}(V) \in R_2$. These balls are disjoint if $V \cap LC = \emptyset$ (Fig. 1.75, right panel).

Many of the considerations made above, for 1-dimensional and 2-dimensional noninvertible maps, can be generalized to n-dimensional ones, even if their visualization becomes more difficult. First of all, from the definition of critical set it is clear that the relation $CS = T(CS_{-1})$ holds in any case. Moreover, the points of CS_{-1} where the map is continuously differentiable are necessarily points where the Jacobian determinant vanishes:

$$CS_{-1} \subseteq J_0 = \{p \in \mathbb{R}^n \mid \text{Det}\,\mathbf{J}(p) = 0\} \tag{1.98}$$

In fact, in any neighborhood of a point of CS_{-1} there are at least two distinct points which are mapped by T in the same point. Accordingly, the map is not locally invertible in points of CS_{-1}, and (1.98) follows from the implicit function theorem. This property provides an easy method to compute the critical set for continuously differentiable maps: from the expression of the jacobian determinant one computes the locus of points at which it vanishes, then the set obtained after an application of the map to these points is the critical set CS.

A problem that often arises in the study of nonlinear dynamical systems concerns the existence of several attracting sets, each with its own basin of attraction. In this case the dynamic process becomes path-dependent, i.e., which kind of long run dynamics characterizes the system depends on the starting condition. Another important problems in the study of applied dynamical systems is the delimitation of a bounded region of the state space where the system dynamics are ultimately trapped, despite of the complexity of the long-run time patterns. This is an useful information, even more useful than a detailed description of step by step time evolution.

Both these questions require an analysis of the global dynamical properties of the dynamical system, that is, an analysis which is not based on the linear approximation of the map. When the map T is noninvertible, its global dynamical properties can be characterized by using the formalism of critical sets, by which the folding action associated with the application of the map, as well as the "unfolding" associated with the action of the inverses, can be described. Loosely speaking, the repeated application of a noninvertible map repeatedly folds the state space along the critical sets and their images, and often this allows one to define a bounded region where asymptotic dynamics are trapped. As some parameter is varied, global bifurcations that cause sudden qualitative changes in the properties of the attracting sets can be detected by observing contacts of critical curves with invariant sets. Instead, the repeated application of the inverses "repeatedly unfold" the state space, so that a neighborhood of an attractor may have preimages far from it, thus giving rise to complicated topological structures of the basins, that may be formed by the union of several (even infinitely many) nonconnected portions.

1.3.3.2 Critical Sets and the Delimitation of Trapping Regions

Portions of the critical set CS and its images $CS_k = T^k(CS)$ can be used to obtain the boundaries of trapping regions where the asymptotic dynamics of the iterated points of a noninvertible map are confined. This has already been explained for the logistic map in Sect. 1.3.1.4, where we have shown that, for $3 < \mu < 4$, starting from an initial condition inside the interval $[c_1, c]$, with $c_1 = f(c)$, no images can be obtained out of this interval, i.e., the interval formed by the critical point c and its rank-1 image c_1 is trapping. Moreover, any trajectory generated from an initial condition in (0, 1), enters $[c_1, c]$ after a finite number of iterations. This is expressed by saying that the interval $[c_1, c]$ is *absorbing*. Examples have been shown in Figs. 1.48 and 1.50.

In general, for an n-dimensional map, an *absorbing region* $\mathscr{A}$ *(intervals in R, areas in* R^2*, volumes in* R^3*, ...)* is defined as a bounded set whose boundary is given by portions of the critical set CS and its images of increasing order $CS_k = T^k\,(CS)$, such that a neighborhood $U \supset \mathscr{A}$ exists whose point enter $\mathscr{A}$ after a finite number of iterations and then never escape it, since $T(\mathscr{A}) \subseteq \mathscr{A}$, i.e., $\mathscr{A}$ is trapping. Loosely speaking, we can say that the iterated application of a noninvertible map, folding and folding again the space, defines trapping regions bounded by critical sets of increasing order.

Sometimes, smaller absorbing regions are nested inside a larger one, as it was illustrated for the logistic map (1.52), as shown in Fig. 1.50, where inside the absorbing interval $[c_1, c]$ a trapping subset is obtained by higher rank images of the critical point, given by $\mathscr{A} = [c_1, c_3] \cup [c_2, c]$.

Inside an absorbing region one or more attractors may exist. However, if a chaotic attractor exists which fills up a whole absorbing region then the boundary of the chaotic attractor is formed by portions of critical sets. To better illustrate this point, we also give a two-dimensional example, obtained by using the map (1.94). In Fig. 1.76a, a chaotic trajectory is shown, and in Fig. 1.76b its outer boundary is obtained by the union of a segment of LC and three iterates $LC_i = T^i(LC)$, $i = 1, 2, 3$.

A practical procedure can be outlined in order to obtain the boundary of an absorbing area (although it is difficult to give a general method). Starting from a portion of LC_{-1}, approximately taken in the region occupied by the area of interest, its images by T of increasing rank are computed until a closed region is obtained. When such

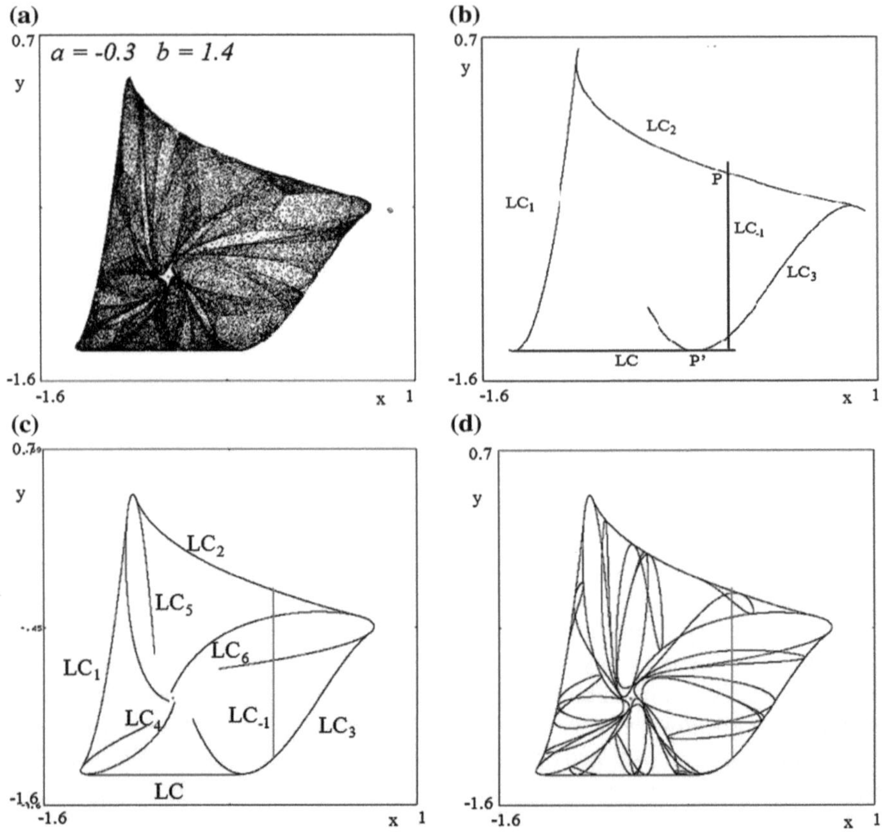

Fig. 1.76 Delimitation of a chaotic area of the map (1.94) by portions of critical curves of increasing rank

a region is mapped into itself, then it is an absorbing area $\mathscr{A}$. The length of the initial segment is to be taken, in general, by a trial and error method, although several suggestions are given in the books in the bibliography. Once an absorbing area $\mathscr{A}$ is found, in order to see if it is invariant or not the same procedure must be repeated by taking only the portion

$$\gamma = \mathscr{A} \cap LC_{-1} \tag{1.99}$$

as the starting segment. Then one of the following two cases occurs:

(**case I**) the union of m iterates of γ (for a suitable m) covers the whole boundary of $\mathscr{A}$; in which case $\mathscr{A}$ is an invariant absorbing area, and

$$\partial \mathscr{A} \subset \bigcup_{k=1}^{m} T^k(\gamma) \tag{1.100}$$

(**case II**) no natural m exists such that $\bigcup_{i=1}^{m} T^i(\gamma)$ covers the whole boundary of $\mathscr{A}$; in which case $\mathscr{A}$ is not invariant but strictly mapped into itself. An invariant absorbing area is obtained by $\cap_{n>0}T^n(\mathscr{A})$ (and may be obtained by a finite number of images of $\mathscr{A}$).

The application of this procedure to the problem of the delimitation of the chaotic area of Fig. 1.76a by portions of critical curves suggests us, on the basis of Fig. 1.76b, to take a smaller segment γ and to take an larger number of iterates in order to obtain also the inner boundary. The result is shown in Fig. 1.76c, where by four iterates we get the outer boundary. By a few more iterates also the inner boundary of the chaotic area is get, as shown in Fig. 1.76d. As it can be clearly seen, and as clearly expressed by the strict inclusion in (1.100), the union of the images also include several arcs internal to the invariant area $\mathscr{A}$. Indeed, the images of the critical arcs which are mapped inside the area play a particular role, because these curves represent the "foldings" of the plane under forward iterations of the map, and this is the reason why these inner curves often denote the portions of the region which are more frequently visited by a generic trajectory inside it (compare Fig. 1.76a, d). This is due to the fact that points close to a critical arc LC_i, $i \geq 0$, are more frequently visited, because there are several distinct parts of the invariant area which are mapped in the same region (close to LC_i).

1.3.3.3 Critical Sets and the Creation of Nonconnected Basins

In the case of noninvertible maps, the multiplicity of preimages may lead to basins with complex structures, such as multiply connected or nonconnected sets, sometimes formed by infinitely many nonconnected portions. As already stressed in Sect. 1.3.1.5, in the context of noninvertible maps it is useful to define the *immediate basin* $\mathscr{B}_0(A)$, of an attracting set A, as the widest connected component of the basin which contains A. Then the total basin can be expressed as

$$\mathscr{B}(A) = \bigcup_{n=0}^{\infty} T^{-n}(\mathscr{B}_0(A))$$

where $T^{-n}(x)$ represents the set of all the rank-n preimages of x, i.e., the set of points which are mapped in x after n iterations of the map T. The backward iteration of a noninvertible map *repeatedly unfolds* the phase space, and this implies that the basins may be nonconnected, i.e., formed by several disjoint portions.

Also in this case, we have already given an example of this property for by using a one-dimensional map, where in Fig. 1.59 the graph of a $Z_1 - Z_3 - Z_1$ noninvertible map is shown, Z_3 being the portion of the codomain bounded by the relative minimum value $c_{\min}$ and relative maximum value $c_{\max}$. In the situation shown in Fig. 1.59 there are three attractors, and after the global bifurcation where $c_{\min} = q$, the portion $(c_{\min}, q)$ enters Z_3, so new preimages $f^{-k}(c_{\min}, q)$ appear with $k \geq 1$. These preimages constitute an infinite (countable) set of nonconnected portions of $\mathscr{B}(r)$ nested inside $\mathscr{B}(A)$, represented by the thick portions of the diagonal in Fig. 1.59, bounded by the infinitely many preimages of any rank, say q_{-k}, $k \in \mathbb{N}$, of q, that accumulate in a left neighborhood of the fixed point z. In fact, as z is a repelling fixed point for the forward iteration of f, it is an attracting fixed point for the backward iteration of the same map. So, the contact between the critical point $c_{\min}$ and the basin boundary q marks the transition from simple connected to nonconnected basins. Similar global bifurcations, due to contacts between critical sets and basin boundaries, also occur in higher dimensional maps. In fact, if a parameter variation causes a crossing between a basin boundary and a critical set which separates different regions Z_k so that a portion of a basin enters a region where a larger number of inverses is defined, then new components of the basin may suddenly appear at the contact. However, for maps of dimension greater than 1, such kinds of bifurcations can be very rarely studied by analytical methods, since the analytical equations of such singularities are not known in general. Hence such studies are mainly performed by geometric and numerical methods.

1.3.3.4 Some Economic Examples

In this section we show how the global properties of noninvertible two-dimensional maps can be used in the study of discrete dynamic models in economics. In particular we will see the practical usage of critical curves to detect global bifurcations that change the qualitative structure of the basins of attraction and how critical curves and their images are employed to bound trapping regions where asymptotic dynamics are confined.

Global Properties of the Cournot Duopoly Model with Linear Demand and Gradient Dynamics We re-consider the duopoly model described in Sect. 1.3.2.3, and we perform a deeper analysis of its global dynamic properties (see, e.g., [10]). The map (1.74) is a noninvertible map of the plane, that is, starting from some non-negative initial production strategy (q_{10}, q_{20}) the iteration of (1.74) uniquely defines the trajectory $(q_1(t), q_2(t)) = T^t(q_{10}, q_{20})$, $t = 1, 2, \ldots$, whereas the backward iteration of (1.74) is not uniquely defined. In fact a point (q_1', q_2') of the plane may have several preimages, obtained by solving the fourth degree algebraic system (1.74) with respect to q_1 and q_2. In order to understand the structure of the critical curves LC and consequently the subdivision of the phase plane into zones Z_k with k preimages, we start from LC_{-1}, that for a differentiable map like (1.74), according to (1.98), is given by the locus of points where the Jacobian determinant vanishes. From the expression of J given in (1.77), the condition $\text{Det}J = 0$ becomes

$$q_1^2 + q_2^2 + 4q_1q_2 - \alpha_1 q_1 - \alpha_2 q_2 + \beta = 0$$

where

$$\alpha_i = \frac{4(1 + v_j(a - c_j)bv_i) + 1 + v_i(a - c_i)bv_j}{4b^2 v_1 v_2} \quad ; \quad i = 1, 2 \, ; \; j \neq i$$

and

$$\beta = \frac{(1 + v_1(a - c_1)bv_1)(1 + v_2(a - c_2)bv_2)}{4b^2 v_1 v_2}.$$

This is an hyperbola in the plane (q_1, q_2) with symmetry centre in the point $((2\alpha_2 - \alpha_1)/3, (2\alpha_1 - \alpha_2)/3)$ and asymptotes of angular coefficients $\left(-2 \pm \sqrt{3}\right)$. Thus LC_{-1} is formed by two branches, denoted by $LC_{-1}^{(a)}$ and $LC_{-1}^{(b)}$ in Fig. 1.77. This implies that also LC is the union of two branches, denoted by $LC^{(a)} = T(LC_{-1}^{(a)})$ and $LC^{(b)} = T(LC_{-1}^{(b)})$. Each branch of the critical curve LC separates the phase plane of T into regions whose points possess the same number of distinct rank-1 preimages. In the case of the map (1.74) $LC^{(b)}$ separates the region Z_0, whose points have no preimages, from the region Z_2, whose points have two distinct rank-1 preimages, and $LC^{(a)}$ separates the region Z_2 from Z_4, whose points have four distinct preimages. In fact, it is possible to show (see below) that the point $\left(q_1', q_2'\right) = (0, 0)$ has four preimages obtained by solving the algebraic system (1.74) with respect to the unknowns (q_1, q_2), hence $(0, 0) \in Z_4$. The other zones are classified by remembering that any branch of LC is characterized by the merging (and disappearance) of two preimages.

In order to study the action of the multivalued inverse relation T^{-1} it is useful to consider a region Z_k of the phase plane as the superposition of k sheets, each associated with a different inverse. Such a representation is known as *foliation* of the plane. Different sheets are connected by folds joining two sheets, and the projections of such folds on the phase plane are arcs of LC. The foliation associated with the

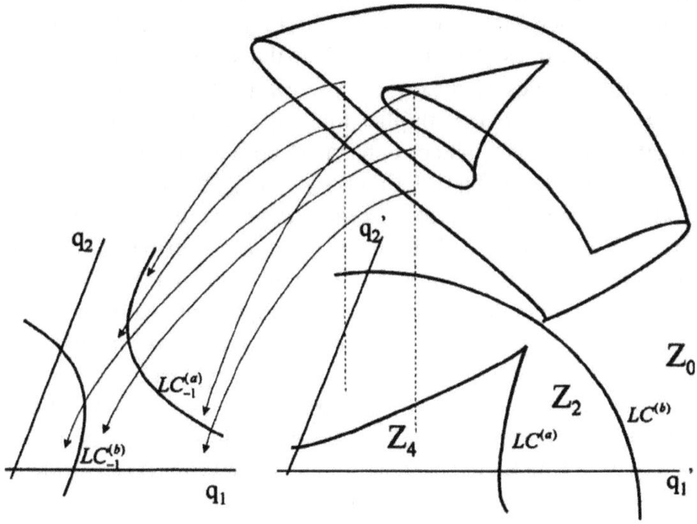

Fig. 1.77 Riemann foliation for the map (1.74)

map (1.74) is qualitatively represented in Fig. 1.77. It can be noticed that the cusp point of *LC* is characterized by three merging preimages at the junction of two folds.

An important property of the map (1.74) is that each coordinate axis $q_i = 0,\ i = 1, 2$, is trapping, that is, mapped into itself, since $q_i = 0$ gives $q_i' = 0$ in (1.74). This means that starting from an initial condition on a coordinate axis (*monopoly case*) the dynamics is confined in the same axis for each t, governed by the restriction of the map T to that axis. Such a restriction is given by the following one-dimensional map, obtained from (1.74) with $q_i = 0$

$$q_j = (1 + v_j(a - c_j))q_j - 2bv_jq_j^2 \qquad j \neq i\ . \tag{1.101}$$

This map is conjugate to the standard logistic map $x' = \mu x\,(1 - x)$ through the linear transformation

$$q_j = \frac{1 + v_j(a - c_j)}{2bv_j}x \tag{1.102}$$

from which we obtain the relation

$$\mu = 1 + v_j(a - c_j)\ . \tag{1.103}$$

This means that the dynamics of (1.101) can be obtained from the well known dynamics of (1.52).

Another important feature of the map (1.74) is that it can generate unbounded (i.e., divergent) trajectories (this can be also expressed by saying that (1.74) has an attracting set at infinite distance). In fact, unbounded (and negative) trajectories are obtained if the initial condition is taken sufficiently far from the origin, i.e., in a suitable neighborhood of infinity, since if $q_{i0} > (1 + a - c_i)\,/(bv_i),\ i = 1, 2$, then the first iterate of (1.74) gives negative values $q_i' < 0,\ i = 1, 2$, so that the successive iterates give negative and decreasing values because $q_i' = q_i + v_iq_i\left(a - c_i - 2bq_i - bq_j\right) < q_i$ being $(a - c_i) > 0$ if (1.76) hold. This implies that any attractor at finite distance cannot be globally attracting in $\mathbb{R}_+^2$, since its basin of attraction cannot extend out of the rectangle $[0, (1 + a - c_1)\,/(bv_1)] \times [0, (1 + a - c_2)\,/\,(bv_2)]$.

In the following we call *attractor at finite distance*, denoted by $\mathscr{A}$, a bounded attracting sets (which may be the Nash equilibrium E_*, a periodic cycle or some more complex attractor around E_*) in order to distinguish it from the limit sets at infinite distance, i.e., the unbounded trajectories, which represent exploding (or collapsing) evolutions of the duopoly system. We denote by $\mathscr{B}(\mathscr{A})$ the basin of attraction of an attractor $\mathscr{A}$, defined as the open set of points (q_1, q_2) of the phase plane whose trajectories $T^t(q_1, q_2)$ have limit sets belonging to $\mathscr{A}$ as $t \to +\infty$. We also denote by $\mathscr{B}\,(\infty)$ the *basin of infinity*, defined as the set of points which generate unbounded trajectories. Let $\mathscr{F}$ be the boundary (or frontier) separating $\mathscr{B}\,(\mathscr{A})$ from $\mathscr{B}\,(\infty)$. An exact determination of $\mathscr{F}$ is the main goal of this section. Indeed, this boundary may be rather complex, as evidenced by the numerical results shown in Fig. 1.78.

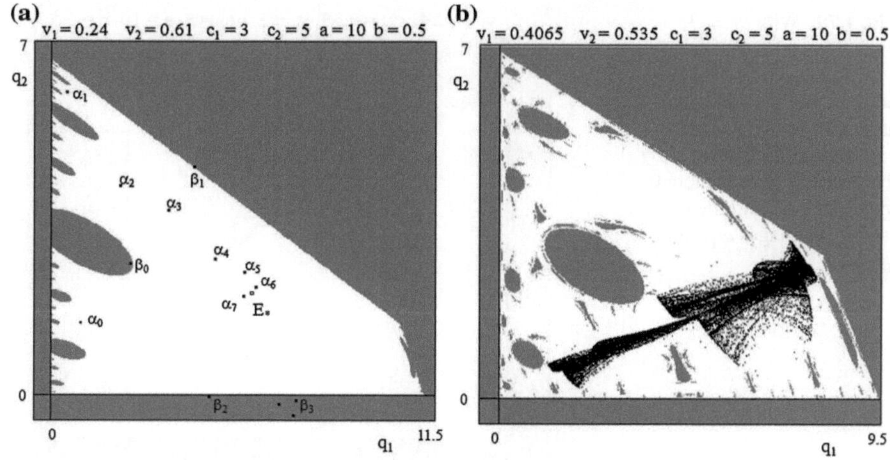

Fig. 1.78 Numerical representation of the basins of attraction for the duopoly map. The two figures are obtained by taking a grid of initial conditions (q_{10}, q_{20}) and generating, for each of them a numerically computed trajectory. If the trajectory is diverging then a *grey dot* is painted in the point corresponding to the initial condition, otherwise a *white dot* is painted. **a** The *white* region represents the basin of attraction of the Nash equilibrium, which is the only attractor at finite distance for that set of parameters; **b** the attractor is a chaotic attractor surrounding the unstable Nash equilibrium

In Fig. 1.78a the attractor at finite distance is the Nash equilibrium E_*, and its basin of attraction is represented by the white area, whereas the grey-shaded area represents the basin of infinity. In the situation shown in Fig. 1.78a the boundary separating $\mathscr{B}(\mathscr{A})$ from $\mathscr{B}(\infty)$ has a fractal structure, as will be explained below. In Fig. 1.78b the bounded attractor $\mathscr{A}$ is a chaotic set, with a multiply connected (or connected with holes) basin of attraction. The same property can be expressed by saying that $\mathscr{B}(\infty)$ is a nonconnected set, with nonconnected regions given by the holes inside $\mathscr{B}(\mathscr{A})$. In this situation there is a great uncertainty about the long-run behavior of a given adjustment process, since a small change in the initial strategy of the game may cause a crossing of $\mathscr{F}$.

The frontier $\mathscr{F} = \partial\, \mathscr{B}(\mathscr{A}) = \partial\, \mathscr{B}(\infty)$ behaves as a repelling line for the points near it, since it acts as a watershed for the trajectories of the map T. Points belonging to $\mathscr{F}$ are mapped into $\mathscr{F}$ both under forward and backward iteration of T, that is, the frontier is invariant for application of T and T^{-1}. More exactly $T(\mathscr{F}) \subseteq \mathscr{F}$, $T^{-1}(\mathscr{F}) = \mathscr{F}$. This implies that if a saddle-point, or a saddle-cycle, belongs to $\mathscr{F}$, then $\mathscr{F}$ must also contain all the preimages of such singularities, and it must also contain the whole stable manifold W^s. In order to understand how complex basin boundaries, like those shown in Fig. 1.78, are obtained, we start from a situation in which $\mathscr{F}$ has a simple shape, and then we study the sequence of bifurcations that cause the main qualitative changes in the structure of the basin boundaries as some parameter is varied. Such bifurcations, typical of noninvertible maps, can be characterized by contacts of the basin boundaries with the critical curves.

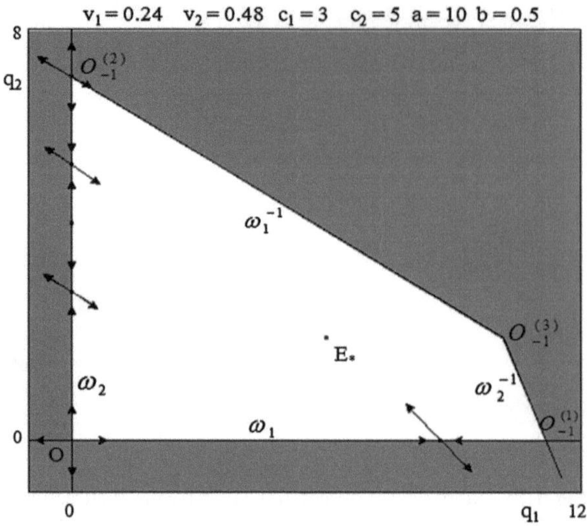

Fig. 1.79 With $c_1 = 3$, $c_2 = 5$, $a = 10$, $b = 0.5$, $v_1 = 0.24$, $v_2 = 0.48$, the boundary of the basin of attraction of the Nash equilibrium E_* is formed by the invariant axes, denoted by ω_1 and ω_2, and their rank-1 preimages ω_1^{-1} and ω_2^{-1}

Indeed, for the parameters' values used to obtain Fig. 1.79, an exact determination of the boundaries separating the basin of E_* from that of infinity can be obtained. In fact the saddle fixed points (or the saddle-cycles, if (1.80) or (1.81) no longer hold) located on the coordinate axes belong to $\mathcal{F}$, and also the invariant coordinate axes ω_1 and ω_2, which form the local stable manifold (or inset) of the saddles, are part of $\mathcal{F}$. These axes behave as repelling lines because the unstable manifolds (or outsets) of the saddles are transverse to the axes, each of them having a branch pointing toward E_* and the opposite branch going to infinity (see Fig. 1.79). The other parts of $\mathcal{F}$ can be obtained by taking all the preimages of these invariant axes, in order to obtain the whole stable sets of the saddles

$$\mathcal{F} = \left(\cup_{n=0}^{\infty} T^{-n}(\omega_1)\right) \bigcup \left(\cup_{n=0}^{\infty} T^{-n}(\omega_2)\right) \tag{1.104}$$

The map T, defined in (1.74), is a noninvertible map. In fact, if we consider a generic point $P = (0, p)$ of the q_2 axis, its preimages are the real solutions of the algebraic system obtained from (1.74) with $(q_1', q_2') = (0, p)$:

$$\begin{cases} q_1\left[1 + v_1(a - c_1) - 2bv_1q_1 - bv_1q_2\right] = 0\ , \\ (1 + v_2(a - c_2))\, q_2 - 2bv_2q_2^2 - bv_2q_1q_2 = p\ . \end{cases} \tag{1.105}$$

From the first of (1.105) we obtain $q_1 = 0$ or

$$1 + v_1(a - c_1) - 2bv_1q_1 - bv_1q_2 = 0 \tag{1.106}$$

which means that if the point P has preimages, then they must be located either on the same invariant axis or on the line of equation (1.106). With $q_1 = 0$ the second equation becomes a second degree algebraic equation which has two distinct, coincident or no real solutions if the discriminant

$$(1 + v_2 (a - c_2))^2 - 8bv_2p \tag{1.107}$$

is positive, zero or negative respectively. A similar conclusion holds if (1.106) is used to eliminate a state variable in the first equation of (1.105). From this we can deduce that the generic point P of the q_2 axis can have no preimages or two preimages on the same axis (which are the same obtained by the restriction (1.101) of T to the axis q_2) or four preimages, two on the same axis and two on the line of equation (1.106). This implies that the set of the rank-one preimages of the q_2 axis belongs to the same axis and to the line (1.106). Following the same arguments we can state that the other invariant axis, q_1, has preimages on itself and on the line of equation

$$1 + v_2(a - c_2) - bv_2q_1 - 2bv_2q_2 = 0 \; . \tag{1.108}$$

It is straightforward to see that the origin $O = (0, 0)$ has always 4 preimages, located at the intersection of the lines (1.106) and (1.108) (see Fig. 1.79). In the situation, shown in Fig. 1.79, the lines (1.106) and (1.108), labelled by ω_2^{-1} and ω_1^{-1} respectively, together with the coordinate axes, labelled by ω_2 and ω_1, delimitate a bounded region of the strategy space (q_1, q_2) which is exactly the basin of attraction of E_*.

These four sides, given by the segments $OO_{-1}^{(1)}$ and $OO_{-1}^{(2)}$ of the coordinate axes and their rank-one preimages, constitute the whole boundary $\mathcal{F}$ because no preimages of higher rank exist, since ω_{-1}^{1} and ω_{-1}^{2} belong to the region Z_0 of the plane whose points (q_1', q_2') have no preimages, i.e., the fourth degree algebraic system has no real solutions. This fact can be characterized through the study of the critical curves of the noninvertible map (1.74). As we have seen, since the map T is continuously differentiable, the critical curve LC_{-1} is the locus of points in which the determinant of $J(q_1, q_2)$, given in (1.77), vanishes, and the critical curve LC, locus of points having two coincident rank-one preimages, can be obtained as the image, under T, of LC_{-1}. For the map (1.74) LC_{-1} is formed by the two branches of an hyperbola, denoted by $LC_{-1}^{(a)}$ and $LC_{-1}^{(b)}$ in Fig. 1.80, thus also $LC = T(LC_{-1})$ consists of two branches, $LC^{(a)} = T(LC_{-1}^{(a)})$ and $LC^{(b)} = T(LC_{-1}^{(b)})$, represented by the thicker curves of Fig. 1.80a. These two branches of LC separate the phase plane into 3 regions, denoted by Z_0, Z_2 and Z_4, whose points have 0, 2 and 4 distinct rank-1 preimages respectively. It can be noticed that, as already stressed above, the origin always belongs to the region Z_4. It can also be noticed that the line LC_{-1} intersects the axis q_j, $j = 1, 2$, in correspondence to the critical point c_{-1} of the restriction (1.101) of T to that axis.

The simple shape that the frontier $\mathcal{F}$ assumes for values of the parameters like those used in Fig. 1.80a, where the basin of attraction of E_* is a simply connected set, is due to the fact that the preimages of the invariant axes, denoted in Fig. 1.80a

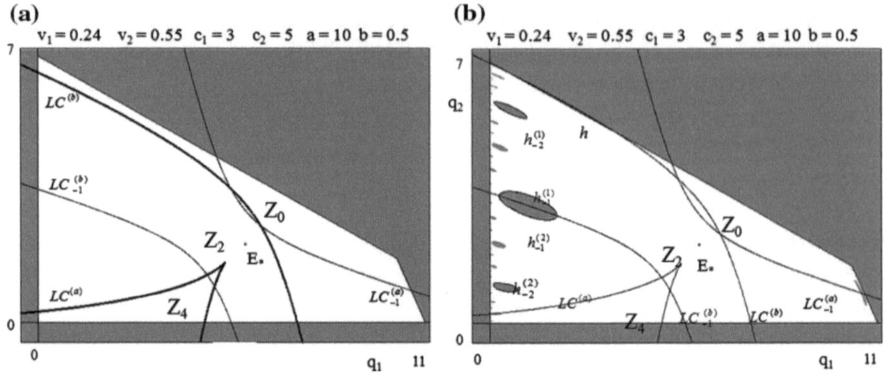

Fig. 1.80 Graphical representation of the basin of attraction of the Nash equilibrium (*white* region) and the basin $\mathscr{B}(\infty)$ of unbounded trajectories (*grey* region) together with the basic critical curve LC_{-1} and of critical curve LC (represented by *heavy lines*). The values of parameters c_1, c_2, a, b are the same as in Fig. 1.78, with **a** $v_1 = 0.24$ and $v_2 = 0.55$; **b** $v_1 = 0.24$ and $v_2 = 0.596$ (just after the contact of LC with ω_1^{-1})

by ω_i^{-1}, $i = 1, 2$, are entirely included inside the region Z_0, so that no preimages of higher rank exist. The situation is different when the values of the parameters are such that some portions of these lines belong to the regions Z_2 or Z_4. In this case preimages of higher order of the invariant coordinate axes are obtained, which form new arcs of the frontier $\mathscr{F}$, so that its shape becomes more complex. The switch between these two qualitatively different situations can be obtained by a continuous variation of some parameters of the model, and determines a global (or contact) bifurcation. The occurrence of these global bifurcations can be revealed by the study of critical curves. In order to illustrate this, in the following of this section we fix the marginal costs and the parameters of the demand function at the parameters' values $c_1 = 3$, $c_2 = 5$, $a = 10$, $b = 1/2$, and we vary the values of the speeds of adjustment v_1 and v_2. If, starting from the parameters' values used to obtain the simple basin structure of Fig. 1.80a, the parameter v_2 is increased, the two branches of the critical curve LC move upwards. The first global bifurcation of the basin occurs when the branch of LC which separates the regions Z_0 and Z_2 becomes tangent to $\mathscr{F}$, that is, to one of the lines (1.108) or (1.106). In Fig. 1.80b it can be seen that just after the bifurcation value of v_2, at which $LC^{(b)}$ is tangent to the line ω_1^{-1} of equation (1.108), a portion of $\mathscr{B}(\infty)$, say H_0 (bounded by the segment h of ω_1^{-1} and LC) that before the bifurcation was in region Z_0, enters inside Z_2. The points belonging to H_0 have two distinct preimages, located at opposite sides with respect to the line LC_{-1}, with the exception of the points of the curve $LC^{(b)}$ inside $\mathscr{B}(\infty)$ whose preimages, according to the definition of LC, merge on $LC_{-1}^{(b)}$. Since H_0 is part of $\mathscr{B}(\infty)$ also its preimages belong to $\mathscr{B}(\infty)$. The locus of the rank-1 preimages of H_0, bounded by the two preimages of h, is composed by two areas joining along LC_{-1} and forms a *hole* (or lake) of $\mathscr{B}(\infty)$ nested inside $\mathscr{B}(E_*)$. This is the largest hole appearing in Fig. 1.80b, and is called the *main hole*. It lies entirely inside region Z_2, hence it has 2 preimages, which are smaller holes bounded by preimages of rank 3 of the q_1 axis. Even these are both inside Z_2. So each of them has two further preimages inside Z_2, and so on. Now the boundary $\mathcal{F}$ is given by the union of an external part, formed by the coordinate axes and their rank-1 preimages (1.108) and (1.106), and the boundaries of the holes, which are sets of preimages of higher rank of the q_1 axis. Thus the global bifurcation just described transforms a *simply connected* basin into a *multiply connected* one, with a countable infinity of holes, called *arborescent sequence of holes*, inside it.

As v_2 is further increased LC continues to move upwards and the holes become larger. This fact causes a sort of predictability loss, since a greater uncertainty is obtained with respect to the destiny of games starting from an initial strategy falling in zone of the holes. If v_2 is further increased a second global bifurcation occurs when LC crosses the q_2 axis at $O_{-1}^{(2)}$. This happens when $v_2 = 3/(a - c_2)$, as in Fig. 1.81a. After this bifurcation all the holes reach the coordinate axis q_2, and the infinite contact zones are the intervals of divergence of the restriction (1.101), which are located around the critical point and all its preimages under (1.101) (compare Fig. 1.81a with Fig. 1.58). After this bifurcation the basin $\mathcal{B}(E_*)$ becomes simply connected again, but its boundary $\mathcal{F}$ has now a fractal structure, since its shape, formed by infinitely many peninsulas, has the self-similarity property.

The sequence of pictures shown in Fig. 1.81 is obtained with $v_1 = 0.24$ (as in Fig. 1.80) and increasing values of v_2. Along this sequence the point (v_1, v_2) reaches, in the plane of adjustment speeds, the line of flip bifurcations. When this line is crossed the Nash equilibrium E_* becomes a repelling saddle point, and an attracting cycle of period 2, say $\mathcal{C}_2$, is created near it (as in Fig. 1.81b). The flip bifurcation opens a cascade of period doublings, which creates a sequence of attracting cycles of

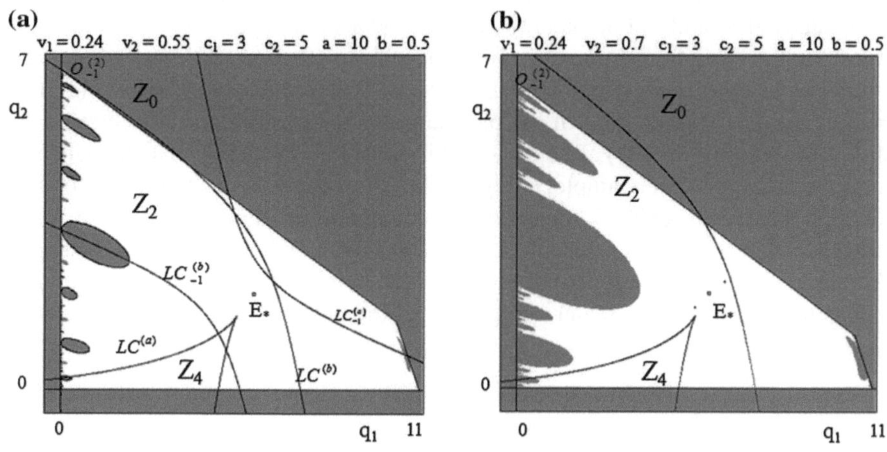

Fig. 1.81 Numerical simulations of the duopoly map, obtained with fixed parameters $c_1 = 3$, $c_2 = 5$, $a = 10$, $b = 0.5$, $v_1 = 0.24$, and increasing values of v_2

period 2^n followed by the creation of chaotic attractors, which may be cyclic chaotic areas.

It is worth noting that in general there are no relations between the bifurcations which change the qualitative properties of the basins and those which change the qualitative properties of the attractor at finite distance. In other words, we may have a simple attractor, like a fixed point or a cycle, with a very complex basin structure, or a complex attractor with a simple basin. Both these sequences of bifurcations, obtained by increasing the speeds of adjustment v_i, cause a loss of predictability. After the local bifurcations the myopic duopoly game no longer converges to the global optimal strategy, represented by the Nash equilibrium E_*, and even if the game starts from an initial strategy very close to E_* the duopoly system goes towards a different attractor, which may be periodic or aperiodic. These bifurcations cause in general a loss of predictability about the asymptotic behavior of the duopoly system: for example, in the sequence shown in Fig. 1.81 the situation of convergence to the unique Nash equilibrium, like in the static Cournot game, is replaced by asymptotic convergence to a periodic cycle with predictable output levels, and then by a cyclic behavior with output levels which are not well predictable since they fall inside cyclic chaotic areas, and, finally, by a situation of erratic behavior, inside a large area of the strategy space, with no apparent periodicity. Instead, the global bifurcations of the basin boundaries cause an increasing uncertainty with respect to the destiny of a duopoly game starting from a given initial strategy since a small change in the initial condition of the duopoly, or a small exogenous shock during the adjustment process, may cause a great modification about the long-run behavior of the system. Similar bifurcation sequences can also be obtained by increasing the parameter v_1 with a fixed value of v_2. In this case a contact between LC and ω_2^{-1}, rank-one preimage of the q_2 axis, gives the first bifurcation that transforms the basin $\mathcal{B}(\mathcal{A})$ from a simply connected into a multiply connected set, with holes near the q_1 axis. Situations with values of v_1 and v_2 both near the critical values $v_i = 3/(a - c_i)$, $i = 1, 2$, can give complex basin boundaries near both the coordinate axes, with two arborescent sequences of holes, generated by contacts of LC with the lines (1.106) and (1.108). In any case, the computation of the preimages of the coordinate axes allows us to obtain, according to (1.104), the exact delimitation of the basin boundary also in these complex situations. For example, in Fig. 1.82 the preimages of the q_1 axis, up to rank-six, are represented for the same set of parameters as that used in Fig. 1.78b. It can be noticed that some preimages of rank five and six bound holes that enter the region Z_4, thus giving a faster exponential growth of the number of higher order preimages. This is the cause for the greater complexity of the basin boundary which is clearly visible in Fig. 1.78b.

A Duopoly Game with Multiple Nash Equilibria Games played by rational players with complete information sets are typically one-shot games: each player knows the complete payoffs' structure of the game and that other players are rational as well so that, having complete information, each player is able to forecast the choices of other players. Thus, the game is studied identifying the so-called "solution concepts", such as Nash equilibria. In fact, if each player is assumed to have all such

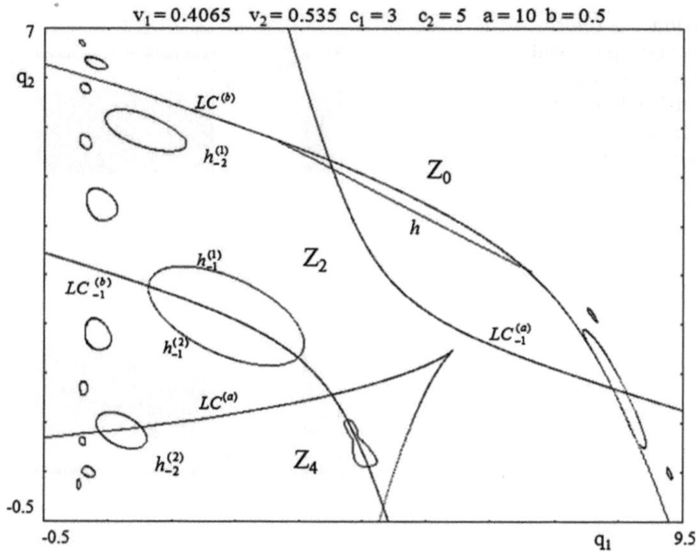

Fig. 1.82 Preimages of the q_1 axis, up to rank 6, obtained with the same set of parameters as those used in Fig. 1.78b

information and computational skills to solve the optimization problem obtained by means of the rationality assumption (expressed as maximization of individual utility) then everybody will choose a Nash equilibrium. However, agents are sometimes neither so astute nor informed, and they behave following adaptive methods, such as learning-by-doing or trial-and-error practices. Sometimes agents do not optimize at all, just following rough rules of thumb. This leads players to replace one-shot optimal decisions with repeated myopic or adaptive decisions, in other words to a dynamic process that may or may not converge to a Nash equilibrium, provided it is an equilibrium point of the dynamical system as well. Moreover, when a game has several Nash equilibrium points represented by equilibrium points of the dynamical system, then the step-by-step dynamic process may act as a selection device, i.e., the stability of the equilibria suggests which of them will prevail in the long-run. And if several equilibrium points are stable, then the study of their basins of attraction will give information about the path dependence, i.e., how the convergence will depend on historical accidents (represented by exogenous shifts of initial conditions).

As an example let us consider a case where the reaction curves are second degree functions, in the form of standard logistic maps $R_i(q_j) = \mu_i q_j \left(1 - q_j\right)$. They can be obtained by assuming a linear demand $p = a - b(q_1 + q_2)$ and cost functions with externalities: $C_i = C_i(q_i, q_j) = d + aq_i - b(1 + 2\mu)q_i q_j + 2b\mu q_i q_j^2$ (see [22])

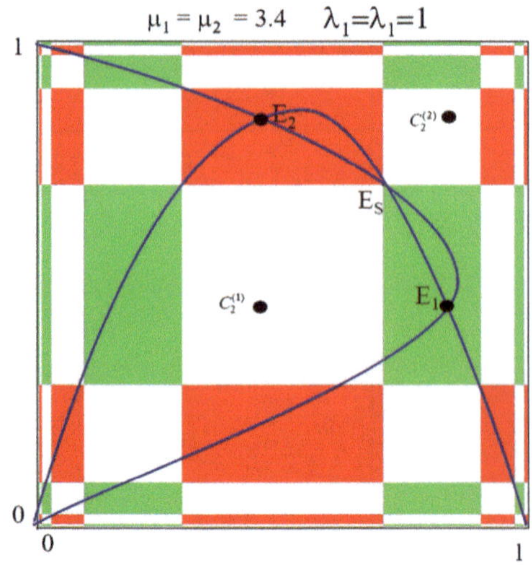

Fig. 1.83 Attractors (*black dots*) and basins (represented by *different colors*) for the duopoly model (1.109) with $\lambda_1 = \lambda_2 = 1$, i.e., the case of best reply with Näive expectations

The adaptive adjustment with inertia becomes:

$$
\begin{aligned}
q_1(t+1) &= (1-\lambda_1)\, q_1(t) + \lambda_1 \mu_1 q_2(t)\, (1 - q_2(t)) \\
q_2(t+1) &= (1-\lambda_2)\, q_2(t) + \lambda_2 \mu_2 q_1(t)\, (1 - q_1(t))
\end{aligned}
\tag{1.109}
$$

As noticed above, this model reduces to the repeated game with best reply and naïve expectations if $\lambda_1 = \lambda_2 = 1$. Several coexisting stable Nash equilibria can be obtained for certain sets of parameters, as well as other more complicated coexisting attractors, such as stable cycles or chaotic attractors. An exemplary case is shown in Fig. 1.83, where two stable Nash equilibria coexist with a stable cycle of period 2, each with its own basin of attraction. The basins are multiply connected, i.e., besides the immediate basin several (really infinitely many) nonconnected portions exist that accumulate along the outer boundary of the phase space (see [9]). As stressed in Sect. 1.3.3, such a situation can only occur in the case of noninvertible maps.

In order to reduce the number of parameters in our model, we will assume that $\mu_1 = \mu_2 = \mu$. Under this assumption the fixed points can be analytically computed as follows. Two fixed points always exist, given by $O = (0,0)$ and $S = (1 - 1/\mu, 1 - 1/\mu)$. For $\mu > 1$, S represents a Nash equilibrium of the duopoly game, at which the two firms produce the same quantities. Moreover, two further Nash equilibria, given by

$$
E_1 = \left(\frac{\mu + 1 + \sqrt{(\mu+1)(\mu-3)}}{2\mu}, \frac{\mu + 1 - \sqrt{(\mu+1)(\mu-3)}}{2\mu} \right) \tag{1.110}
$$

and

$$E_2 = \left(\frac{\mu + 1 - \sqrt{(\mu+1)(\mu-3)}}{2\mu}, \frac{\mu + 1 + \sqrt{(\mu+1)(\mu-3)}}{2\mu} \right), \tag{1.111}$$

are created at $\mu = 3$, and for $\mu > 3$ they are located in symmetric positions with respect to the diagonal Δ of equation $q_1 = q_2$. Each of them represents a Nash equilibrium, characterized by different quantities produced by two firms. In the presence of multiple Nash equilibria the problem of equilibrium selection arises. The following result holds (see also Fig. 1.84).

Proposition 1.3 (Local Stability and Bifurcations, Homogeneous Players [7]) *Let* $\mu_1 = \mu_2 = \mu$ *and* $\lambda_1 = \lambda_2 = \lambda$. *Then*

(i) *For* $0 < \mu < 1$ *the fixed point* $O = (0, 0)$ *is a stable node, for* $1 < \mu < 2/\lambda - 1$ *it is a saddle point, with unstable set along* Δ *and local stable set which crosses through* O *perpendicular to* Δ*, and for* $\mu > 2/\lambda - 1$ *it is an unstable node;*
(ii) *For* $1 < \mu < 3$ *the fixed point* $S = (1 - 1/\mu, 1 - 1/\mu)$ *is a stable node, for* $3 < \mu < 1 + 2/\lambda$ *it is a saddle point, with local stable set along* Δ *and unstable set which crosses through* S *perpendicular to* Δ*, and for* $\mu > 1 + 2/\lambda$ *it is an unstable node;*
(iii) *The fixed points* E_i, $i = 1, 2$, *given in (1.110) and (1.111), are created at* $\mu = 3$ *through a pitchfork bifurcation of* S*, and are stable nodes for* $3 < \mu < 1 + \sqrt{5}$*, stable foci for* $1 + \sqrt{5} < \mu < 1 + \sqrt{4 + 2/\lambda}$ *and at* $\mu = 1 + \sqrt{4 + 2/\lambda}$ *they become unstable foci through a Neimark-Sacker bifurcation.*

Proof The Jacobian matrix of (1.109) is

$$J(q_1, q_2; \mu, \lambda) = \begin{bmatrix} 1 - \lambda & \lambda\mu(1 - 2q_2) \\ \lambda\mu(1 - 2q_1) & 1 - \lambda \end{bmatrix}. \tag{1.112}$$

In the points of the diagonal Δ of equation $q_1 = q_2$ on which both O and S are located, the matrix (1.112) assumes the structure

$$DT(x, x; \lambda, \mu) = \begin{bmatrix} 1 - \lambda & \lambda\mu(1 - 2x) \\ \lambda\mu(1 - 2x) & 1 - \lambda \end{bmatrix}. \tag{1.113}$$

Such a matrix has real eigenvalues. In particular, in O the eigenvalues are:

$$z_{\parallel}(O) = 1 + \lambda(\mu - 1) \quad \text{with eigenvector} \quad \mathbf{r}_{\parallel} = (1, 1) \quad \text{along } \Delta \tag{1.114}$$

and

$$z_{\perp}(O) = 1 - \lambda(\mu + 1) \quad \text{with eigenvector} \quad \mathbf{r}_{\perp} = (1, -1) \quad \text{perpendicular to } \Delta\,. \tag{1.115}$$

In the fixed point S we have

$$z_{\parallel}(S) = 1 + \lambda(1-\mu) \quad \text{and} \quad z_{\perp}(S) = 1 + \lambda(\mu - 3) \ .$$

So, the fixed point O is locally asymptotically stable (a stable node) in the region

$$\Omega_2(O) = \{(\mu, \lambda) \in \Omega_2 \mid \mu < 1\} \ . \tag{1.116}$$

Analogously, since $z_{\parallel}(S) \in (-1, 1)$ for $0 < \lambda(\mu - 1) < 2$ and $z_{\perp}(S) \in (-1, 1)$ for $-2 < \lambda(\mu - 3) < 0$, the fixed point S is locally asymptotically stable (a stable node) in the region

$$\Omega_2(S) = \{(\mu, \lambda) \in \Omega_2 \mid 1 < \mu < 3\} \ . \tag{1.117}$$

At $\mu = 1$, $O \equiv S$ and a *transcritical* (or stability exchange) bifurcation occurs at which the two fixed points exchange their stability property along Δ: for $\mu < 1$, just before the bifurcation, O is a stable node and S is a saddle, with local stable set along Δ, and for $\mu > 1$, just after the bifurcation, O is a saddle, with unstable set along Δ, and S is a stable node.

At $\lambda(\mu + 1) = 2$ a *period doubling* (or flip) bifurcation of O occurs which creates a cycle of period 2 along the invariant manifold associated with $z_{\perp}(O)$. For $\lambda \in (0, 1)$ this bifurcation occurs for $\mu > 1$, i.e., when O is a saddle, hence at the flip bifurcation O becomes an unstable node and a saddle cycle of period two is created, with stable set along the direction associated with $z_{\perp}(O)$.

At $\mu = 3$ a *pitchfork bifurcation* occurs at which the fixed point S becomes a saddle point with unstable set in the direction transverse to Δ, and the fixed points E_1 and E_2 are created. At $\lambda(\mu - 1) = 2$ a flip bifurcation along Δ occurs at which S becomes a repelling node and a saddle cycle of period 2 is created along Δ, with stable set along Δ and unstable set transverse to it.

The Jacobian matrix (1.112) computed at the two fixed points E_1 and E_2 which exist for $\mu > 3$, respectively, assume the forms

$$J(E_1; \lambda, \mu) = \begin{bmatrix} 1-\lambda & -\lambda(1 - \sqrt{(\mu+1)(\mu-3)}) \\ -\lambda(1 + \sqrt{(\mu+1)(\mu-3)}) & 1-\lambda \end{bmatrix},$$

$$J(E_2; \lambda, \mu) = \begin{bmatrix} 1-\lambda & -\lambda(1 + \sqrt{(\mu+1)(\mu-3)}) \\ -\lambda(1 - \sqrt{(\mu+1)(\mu-3)}) & 1-\lambda \end{bmatrix}.$$

Hence, E_1 and E_2 have the same characteristic equation with $\mathrm{Tr} = 2(1-\lambda)$ and $\mathrm{Det} = (1-\lambda)^2 - \lambda^2 \left(4 + 2\mu - \mu^2\right)$. Being $\mathrm{Tr}^2 - 4\mathrm{Det} = 4\lambda^2 \left(4 + 2\mu - \mu^2\right)$, the eigenvalues are real for $\mu \leq 1 + \sqrt{5}$ and are given by

$$z_1 = 1 - \lambda - \lambda\sqrt{4 + 2\mu - \mu^2} \quad \text{and} \quad z_2 = 1 - \lambda + \lambda\sqrt{4 + 2\mu - \mu^2} \ .$$

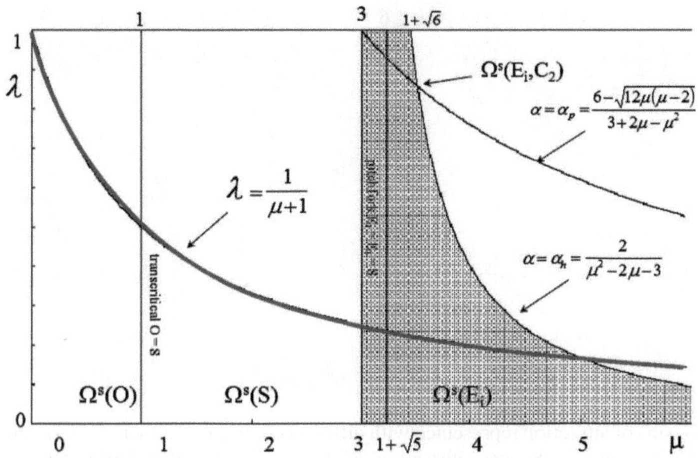

Fig. 1.84 Stability regions and bifurcation curves in the parameters' plane $(\mu.\lambda)$ for the model (1.109) in the homogenous case

For $\mu > 1 + \sqrt{5}$ the eigenvalues are complex and are given by

$$z_1 = 1 - \lambda - i\lambda\sqrt{\mu^2 - 2\mu - 4} \quad \text{and} \quad z_2 = 1 - \lambda + i\lambda\sqrt{\mu^2 - 2\mu - 4}\,.$$

In the parameters space Ω_2 the region of stability of E_i is

$$\Omega_2(E_i) = \left\{(\mu, \lambda) \in \Omega_2 | \mu > 3 \text{ and } \lambda\left(\mu^2 - 2\mu - 3\right) < 2\right\} \tag{1.118}$$

At $\lambda\left(\mu^2 - 2\mu - 3\right) = 2$ the eigenvalues exit the unit circle, so that the fixed points are transformed from stable to unstable foci through a supercritical Neimark-Sacker bifurcation at which two stable closed orbits are created around the two Nash equilibria E_1 and E_2. The rigorous proof of the occurrence of a supercritical Hopf bifurcation requires the evaluation of some long expressions involving derivatives of the map up to order three. We claim numerical evidence for the existence of a stable closed orbit around the unstable focus after the bifurcation (see Fig. 1.85). □

In Fig. 1.84, the red line of equation $\lambda = /(\mu + 1)$ represents a global bifurcation curve at which the basins change their topological structure from simply to multiply connected, i.e., connected with holes, according to the following proposition (that we give without a proof).

Proposition 1.4 (Global Bifurcation of the Basins, Homogeneous Players) *If* $\mu_1 = \mu_2 = \mu$, $\lambda_1 = \lambda_2 = \lambda$ *and* $(\mu, \lambda) \in \widetilde{\Omega}_2(E_i)$, *the bounded trajectories of (1.109) converge to one of the stable Nash equilibria* E_1 *or* E_2, *given by (1.110) and (1.111), respectively, and the common boundary which separates the basin* $\mathcal{B}(E_1)$ *from the basin* $\mathcal{B}(E_2)$ *is given by the stable set* $W^s(S)$ *of the saddle point* S. *If* $\lambda(\mu + 1) < 1$,

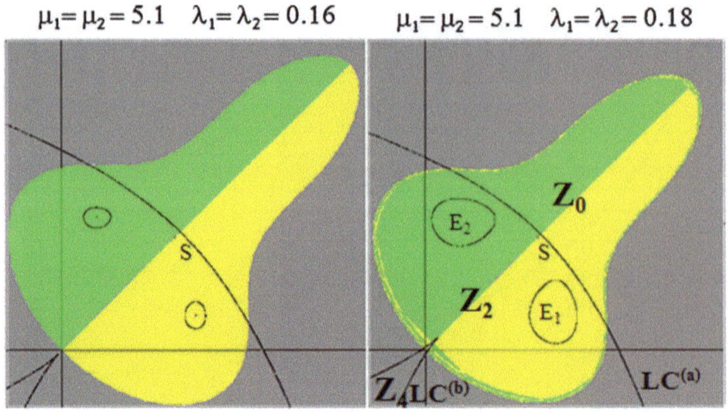

Fig. 1.85 Basins of attraction represented with different colors, and attractors of the model (1.109). *Left* Just after the Neimark-Sacker bifurcation. *Right* Just after the global bifurcation

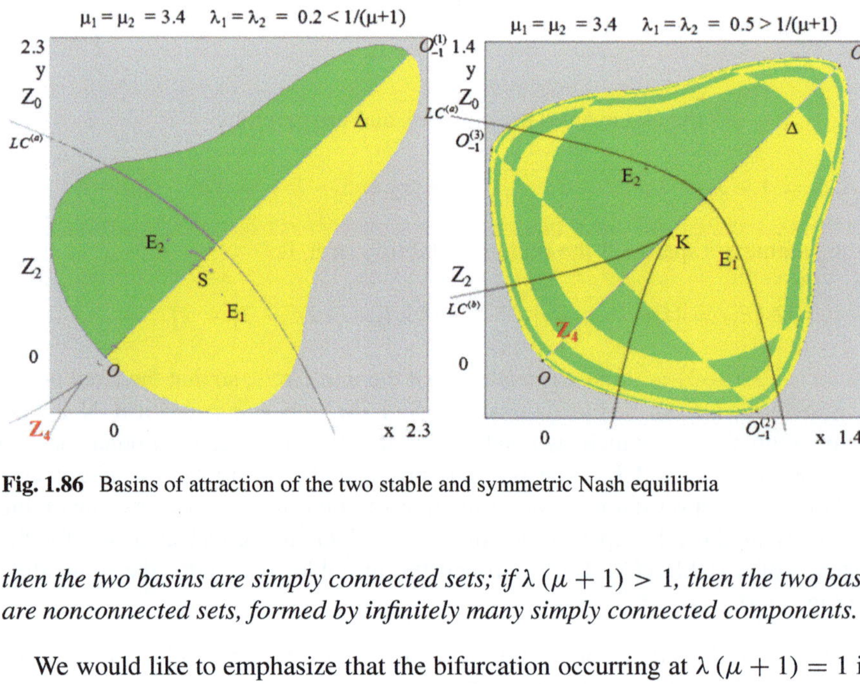

Fig. 1.86 Basins of attraction of the two stable and symmetric Nash equilibria

then the two basins are simply connected sets; if $\lambda\,(\mu+1) > 1$, *then the two basins are nonconnected sets, formed by infinitely many simply connected components.*

We would like to emphasize that the bifurcation occurring at $\lambda\,(\mu+1) = 1$ is a global bifurcation, i.e., it cannot be revealed by a study of the linear approximation of the dynamical system. The occurrence of such a bifurcation has been characterized by a contact between the stable set of S and a critical curve LC, i.e., a *contact* (or *global*) bifurcation (Fig. 1.86).

The occurrence of the bifurcation, which transforms the basins from simply connected to nonconnected, causes a loss of predictability about the long-run evolution of this Cournot game starting from given initial quantities of the two players. In fact, in contrast to what happens in the case of simply connected basins, when the basins are no longer simply connected, the adjustment dynamic starting with $q_1(0) > q_2(0)$ may lead to convergence to either of the Nash equilibria. Furthermore, if the initial quantities are sufficiently far away from a Nash equilibrium, for example near the boundary $\partial\mathscr{B}$ of $\mathscr{B}$, then the presence of the infinitely many components of both basins causes a sort of sensitivity with respect to these initial conditions. Even a very small perturbation of the starting point of the Cournot game may lead to a crossing of the boundary that separates the two basins, with consequent convergence to a different Nash equilibrium.

We now turn to the case of heterogeneous behavior, and assume that $\lambda_1 \neq \lambda_2$ holds. Although we get the same Nash equilibria since the fixed points do not depend on the speeds of adjustment, the eigenvalues of the Jacobian matrix of the map (1.109) depend on both of the parameters λ_1 and λ_2. Furthermore, note that the diagonal Δ is no longer trapping. The following proposition defines the stability regions for each Nash equilibrium in the three-dimensional parameters space $\Omega_3 = \{(\mu, \lambda_1, \lambda_2) \in \mathbb{R}^3 | \mu > 0, 0 \leq \lambda_1 \leq 1, 0 \leq \lambda_2 \leq 1\}$.

Proposition 1.5 (Local Stability, Heterogeneous Behavior [7])
Let $\mu_1 = \mu_2$. *Then*

(i) The fixed point $O = (0, 0)$ *is*

- *a stable node for* $0 < \mu < 1$;
- *a saddle point for* $1 < \mu < \sqrt{1 + (4 - 2(\lambda_1 + \lambda_2))/(\lambda_1\lambda_2)}$;
- *an unstable node for* $\mu > \sqrt{1 + (4 - 2(\lambda_1 + \lambda_2))/(\lambda_1\lambda_2)}$.

(ii) The fixed point $S = (1 - 1/\mu, 1 - 1/\mu)$ *is*

- *a stable node for* $1 < \mu < 3$;
- *a saddle point for* $3 < \mu < 2 + \sqrt{1 + (4 - 2(\lambda_1 + \lambda_2))/(\lambda_1\lambda_2)}$;
- *an unstable node for* $\mu > 2 + \sqrt{1 + (4 - 2(\lambda_1 + \lambda_2))/(\lambda_1\lambda_2)}$.

(iii) The fixed points E_i, $i = 1, 2$, *given in (1.110) and (1.111) are created at* $\mu = 3$ *through a pitchfork bifurcation of S and*

- *for* $3 < \mu < 1 + \sqrt{9/2 + \lambda_1/(4\lambda_2) + \lambda_2/(4\lambda_1)}$ *are stable nodes;*
- *for* $1 + \sqrt{9/2 + \lambda_1/(4\lambda_2) + \lambda_2/(4\lambda_1)} < \mu < 1 + \sqrt{4 + 1/\lambda_1 + 1/\lambda_2}$ *are stable foci;*
- *at* $\mu = 1 + \sqrt{4 + 1/\lambda_1 + 1/\lambda_2}$ *become unstable foci through a Neimark-Sacker bifurcation.*

Proof The analysis of the local stability of a fixed point is obtained through the localization of the eigenvalues of the Jacobian matrix in the complex plane, where the Jacobian

$$(q_1, q_2) = \begin{bmatrix} 1 - \lambda_1 & \lambda_1\mu(1 - 2q_2) \\ \lambda_2\mu(1 - 2q_1) & 1 - \lambda_2 \end{bmatrix}$$

computed at the corresponding fixed point has to be considered. The stability conditions

$$P(1) = 1 - \text{Tr} + \text{Det} > 0\,, \quad P(-1) = 1 + \text{Tr} + \text{Det} > 0\,, \quad 1 - \text{Det} > 0$$

at the fixed point $O = (0, 0)$ become

$$\text{Tr}^2 - 4\text{Det} = (\lambda_1 - \lambda_2)^2 + 4\lambda_1\lambda_2\mu^2 > 0 \quad \forall\, (\mu, \lambda) \in \Omega_3\,,$$

$$P(1) = \lambda_1\lambda_2\,(1 + \mu)\,(1 - \mu) > 0 \quad \text{for} \quad \mu < 1\,,$$

$$P(-1) = 4 - 2\,(\lambda_1 + \lambda_2) + \lambda_1\lambda_2\left(1 - \mu^2\right) > 0 \quad \text{for} \quad \mu < \sqrt{1 + 2\frac{2 - (\lambda_1 + \lambda_2)}{\lambda_1\lambda_2}}\,.$$

At the fixed point $S = (1 - 1/\mu, 1 - 1/\mu)$ we have

$$\text{Tr}^2 - 4\text{Det} = \lambda_1^2 + \lambda_2^2 + 14\lambda_1\lambda_2 + 4\lambda_1\lambda_2\mu\,(\mu - 4) \geq (\lambda_1 - \lambda_2)^2 \geq 0\,,$$

being $\mu\,(\mu - 4) \geq -4$. So, the eigenvalues are always real at the fixed point S, and the stability conditions reduce to

$$P(1) = \lambda_1\lambda_2\left(-\mu^2 + 4\mu - 3\right) > 0 \quad \text{for} \quad 1 < \mu < 3\,,$$

$$P(-1) = \lambda_1\lambda_2\mu^2 - 4\lambda_1\lambda_2\mu + 3\lambda_1\lambda_2 + 2\,(\lambda_1 + \lambda_2) - 4 > 0$$
$$\text{for} \quad \mu < 2 + \sqrt{1 + 2\frac{2 - (\lambda_1 + \lambda_2)}{\lambda_1\lambda_2}}\,.$$

Hence, O is locally asymptotically stable (a stable node) in the region

$$\Omega_3(O) = \{(\mu, \lambda_1, \lambda_2) \in \Omega_3 | \mu < 1\}\,,$$

and S is locally asymptotically stable (a stable node) in the region

$$\Omega_3(S) = \{(\mu, \lambda_1, \lambda_2) \in \Omega_3 | 1 < \mu < 3\}\,.$$

At $\mu = 1$ a transcritical bifurcation occurs at which O and S exchange stability, at $\mu = 3$ a pitchfork bifurcation of S occurs at which the fixed points E_1 and E_2 are created. The main difference with respect to the homogeneous case lies in the fact that the eigendirections associated with the fixed points are no longer parallel and perpendicular to Δ, and Δ is no longer invariant.

At $\mu = \sqrt{1 + 2(2 - (\lambda_1 + \lambda_2))/(\lambda_1\lambda_2)} > 1$ a flip bifurcation of O occurs at which O is transformed from saddle to unstable node, and a saddle cycle of period 2 is created with stable set through O.

At $\mu = 2 + \sqrt{1 + 2(2 - (\lambda_1 + \lambda_2))/(\lambda_1\lambda_2)} > 3$ a flip bifurcation of S occurs at which S is transformed from saddle to unstable node, and a saddle cycle of period 2 is created with stable set through S.

The Jacobian matrix computed at the two fixed points E_1 and E_2 assumes, respectively, the forms

$$J(E_1; \mu, \lambda_1, \lambda_2) = \begin{bmatrix} 1 - \lambda_1 & -\lambda_1\left(1 - \sqrt{(\mu+1)(\mu-3)}\right) \\ -\lambda_2\left(1 + \sqrt{(\mu+1)(\mu-3)}\right) & 1 - \lambda_2 \end{bmatrix}$$

and

$$J(E_2; \mu, \lambda_1, \lambda_2) = \begin{bmatrix} 1 - \lambda_1 & -\lambda_1\left(1 + \sqrt{(\mu+1)(\mu-3)}\right) \\ -\lambda_2\left(1 - \sqrt{(\mu+1)(\mu-3)}\right) & 1 - \lambda_2 \end{bmatrix}.$$

It is easy to see that, like in the homogeneous case, E_1 and E_2 have the same characteristic equation, with $\mathrm{Tr} = 2 - \lambda_1 - \lambda_2$ and $\mathrm{Det} = 1 - \lambda_1 - \lambda_2 + \lambda_1\lambda_2(\mu+1)(\mu-3)$.

The fixed points E_i are transformed from stable nodes into stable foci when

$$\mathrm{Tr}^2 - 4\mathrm{Det} = -4\lambda_1\lambda_2\mu^2 + 8\lambda_1\lambda_2\mu + 14\lambda_1\lambda_2 + \lambda_1^2 + \lambda_2^2 = 0\,,$$

i.e., at $\mu = 1 + \sqrt{\frac{9}{2} + \frac{\lambda_1}{4\lambda_2} + \frac{\lambda_2}{4\lambda_1}}$.

Since

$$P(1) = \lambda_1\lambda_2(\mu+1)(\mu-3) > 0 \quad \text{for} \quad \mu > 3\,,$$

$$P(-1) = 4 - 2(\lambda_1 + \lambda_2) + \lambda_1\lambda_2(\mu+1)(\mu-3) > 0 \quad \text{for} \quad \mu > 3\,,$$

the stability conditions for E_i, $i = 1, 2$, reduce to

$$\mathrm{Det} - 1 = \lambda_1\lambda_2\mu^2 - 2\lambda_1\lambda_2\mu - 3\lambda_1\lambda_2 - \lambda_1 - \lambda_2 < 0\,.$$

Hence, in the parameters space Ω_3 the region of stability of E_i is

$$\Omega_3(E_i) = \left\{ (\mu, \lambda_1, \lambda_2) \in \Omega_3 \mid \mu > 3 \text{ and } \mu < 1 + \sqrt{4 + \frac{\lambda_1 + \lambda_2}{\lambda_1\lambda_2}} \right\}.$$

The equation $\mu = 1 + \sqrt{4 + (\lambda_1 + \lambda_2)/(\lambda_1\lambda_2)}$ defines a bifurcation surface in Ω_3 through related to a supercritical Neimark-Sacker bifurcation occurs, at which the fixed points E_1 and E_2 are transformed from stable to unstable foci and a stable closed invariant curve is created around them. □

From a comparison of the two propositions on local stability given above, it appears that the influence of heterogeneous behavior on the stability of the Nash equilibria is not too strong. However, in the case of coexisting stable Nash equilibria, an important question concerns the delimitation of their basins of attraction and

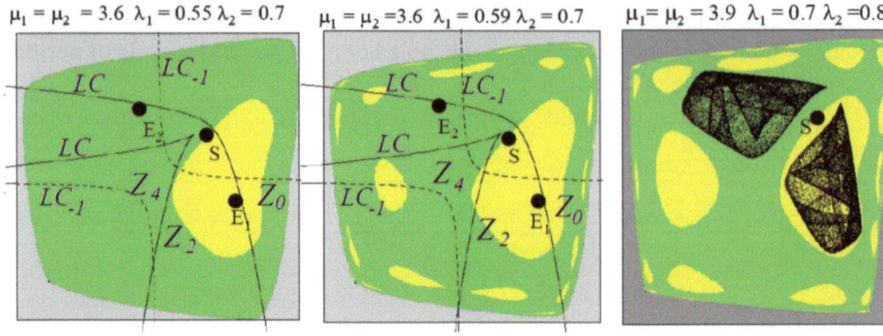

Fig. 1.87 Basins of attraction in the case of heterogeneous players

the global bifurcations that cause qualitative modifications of their boundaries. In fact, due to the heterogeneous behavior of the two competing firms, the symmetry properties of the dynamical system which allowed us to obtain a simple analytical expression of the global bifurcation given in the proposition stated above no longer hold. Hence, the occurrence of contact bifurcations can only be revealed numerically. This is illustrated in Fig. 1.87, where in the left panel a contact between the boundary of the basin of E_1 (formed by the stable set of the saddle point S) and the critical curve LC that separates Z_2 from Z_4 is shown. The portion of the basin of E_1 that enters Z_4 after the contact generates new preimages that give rise to a sequence of nonconnected portion of the basin, as shown in the central panel of Fig. 1.87. However, as the equation of the boundary is not known in this case, an analytic computation of the values of the parameters at which the contact occurs is not possible. This is an usual occurrence, as the analytical expressions of the stable sets (that bound the basins) as well as the analytic expressions of the critical curves, are very rarely known.

1.4 An Introduction to Optimal Control in Continuous Time

Here we provide a very brief introduction to optimal control problems in continuous time. The following part does not aim at giving a complete nor mathematically detailed analysis of the topic, but just a non-rigorous discussion on the theory of optimal control, some connections to the qualitative theory of dynamical systems and an overview on some applications in economics. We refer the reader to a more complete treatment in the bibliography.

In many economic application, it is necessary to solve a problem of constrained maximization, i.e., finding the values of variables such that a given function is assumes its maximum value given some constrains, generally expressed in terms of equations to be satisfied. However, it is often useful to consider maximization of a given functional that depends on dynamic variables constrained to follow the trajectories of a given dynamical system. In the more standard case, this problem assumes the form:

$$\max_{u(t)\in A(t)} \int_0^T f(x(t), u(t), t)\mathrm{d}t + F(x(T), T) \tag{1.119}$$

such that:

$$\begin{cases} \dot{x} = g(x(t), u(t), t) \\ x(0) = x_0 \end{cases}$$

and with one of the following terminal conditions: (1.120)

$$\text{(a) } x(T) \text{ free} \qquad \text{(b) } x(T) = x_T \qquad \text{(c) } x(T) \geq x_T \tag{1.121}$$

where:

- 0 is the *initial time*;
- $T \in (0, +\infty]$ is the *terminal time*;
- $x(t)$ is the *state variable* of the system, whose dynamics is determined by the following differential equation:
- $\dot{x} = g(x(t), u(t), t)$ is called the *state equation* or the *dynamics*;
- $u(t)$ is the *control*, to be determined to maximize the previous integral;
- $f(x(t), u(t), t)$ is the *instantaneous payoff*;
- $A(t)$ is the *constraint set* on the control, specifying that at each time t the control $u(t)$ must belong to the set $A(t)$;
- $F(x(T), T)$ is the *terminal payoff* (or scrap value or salvage value).

When $T = +\infty$, no condition is usually imposed on the state $x(t)$. In some cases, however, it is required that

$$\lim_{t\to+\infty} x(t) \geq \overline{x}$$

Functions $f(., ., .)$, $g(., ., .)$ and $F(., .)$ are assumed continuously differentiable.

Often in economic applications, in the problem (1.119) it is $f(x(t), u(t), t) = f(x(t), u(t))$ and $g(x(t), u(t))$, i.e., the performance criterion and the differential equations do not depend directly on time.[9] In this case, the optimal control problem (1.119) is called *autonomous*. In many economic application it is also $T = +\infty$ so that no scrap value is included in the problem, i.e., $F(x(T), T) = 0$. We will briefly

[9]More precisely, in economic applications usually the performance criterion is given in the form $f(x(t), u(t), t) = \mathrm{e}^{-\delta t}h(x(t), u(t))$, i.e., it depends directly on time but only through a discount term. This point is more extensively discussed below.

review some other formulations of the basic problem, such as the formulation of the problem with discount, which is of primary importance in economics.

The control $u(t)$ represents a choice variable that the agent can set continuously as long it is in the constraint set $A(t)$. Often this constraint on the control is imposed for reasons of feasibility. For instance, if the control $u(t)$ represents the consumption of a good, this can not be negative, so that $u(t) \geq 0$ for all t. In addition, some upper bound of consumption can be given, for instance through a budget constraint at time t; denoting by $B(t)$ the total budget of the agent, the constraint set $A(t)$ becomes $0 \leq u(t) \leq B(t), \forall t \in [0, T]$.

Any function $u(t)$, with $u(t) \in A(t)$ for all $t \in [0, T]$, such that $u(t)$ is (piecewise) continuous represents an *admissible* control. Consider that we fix a particular admissible control, for instance $u(t) = \overline{u}(t)$. Then, for this choice of the control, $\dot{x} = g(x(t), \overline{u}(t), t)$ is a (first order) differential equation that, together with the initial condition $x(0) = x_0$, determines entirely the trajectory of the state, i.e., the value of $x(t)$ for all $t \in [0, T]$, which is called an *admissible* path provided that the constraints on the final state are met (e.g., $x(T) = x_T$ or $x(T) \geq x_T$). Thus, with this choice of $\overline{u}(t)$, the state and, consequently, the value of the performance criterion $f(x(t), \overline{u}(t), t)$ is univocally determined. Since we are interested in maximizing the definite integral of the performance criterion over the interval $[0, T]$, the optimal control problem (1.119) consists in selecting, among all admissible controls $u(t) \in A(t)$, the control $u^*(t)$ that maximizes the value of this definite integral plus, if present, the terminal payoff. Such an admissible control is called an '*optimal control*'. The corresponding state $x^*(t)$ obtained as the solution of the *Cauchy problem* (differential equation with an initial condition)

$$\begin{cases} \dot{x} = g(x(t), u^*(t), t) \\ x(0) = x_0 \end{cases}$$

is called an *optimal trajectory* or *optimal path*.

Here we state the main results to find a candidate solution to the optimal control problem (1.119), i.e., an optimal control and the corresponding optimal path. The most important necessary conditions are *Bellman's Optimality principle* (or *Dynamic Programming principle*) and *Pontryagin's maximum principle*. In the following Section we provide a simplified derivation of these results.

For defining Bellman's optimality principle, define the following function $V(x, t) : \mathbb{R} \times \mathbb{R} \to \mathbb{R}$

$$V(x, t) = \max_{u(t) \in A(t)} \int_t^T f(x(s), u(s), s) ds + F(x(T), T) \tag{1.122}$$

called the *value function*. $V(x, t)$ represents the maximum (more precisely the sup) possible value attainable starting at time t with initial state x, as also explained below.

Proposition 1.6 (Bellman's Optimality Principle) *If $V(x,t)$ is differentiable in t and x, then it solves the following (Partial) Differential Equation*

$$-\frac{\partial V}{\partial t} = \max_{u(t)\in A(t)} \left[f(x(t), u(t), t) + \frac{\partial V}{\partial x} g(x(t), u(t), t) \right]$$

with terminal condition $V(x(T), T) = F(x(T), T)$.

For stating Pontryagin's maximum principle, define the Hamiltonian function

$$H := H(x, u, \lambda, t) = f(x, u, t) + \lambda g(x, u, t) \ , \qquad (1.123)$$

where $\lambda = \lambda(t)$ is called the costate variable.

Proposition 1.7 (Pontryagin's Maximum Principle) *If $u^*(t) = u^*$ is an optimal control and $x^*(t) = x^*$ is the corresponding optimal path for the problem (1.119), then there exists a costate variable $\lambda^*(t) = \lambda^*$ such that x^*, λ^*, u^* are the solution in $[0, T]$ of the following problem:*

$$\begin{cases} \dot{x} = g(x^*, u^*, t) = H_\lambda \ \textit{(state equation)} \\ \dot{\lambda} = -\left[f_x(x^*, u^*, t) + \lambda g_x(x^*, u^*, t) \right] = -H_x \ \textit{(costate equation)} \\ u^* = \arg\max H(x^*, u, \lambda^*, t) (\textit{ maximum principle}) \\ x^*(0) = x_0 \ \textit{(initial condition)} \\ (a)\ \lambda^*(T) = \frac{\partial}{\partial x} F(x_T, T) \ \textit{(transversality condition when } x(T) \textit{ free)} \\ \quad or \\ (b)\ \lambda^*(T) \geq \frac{\partial}{\partial x} F(x_T, T) \ \textit{(transversality condition when } x(T) \geq x_T) \end{cases}$$

Notice that when $x(T) = x_T$ no transversality condition is imposed. Moreover, when $F(x_T, T) = 0$, i.e., without scrap value, the transversality condition when $x(T)$ is free reduces to $\lambda^*(T) = 0$. Analogously, when $F(x_T, T) = 0$ and $x(T) \geq x_T$, the transversality condition is $\lambda^*(T) \geq 0$. Before closing this section, we would like to provide some important remarks.

The first remark concerns admissible controls. In particular, notice that continuity of the control $u(t)$ is not assumed, as we defined an admissible control as a piecewise continuous function that belongs to the constraint set $A(t)$ for all t. In some cases, such as when the Hamiltonian is linear in the control, it turns out that the optimal control can be (jump) discontinuous. The times at which a jump in the control occurs are called *switching points*. Whenever this happens, the state equation $\dot{x}$ may have a different RHS (right hand side), and, consequently, the state can have a kink point. Nevertheless, the Maximum Principle continues to hold for all the points where the control $u(t)$ is continuous and the costate equation holds whenever the control is continuous. It can be shown that the Hamiltonian is continuous even at the switching points. Also in dynamic programming, an admissible control is a function with a finite number of jump discontinuities. We provide some examples below.

The second remark is related to the formulation of the problem we presented. In some cases, the terminal time T is unspecified, but it is a variable of the problem. A typical example is the so-called *time-optimal* control, for which one wants to find the smallest time such that the state of the system reaches a given point starting from a given initial condition. This problem can be written in standard form and analyzed with the principles that we described above.

Another remark concerns the possible constraints that can be part of the problem. The most common kinds are the *mixed inequality constraints*, where it is required that for each $t \in [0, T]$ inequalities of the form $q(x(t), u(t), t) \geq 0$ or the more involved "pure-state" constraints of the form $s(x(t), t) \geq 0$ hold. The maximum principle can be reformulated to deal with these cases. Given the introductory aim of this section, we do not enter the details here but we refer to [31] for a comprehensive overview.

In the next two Sections, which can be skipped for the first reading, we provide a justification for these results.

1.4.1 Bellman's Optimality Principle: The Hamilton-Jacobi-Bellman Equation

The main tools at hand for solving an optimal control problem are Bellman's optimality principle and Pontryagin's maximum principle.

Let us first try to find a necessary condition for an optimal control. In other words, we try to answer the following questions: How does an optimal control look like? What are the main properties that an optimal control should possess?

A very clear answer is given in the famous optimality principle, which Bellman himself describes with these words in his book [5]:

> An optimal policy has the property that whatever the initial state and initial decision are, the remaining decisions must constitute an optimal policy with regard to the state resulting from the first decision.[10]

For an everyday life example, suppose that a marathon runner has to run 42 km. If the marathon runner divided into two (not necessarily equals) parts the run and used her energies to take the first part of the race at full speed, then she would no longer have necessary energy for the second part of the run. Clearly it does not make sense to divide the entire path into two parts and maximize over the first part: the final outcome would not be the optimal one and she probably would not finish the race. However, if the runner divided the way into two parts, then she would run the second part of the journey at best, given the energy left over from the first part of the route. In other words, *the second part of an optimal path must be optimal.*

Now let us try to describe the optimality principle in mathematical terms, without providing rigorous proofs. Suppose that an optimal control exists and it is used in solving (1.119). Then the objective in (1.119) becomes a number, since it is the sum

[10]The word "policy" is nowadays substituted with the most common "control".

of the definite integral (an area) in (1.119) and the terminal payoff. Let us denote this number by $V(x_0, 0)$, emphasizing that the value of the definite integral in (1.119) plus the scrap value depends only on the initial state of the system $x(0) = x_0$, and not on the control if it has been chosen to be an optimal one. Assume that the value function (1.122) is well definite (e.g., the integral converges). The value function (1.122) returns the value of the integral plus the terminal payoff, for a generic initial time t and initial state x in the optimal path, i.e., $x = x^*(t)$. The optimality principle implies that the value function must satisfy the following condition: if at time t the state is $x^*(t)$ (a point of the optimal path) and the interval $[t, T]$ is split in two parts, say $[t, t + \Delta t]$ and $[t + \Delta t, T]$, then the optimal control must maximize the integral of the instantaneous payoff in the period $[t, t + \Delta t]$ and then, from $t + \Delta t$ onwards to T, it must hold that the value function gives the maximum attainable starting at time $t + \Delta t$ with the updated state $x + \Delta x$, reached through the optimal control during the interval $[t, t + \Delta t]$. In practice, the value function solves a *functional* equation of the form:

$$V(x,t) = \max_{u(s)\in A(s)} \left[\int_t^{t+\Delta t} f(x(s), u(s), s)ds + V(x + \Delta x, t + \Delta t) \right] , \qquad (1.124)$$

where $V(x + \Delta x, t + \Delta t)$ is the value of the "second part" of the optimal path that must be optimal. The functional equation (1.124) can be written as a differential equation, as follows.

By the fundamental theorem of the integral calculus, for a 'small' Δt increment it is

$$\int_t^{t+\Delta t} f(x(s), u(s), s)ds \approx f(x(t), u(t), t)\Delta t \, . \qquad (1.125)$$

If the value function $V(x, t)$ is continuously differentiable, we can approximate it through a Taylor expansion about the point (x, t)

$$V(x + \Delta x, t + \Delta t) \approx V(x,t) + \frac{\partial V}{\partial x}\Delta x + \frac{\partial V}{\partial t}\Delta t \, . \qquad (1.126)$$

By substituting (1.125) and (1.126) into (1.124), we get

$$V(x,t) \approx \max_{u(t)\in A(t)} \left[f(x(t), u(t), t)\Delta t + V(x,t) + \frac{\partial V}{\partial x}\Delta x + \frac{\partial V}{\partial t}\Delta t \right] ,$$

which can be written as

$$0 \approx \max_{u(t)\in A(t)} \left[f(x(t), u(t), t)\Delta t + \frac{\partial V}{\partial x}\Delta x + \frac{\partial V}{\partial t}\Delta t \right] ,$$

since $V(x,t)$ does not depend on $u(t)$ and the $V(x,t)$ terms on the LHS and on the RHS cancel out.

Now if we divide both terms of the last expression by Δt we get

$$0 \approx \max_{u(t)\in A(t)} \left[f(x(t),u(t),t) + \frac{\partial V}{\partial x}\frac{\Delta x}{\Delta t} + \frac{\partial V}{\partial t} \right]$$

which, taking the limit as $\Delta t \to 0$, becomes

$$0 = \max_{u(t)\in A(t)} \left[f(x(t),u(t),t) + \frac{\partial V}{\partial x} g(x(t),u(t),t) + \frac{\partial V}{\partial t} \right], \tag{1.127}$$

where we use the fact that $\lim_{\Delta t\to 0} \frac{\Delta x}{\Delta t} = \dot{x}$, which is the state equation. Moreover, for (1.127) it must hold the boundary condition that

$$V(x(T),T) = F(x(T),T). \tag{1.128}$$

In other words, if the problem starts at the terminal time, then the integral in (1.122) is zero and the value function coincides with the scrap value. Since in (1.127) $\frac{\partial V}{\partial t} = \frac{\partial V(x,t)}{\partial t}$ does not depend on $u(t)$, we can rewrite (1.127) as

$$-\frac{\partial V}{\partial t} = \max_{u(t)\in A(t)} \left[f(x(t),u(t),t) + \frac{\partial V}{\partial x} g(x(t),u(t),t) \right]. \tag{1.129}$$

Technically, (1.127) known as the Hamilton-Jacobi-Bellman (HJB) equation, is a (first order) Partial Differential Equation for the value function. This problem, in general, is very difficult to tackle. In the following, we will provide some examples for which the HJB equation can be written as an ODE or for which the value function can be found starting from some trial functions.

1.4.2 From HJB to Pontryagin's Maximum Principle

Consider the derivative $\frac{\partial V}{\partial x}$ in (1.127), where $x = x^*(t)$. Define the *costate* variable $\lambda(t)$ as follows

$$\lambda(t) := \frac{\partial V}{\partial x} = \frac{\partial V(x^*,t)}{\partial x} = \frac{\partial V(x^*(t),t)}{\partial x}. \tag{1.130}$$

Notice that $\lambda(t)$ represents the derivative of the value function with respect to the state variable at each time.

From (1.129) and the definition of costate variable, it follows that an optimal control maximizes the Hamiltonian function H in (1.123) with respect to u. This important fact is referred to as the "Maximum principle". Observe that (1.127) can be written in terms of the Hamiltonian function (1.123) as follows

$$0 = \max_{u(t)\in A}\left[H(x,u,\lambda,t)+\frac{\partial V}{\partial t}\right] = H(x^*,u^*,\frac{\partial V(x^*,t)}{\partial x},t)+\frac{\partial V(x^*,t)}{\partial t}\ . \quad (1.131)$$

Note that in the RHS of (1.131) the max disappears since we are considering optimal control and optimal state.

Now imagine that the control remains u^* but the state is "perturbed": instead of the optimal state x^* consider the "perturbed" state

$$x = x^* + hv\ ,$$

where $v = v(t)$ is an arbitrary function, which we assume continuous and $h \in \mathbb{R}$. Obviously, for $h = 0$, the perturbed state coincides with the optimal state. For any fixed $t \in [0, T]$ and a fixed $v(t)$, define the function $R(h)$

$$R(h) = H(x,u^*,\frac{\partial V(x,t)}{\partial x},t)+\frac{\partial V(x,t)}{\partial t}\ .$$

Notice that $R(h)$ is a differentiable function of one variable (v is fixed as well as u^*) and it has a maximum point at $h = 0$, being from (1.131)

$$R(0) = H(x^*,u^*,\frac{\partial V(x^*,t)}{\partial x},t)+\frac{\partial V(x^*,t)}{\partial t} \geq H(x,u^*,\frac{\partial V(x,t)}{\partial x},t)+\frac{\partial V(x,t)}{\partial t} = R(h)\ .$$

Since $R(h)$ is differentiable, it must be that $R'(0) = 0$. By the chain rule:

$$\begin{aligned}
R'(h) &= \frac{d}{dh}\left[H(x,u^*,\frac{\partial V(x,t)}{\partial x},t)+\frac{\partial V(x,t)}{\partial t}\right] \quad (1.132)\\
&= \frac{d}{dh}\left[f(x,u^*,t)+V_x(x,t)g(x,u^*,t)+\frac{\partial V(x,t)}{\partial t}\right]\\
&= f_x(x,u^*,t)v+V_x(x,t)g_x(x,u^*,t)v+V_{xx}(x,t)g(x,u^*,t)v+V_{tx}(x,t)v\\
&= \left[f_x(x,u^*,t)+V_x(x,t)g_x(x,u^*,t)+V_{xx}(x,t)g(x,u^*,t)+V_{tx}(x,t)\right]v
\end{aligned}$$

from which

$$R'(0) = \left[f_x(x^*,u^*,t)+V_x(x^*,t)g_x(x^*,u^*,t)+V_{xx}(x^*,t)g(x^*,u^*,t)+V_{tx}(x^*,t)\right]v = 0\ . \quad (1.133)$$

Since v is an arbitrary function, in (1.133) the term in square brackets must be zero.

Now derive $\frac{\partial V(x^*(t),t)}{\partial x}$ in (1.130) with respect to t. Using again the chain rule one obtains

$$\frac{\mathrm{d}V_x}{\mathrm{d}t} = V_{xx}\left(x^*(t),t\right)\dot{x}+V_{xt}\left(x^*(t),t\right) = V_{xx}\left(x^*,t\right)g(x^*,u^*,t)+V_{tx}\left(x^*,t\right)\ . \quad (1.134)$$

Substitute (1.134) in the square bracket term in (1.133) to obtain

$$f_x(x^*, u^*, t) + V_x(x^*, t)g_x(x^*, u^*, t) + \frac{\mathrm{d}V_x}{\mathrm{d}t} = 0$$

which, recalling the definitions of costate in (1.130) and Hamiltonian in (1.123), can be rewritten as

$$\dot{\lambda} = -\frac{\partial H}{\partial x} \,. \tag{1.135}$$

Now consider the terminal condition on the costate, which is referred to as the *transversality* condition. In the simplest case, there is no constraint on the value that the optimal state must assume in T, i.e., $x(T)$ is free. In this case, the scrap value $F(x(T), T)$ in (1.119) is independent on $x(T)$ so that $\frac{\partial}{\partial x}F(x(T), T) = 0$. From (1.128) and from the definition of (1.130) it is

$$0 = \frac{\partial}{\partial x}F(x(T), T) = \frac{\partial}{\partial x}V(x(T), T) = \lambda(T) \,.$$

More generally, if $x(T) = x_T$ is given, then from the same reasoning we obtain the transversality condition

$$\lambda(T) = \frac{\partial}{\partial x}F(x_T, T) \,.$$

Summing up, we obtain the Maximum principle already recalled.

The differential equations for the state and costate variables, together with the boundary conditions $x^*(0)$ and $\lambda^*(T)$, constitute a two-point boundary value problem, where it is specified the initial value of the state and the final value of the costate.

Necessary conditions are important to select possible candidates for the optimal control. In addition, the following sufficient conditions are useful to confirm that a solution candidate is indeed an optimum. We recall below the most important sufficient conditions.

Proposition 1.8 (Mangasarian sufficient condition) *Consider a candidate solution of the optimal control problem (1.119), i.e., an admissible control u^*, the corresponding admissible path x^* and the costate variable λ^*, obtained through Pontryagin's maximum principle.*

- *If the Hamiltonian H in (1.123) is concave in x and u for all $t \in [0, T]$ then u^* is an optimal control and x^* is an optimal path;*
- *If the Hamiltonian H in (1.123) is strictly concave in x and u for all $t \in [0, T]$ then u^* is the unique optimal control and x^* is the unique optimal path.*

An immediate corollary of the previous proposition is the following.

Corollary 1.3 (Mangasarian sufficient condition) *Consider a candidate solution of the optimal control problem (1.119), i.e., an admissible control u^*, the corresponding admissible path x^* and the costate variable λ^*, obtained through Pontryagin's maximum principle. Assume that the instantaneous payoff $f(x, u, t)$ is concave in x and u for all $t \in [0, T]$ and that one of the following conditions holds:*

- *for all $t \in [0, T]$, $g(x, u, t)$ is concave in x and u and $\lambda^* \geq 0$;*
- *for all $t \in [0, T]$, $g(x, u, t)$ is convex in x and u and $\lambda^* \leq 0$;*
- *$g(x, u, t)$ is linear in x and u.*

Then u^* is an optimal control and x^* is an optimal path.

Another useful sufficient condition is based on the concavity of the maximized Hamiltonian H_M, defined as

$$H_M(x, \lambda, t) = \max_u H(x, u, \lambda, t) = \max_u \left[f(x, u, t) + \lambda g(x, u, t)\right] . \qquad (1.136)$$

Proposition 1.9 (Arrow sufficient condition) *Consider a candidate solution of the optimal control problem (1.119), i.e., an admissible control u^*, the corresponding admissible path x^* and the costate variable λ^*, obtained through Pontryagin's maximum principle. If the maximized Hamiltonian H_M in (1.136) is concave in x for all $t \in [0, T]$ then u^* is an optimal control and x^* is an optimal path.*

These sufficient conditions are employed in the examples below.

1.4.3 Some Basic Examples

Example (Basic)

Consider the problem

$$\max_u \int_0^2 \left(x - 2u^2\right) \mathrm{d}t \quad \text{such that:} \quad \begin{cases} \dot{x} = 3 + u \\ x(0) = 5 \end{cases} \qquad (1.137)$$

The Hamiltonian function is

$$H = x - 2u^2 + \lambda\,(3 + u) \ .$$

Notice that the Hamiltonian is concave in state x and control u, so the necessary conditions are also sufficient. We apply the maximum principle to find the optimal control. In this particular case, being the Hamiltonian strictly concave in u and since no constraints on u are imposed, the maximizer can be found through the first order condition:

$$\frac{\partial H}{\partial u} = -4u + \lambda = 0 \rightarrow u^* = \frac{\lambda}{4} \ . \qquad (1.138)$$

The costate equation is

$$\dot{\lambda} = -\frac{\partial H}{\partial x} = -1 \to \lambda(t) = -t + c \,, \tag{1.139}$$

where c is a constant to be determined through the transversality condition, i.e., the value of the costate at terminal time $T = 2$.

Since the final state $x(2)$ is free, the transversality condition becomes $\lambda(2) = 0$. Therefore, the costate is $\lambda(t) = -t + 2$, and the optimal control is $u^*(t) = \lambda(t)/4 = -t/4 + 1/2$. Now, the optimal state path can be obtained by integrating the state equation with the obtained optimal control:

$$\dot{x} = 3 + u^* = -\frac{t}{4} + \frac{7}{2} \to x(t) = -\frac{t^2}{8} + \frac{7}{2}t + 5 \,.$$

The constant (5) in the optimal path has been obtained by the initial condition on the state, see (1.137).

Let us slightly modify problem (1.137) by including a scrap value, for instance consider the objective

$$\max_u \int_0^2 \left(x - 2u^2\right) \mathrm{d}t + 4x(2) \quad \text{such that} \quad \begin{cases} \dot{x} = 3 + u \\ x(0) = 5 \end{cases}$$

Clearly, the Hamiltonian is unchanged as well as conditions (1.138) and (1.139). Now the right transversality condition is $\lambda(T) = \lambda(2) = \frac{\mathrm{d}(4x)}{\mathrm{d}x} = 4$. Through analogous calculations as before we obtain, $\lambda^*(t) = -t + 6$, $u^*(t) = -t/4 + 3/2$ and $x^*(t) = -t^2/8 + 9/2t + 5$.

Let us now try to solve problem (1.137) by dynamic programming. HJB equation for problem (1.137) implies that $V(x, t)$ must solve

$$V_t + \max_u \left[x - 2u^2 + V_x (3 + u)\right] = 0 \,,$$
$$V_t + x + 3V_x + \max_u \left[-2u^2 + uV_x\right] = 0.$$

Maximizing $-2u^2 + uV_x$ with respect to u by setting $\frac{\partial(-2u^2+uV_x)}{\partial u} = 0$, we get that the optimal control satisfies $u = V_x/4$. The HJB equation then becomes

$$V_t + x + 3V_x + \frac{1}{8}(V_x)^2 = 0 \tag{1.140}$$

Usually, it is extremely hard if not impossible to solve in closed form the HJB equation. In this case, we try to obtain a solution starting by a trial function. Consider a function of the form

$$V(x,t) = ax + bxt + ct^3 + dt^2 + et + f\ ,$$

where a, b, c, d, e, f are constant to be determined. Inserting the trial solution in (1.140) it is

$$V(x,t) = (b+1)x + t^2\left(\frac{b^2}{8} + 3c\right) + t\left(\frac{ab}{4} + 3b + 2d\right) + \frac{a^2}{8} + 3a + e\ .$$

Moreover, the transversality condition $V(x,2) = 0$ implies that $(a+2b)x + 8c + 4d + 2e + f = 0$. At this point, the HJB equation is satisfied for all x and t if and only if the following system of equations is satisfied

$$\begin{cases} b+1 = 0\ , \\ \frac{b^2}{8} + 3c = 0\ , \\ \frac{ab}{4} + 3b + 2d = 0\ , \\ \frac{a^2}{8} + 3a + e = 0\ , \\ 8c + 4d + 2e + f = 0\ , \\ a + 2b = 0\ , \end{cases} \tag{1.141}$$

which gives the solution

$$a = 2,\quad b = -1,\quad c = -\frac{1}{24},\quad d = \frac{7}{4},\quad e = -\frac{13}{2},\quad f = \frac{19}{3}$$

that determines the value function

$$V(x,t) = 2x - xt - \frac{1}{24}t^3 + \frac{7}{4}t^2 - \frac{13}{2}t + \frac{19}{3}\ .$$

Notice that $u^*(t) = V_x/4 = (2-t)/4$ coincides with the solution previously obtained through the maximum principle.

In the case of scrap value, the transversality condition requires that $V(x,2) = 4x$. In that case, the last equation of system (1.141) is replaced by $a + 2b - 4 = 0$. We left to the reader to verify that the value function in this case has coefficients

$$a = 6,\quad b = -1,\quad c = -\frac{1}{24},\quad d = \frac{9}{4},\quad e = -\frac{45}{2},\quad f = \frac{109}{3}\ .$$

Example (Bang-bang control) Consider the problem

$$\min_{u\in[-1,1]} \int_0^2 x^2 \mathrm{d}t \quad \text{such that:} \quad \begin{cases} \dot{x} = u\ , \\ x(0) = -2\ . \end{cases}$$

Since the integrand x^2 is the quadrate of the distance between a point x and the origin, the problem can be interpreted as follows: start from $x(0) = -2$ and try to steer x as near to zero as possible. To write the problem as a maximization one, we rewrite the objective as

$$\max_{u \in [-1,1]} \int_0^2 -x^2 \mathrm{d}t \ .$$

The Hamiltonian function is

$$H = -x^2 + \lambda u \ ,$$

so that, applying the Maximum Principle, we obtain

$$\begin{cases} \dot{x} = u \text{ (state equation)} \ , \\ \dot{\lambda} = -H_x = 2x \text{ (costate equation)} \ , \\ u^* = \arg \max_{u \in [-1,1]} H \text{ (maximum principle)} \ , \\ x^*(0) = -2 \text{ (initial condition)} \ , \\ \lambda^*(2) = 0 \text{ (transversality condition when } x(T) \text{ is free)} \ . \end{cases}$$

Notice that Mangasarian's sufficient condition holds, as the instantaneous payoff is strictly concave and the state equation is linear.

H is linear in u, so we must adopt a control of the type

$$u^* = \begin{cases} 1 & \text{if } \lambda > 0 \ , \\ ? & \text{if } \lambda = 0 \ , \\ -1 & \text{if } \lambda < 0 \ . \end{cases} \tag{1.142}$$

In our example, since $x^*(0) = -2 < 0$, it is clear that, in order to steer the system towards zero, we must select $u = 1$. The state-costate system becomes

$$\begin{cases} \dot{x} = 1 \ , \\ \dot{\lambda} = 2x \end{cases}$$

with the previously reminded initial conditions. Integrating the first equation with the initial condition $x^*(0) = -2$ we get $x(t) = t - 2$, which then gives the following equation for the costate

$$\dot{\lambda} = 2x = 2t - 4$$
$$\lambda(t) = t^2 - 4t + 4 = (t-2)^2 \ .$$

Notice that $\lambda(t) > 0$ for all $t \in [0, 2)$, confirming that the initial choice of $u = 1$ was correct. This control is called *bang-bang*, since, of all the possible values of u in the interval $[-1, 1]$, we were only interested in the terminal points, see (1.142), at least for $\lambda \neq 0$. Next example clarifies the presence of the "?" in (1.142).

Example (Singular control) Consider the problem

$$\max_{u \in [-1,1]} \int_0^3 -x^2 \mathrm{d}t \quad \text{such that:} \quad \begin{cases} \dot{x} = u \,, \\ x(0) = -2 \,. \end{cases}$$

The problem is formally identical to the previous one, with the exception that the terminal time is now $T = 3$. Employing the same principle as before, it is clear that it is optimal to steer the system towards the origin. Thus we get that in the interval $[0, 2]$ the optimal trajectory is $x(t) = t - 2$ and $\lambda(t) = (t - 2)^2$. At $T = 2$, it is $x(2) = 0$ and also $\lambda(2) = 0$. In (1.142) we observed that when $\lambda = 0$ the control is undefined. In this example, it is obvious that in the interval $(2, 3]$ the control must be $u = 0$: at $t = 2$ the system has reached the value $x = 0$ and any other control $u \neq 0$ would bring the state away from the origin (remember that the interpretation of the problem is that of minimizing an area). Since in the interval $(2, 3]$ it is $\dot{x} = 0$ it is also $\dot{\lambda} = 2x = 0$, so that $\lambda(t)$ is constant in $(2, 3]$. In order to guarantee that $\lambda(3) = 0$ it must be $\lambda(t) = 0$, for all $(2, 3]$, with also $x(t) = 0$, for all $(2, 3]$. We have thus established that the "?" in (1.142) is indeed $u = 0$. This is a typical example of a *singular* control, since there exists an interval where the Hamiltonian is independent on the control u.

1.4.4 Current Value Formulations

In economics, the typical problem of optimal control assumes the following form

$$\max_{u(t) \in A} \int_0^T \mathrm{e}^{-\delta t} f(x(t), u(t)) \mathrm{d}t + F(x(T), T) \tag{1.143}$$

such that:

$$\begin{cases} \dot{x} = g(x(t), u(t)) \\ x(0) = x_0 \\ \text{and with one of the following terminal conditions:} \\ \text{(a) } x(T) \text{ free} \quad \text{(b) } x(T) = x_T \quad \text{(c) } x(T) \geq x_T \end{cases}$$

where $\delta > 0$ is the discount factor. Remember that in most economic applications it is $T = +\infty$ and, clearly $F(x(t), t) = 0$. In the following, we reformulate the HJB equation and the maximum principle for this specific problem.

First consider the HJB equation in (1.129) for problem (1.143)

$$-\frac{\partial V}{\partial t} = \max_{u(t)\in A(t)} \left[e^{-\delta t} f(x(t), u(t)) + \frac{\partial V}{\partial x} g(x(t), u(t)) \right]$$

and assume that the value function can be written as the product of a function in x times $e^{-\delta t}$:

$$V(x,t) = J(x)e^{-\delta t}$$

from which it is $\frac{\partial V}{\partial t} = -\delta J(x)e^{-\delta t}$ and $\frac{\partial V}{\partial x} = J'(x)e^{-\delta t}$. Thus, the HJB equation becomes

$$\delta J(x)e^{-\delta t} = \max_{u(t)\in A(t)} \left[e^{-\delta t} f(x(t), u(t)) + J'(x)e^{-\delta t} g(x(t), u(t)) \right].$$

Multiplying both sides by $e^{\delta t}$, we obtain the following ODE in the unknown function $J(x)$:

$$\delta J(x) = \max_{u(t)\in A(t)} \left[f(x(t), u(t)) + J'(x) g(x(t), u(t)) \right] . \tag{1.144}$$

It is easily verifiable that the transversality condition $F(x(T), T) = V(x(T), T)$ now becomes

$$J(x(T)) = e^{\delta T} F(x(T), T) .$$

Now we restate the maximum principle for problem (1.143). Consider the corresponding Hamiltonian in (1.123)

$$H(x, u, \lambda, t) = e^{-\delta t} f(x, u) + \lambda g(x, u)$$

and multiply it by $e^{\delta t}$ to obtain the so-called Current-value Hamiltonian H^c:

$$H^c(x, u, \mu, t) = f(x, u) + \mu g(x, u) , \tag{1.145}$$

where $\mu = \mu(t) = \lambda(t)e^{\delta t}$. From $\lambda(t) = \mu(t)e^{-\delta t}$ we obtain that

$$\dot{\lambda} = \dot{\mu} e^{-\delta t} - \delta \mu e^{-\delta t} ,$$

and thus,

$$\dot{\lambda} e^{\delta t} = \dot{\mu} - \delta \mu \tag{1.146}$$

Since the current value Hamiltonian is given by the Hamiltonian times a constant, the optimal control u^* maximizes the current-value Hamiltonian as well as the Hamiltonian; moreover $H^c_\mu = H_\lambda$ so that the state equation can be written as $\dot{x} = H^c_\mu$. Now consider the costate equation

$$\dot{\lambda} = -H_x = -\left[f_x(x^*, u^*) + \lambda g_x(x^*, u^*)\right] \tag{1.147}$$
$$= -\left[e^{-\delta t} f_x(x^*, u^*) + \lambda g_x(x^*, u^*)\right]$$
$$= -\left[e^{-\delta t} f_x(x^*, u^*) + \mu e^{-\delta t} g_x(x^*, u^*)\right] . \tag{1.148}$$

Multiply (1.147) by $e^{\delta t}$ and, considering (1.146), we can write

$$\dot{\mu} - \delta\mu = -\left[f_x(x^*, u^*) + \mu g_x(x^*, u^*)\right] = -H_x^c .$$

It is now possible to restate Pontryagin's maximum principle for problem (1.143).

Proposition 1.10 (Maximum Principle with current-value formulation) *If u^* is an optimal control and x^* is the corresponding optimal path for the problem (1.143), then there exists a costate variable μ^* such that x^*, μ^*, u^* are the solution in* $[0, T]$ *of the following problem:*

$$\begin{cases} \dot{x} = g(x^*, u^*) = H_\mu^c \ (\textit{state equation}) \\ \dot{\mu} = -\left[f_x(x^*, u^*) + \mu^* g_x(x^*, u^*)\right] + \delta\mu = -H_x^c + \delta\mu \ (\textit{costate equation}) \\ u^* = \arg\max H^c(x^*, u, \mu^*) \ (\textit{maximum principle}) \\ x^*(0) = x_0 \ (\textit{initial condition}) \\ (a)\ \mu^*(T) = \frac{\partial}{\partial x} F(x_T, T) \ (\textit{transversality condition when } x(T) \textit{ free}) \\ \quad \textit{or} \\ (b)\ \mu^*(T) \geq \frac{\partial}{\partial x} F(x_T, T) \ ((\textit{transversality condition when } x(T) \geq x_T) \end{cases}$$

Obviously, when the scrap value is zero, in the transversality conditions it is $\frac{\partial}{\partial x} F(x_T, T) = 0$. When $T = +\infty$, the following transversality condition is necessary

$$\lim_{t \to +\infty} e^{-\delta t} H^c(x^*, u^*, \mu^*)$$
$$= \lim_{t \to +\infty} e^{-\delta t} f(x^*, u^*) + e^{-\delta t} \mu^* g(x^*, u^*) = 0. \tag{1.149}$$

In economic applications, the following "simplified" transversality condition is often employed:

$$\lim_{t \to +\infty} \mu^* e^{-\delta t} = 0. \tag{1.150}$$

Notice that condition (1.150) is not necessary, as it can be shown by counterexamples, see [32]. Clearly, if in the considered optimal control problem with infinite time horizon the optimal state converges to an equilibrium value (x^*, u^*), then $\lim_{t \to +\infty} e^{-\delta t} f(x^*, u^*) = 0$ and $\lim_{t \to +\infty} g(x^*, u^*) = 0$ so that (1.150) implies (1.149). We refer to [2] for details on this point.

We conclude by recalling that Mangasarian as well as Arrow sufficient conditions hold. These conditions can be applied, respectively, to the current-value Hamiltonian and to the current-value maximized Hamiltonian H_M^c. In the following example, these two theorems are useful to provide sufficient conditions for optimality.

1.4.5 Economic Examples

Example (Optimal use of a machine). Suppose that you possess a machine whose value at time t is denoted by $x(t)$. For each unit of capital invested in the machine, you obtain a unit of a good, which is then sold in the market at constant price p. The machine depreciates over time at the rate η, but it is possible to reduce depreciation by investing in maintenance. Denote by $u = u(t)$ the instantaneous maintenance activity (or repair effort), which is our control variable. The cost of maintenance is $c(u) = \gamma u^2$. Instantaneous profit is then $px - \gamma u^2$ (revenues less costs). Assume that the total life of the machine is $T > 0$. Indicating by δ the discount factor, the objective is

$$\max_{u \geq 0} \int_0^T e^{-\delta t} \left[px - \gamma u^2 \right] dt \quad \text{such that:} \quad \begin{cases} \dot{x} = -\eta x + u \ , \\ x(0) = x_0 \ , \\ x(T) \ \text{free} \ . \end{cases} \tag{1.151}$$

The current-value Hamiltonian is

$$H^c = px - \gamma u^2 + \mu \left(-\eta x + u \right) \ .$$

The function H^c is concave in x and u, so by the Mangasarian theorem the necessary conditions are also sufficient for an optimal control. Moreover, since H^c is strictly concave in u, we can deduce that, under the assumption that the optimal control is strictly positive, i.e., $u^* > 0$ (see below), the optimal control must satisfy the usual necessary condition for a max: $\frac{\partial H^c}{\partial u} = -2\gamma u^* + \mu = 0$. Thus, H^c is maximized at

$$u^* = \frac{\mu}{2\gamma} \ . \tag{1.152}$$

The costate equation is

$$\dot{\mu} = -\frac{\partial H^c}{\partial x} + \mu\delta = -p + \mu \left(\eta + \delta \right) \tag{1.153}$$

with terminal condition (transversality) $\mu(T) = 0$. Equation (1.153) is a linear ODE with constant coefficients, whose general solution can be easily calculated:

$$\mu(t) = K e^{(\eta+\delta)t} + \frac{p}{\eta + \delta} .$$

Imposing the boundary condition $\mu(T) = 0$, we specify the value of the constant K. The required solution is $\mu(t) = -pe^{(\eta+\delta)(t-T)}/(\eta + \delta) + p/(\eta + \delta)$.

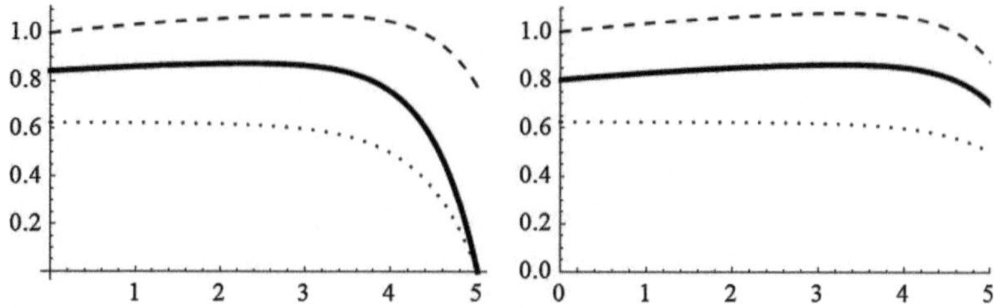

Fig. 1.88 Optimal use of a machine—time evolution of the optimal path $x^*(t)$ (dashed) and optimal control $u^*(t)$ (*black*) and costate $\mu(t)$ (*dotted*) with parameters $\delta = 0.8$; $\eta = 0.8$; $p = 1$; $\gamma = 0.9$; $T = 5$; $x_0 = 1$. **a** without scrap value; **b** with scrap value $S(x) = x/2$

By (1.152), the optimal control is therefore

$$u^*(t) = \frac{1}{2\gamma}\left[-p\frac{e^{(\eta+\delta)(t-T)}}{\eta+\delta} + \frac{p}{\eta+\delta}\right]. \tag{1.154}$$

Observe from (1.154) that, being $t < T$ it is $u^*(t) > 0$ as conjectured above. Finally, the ODE $\dot{x} = -\eta x + u^*$ is linear (but with nonconstant coefficients) and can be solved. Figure 1.88a shows the time evolution of the optimal path $x^*(t)$ (dashed), optimal control $u^*(t)$ (black) and costate $\mu(t)$ (dotted) with parameters $\delta = 0.8$; $\eta = 0.8$; $p = 1$; $\gamma = 0.9$; $T = 5$; $x_0 = 1$.

In Fig. 1.89, typical solutions of the state-costate ODE system are depicted, in the (x, μ) plane:

$$\begin{cases} \dot{x} = -\eta x + \frac{\mu}{2\gamma}\ , \\ \dot{\mu} = -p + \mu\,(\eta+\delta)\ . \end{cases} \tag{1.155}$$

The nullclines are the red lines (obviously $\dot{\mu} = 0$ is the horizontal line). Observe that the equilibrium of the system, obtained by solving the system $\dot{x} = \dot{\mu} = 0$:

$$\left(x^*, \mu^*\right) = \left(\frac{p}{2\gamma\eta(\delta+\eta)}, \frac{p}{\delta+\eta}\right)$$

is a saddle point, as the Jacobian $J = \begin{bmatrix} -\eta & 1/(2\gamma) \\ 0 & \eta+\delta \end{bmatrix}$ has $\mathrm{Tr}(J) = \delta > 0$ and $\mathrm{Det}(J) = -\eta\,(\eta+\delta) < 0$. Below the nullcline $\dot{\mu} = 0$, the optimal combination of state and control is a blue trajectory with the property that it ends exactly in the x-axis at time T. Above the nullcline $\dot{\mu} = 0$, the generic trajectory diverges and can not satisfy the transversality condition $\mu(T) = 0$.

Before ending the example, let us modify it slightly and suppose that at time T the machine has a scrap value, i.e., it can be sold in a second-hand market at price $S(x) = ax$. The problem can be stated as

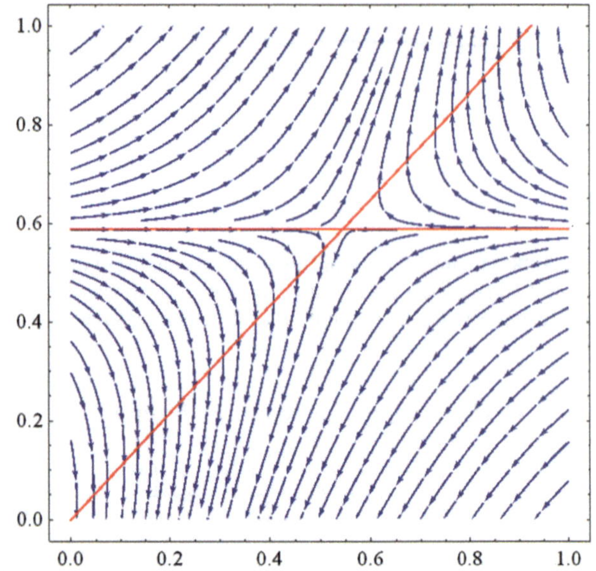

Fig. 1.89 Optimal use of a machine—phase portrait of the optimal state-costate ODEs

$$\max_{u \geq 0} \left[\int_0^T e^{-\delta t} \left[px - \gamma u^2 \right] dt + e^{-\delta t} ax(T) \right]$$

In this case, the ODE (1.155) remains unchanged, whereas only the transversality condition changes and becomes $\mu(T) = \frac{\partial S(x)}{\partial x} = a$. Figure 1.88b shows the relevant quantities with the parameters as in Fig. 1.88a and with scrap value $S(x) = x/2$.

Example (Linear-Quadratic optimal control). Suppose that we want to solve the following problem

$$\min_{u=u(t)} \int_0^{+\infty} e^{-\delta t} \left[ax^2 + bu^2 \right] dt \quad \text{such that:} \quad \begin{cases} \dot{x} = u \ , \\ x(0) = x_0 \ , \end{cases} \tag{1.156}$$

where $a, b > 0$. No constraints on the control are imposed. Being the integrand a quadratic function of state and control and the state equation linear in control, such a problem is often referred to as a *linear-quadratic* (LQ) optimal control problem.

Notice that, since we want to minimize the integral, we can restate the problem in equivalent form by considering the following objective:

$$\max_{u} \int_0^{+\infty} - e^{-\delta t} \left[ax^2 + bu^2 \right] dt \ .$$

Let us start the analysis by characterizing the solution through the maximum principle. The current-value Hamiltonian is given by

$$H^c(x, u, \mu) = -ax^2 - bu^2 + \mu u \,.$$

Clearly Mangasarian's sufficient conditions are verified. Moreover, being H^c strictly concave in u, we can find the optimal control $u^* = u$ by solving the equation of the first order condition

$$\frac{\partial H^c}{\partial u} = -2bu + \mu = 0 \rightarrow u^* = \frac{\mu}{2b} \,.$$

Now consider the costate equation

$$\dot{\mu} = -H_x^c + \delta\mu = 2ax + \delta\mu \,.$$

Summing up, a solution of the problem is given by x^*, u^*, μ^* that solve the following system of linear differential equations

$$\begin{cases} \dot{x} = \frac{\mu}{2b} \,, \\ \dot{\mu} = 2ax + \delta\mu \end{cases} \tag{1.157}$$

with initial condition on the state $x(0) = x_0$ and such that the following transversality condition holds:

$$\lim_{t\to+\infty} \mathrm{e}^{-\delta t} H^c(x^*, u^*, \mu^*) = \lim_{t\to+\infty} -\mathrm{e}^{-\delta t}\left[a\left(x^*\right)^2 + b\left(\frac{\mu^*}{2b}\right)^2\right] + \mathrm{e}^{-\delta t}\frac{(\mu^*)^2}{2b} = 0. \tag{1.158}$$

One particular solution of (1.157) is the fixed point, obtained by solving the system $\dot{x} = \dot{\mu} = 0$. Being a 2×2 system of linear equations with nonzero determinant, this solution is unique and it is the equilibrium $(x, \mu) = (0, 0)$. The Jacobian matrix is given by

$$J = \begin{bmatrix} 0 & \frac{1}{2b} \\ 2a & \delta \end{bmatrix}$$

whose eigenvalues are

$$z_{1,2} = \frac{\delta \pm \sqrt{\delta^2 + 4ab}}{2} \,.$$

Since it is $z_1 = (\delta - \sqrt{\delta^2 + 4ab})/2 < 0 < (\delta + \sqrt{\delta^2 + 4ab})/2 = z_2$, the origin $(0, 0)$ is a saddle point. The equilibrium solution satisfies the transversality condition (1.158) because there μ^* is constant. Thus, when $x(0) = 0$ it is optimal to set $u = 0$ for all t. This is elementary considering again the meaning of problem (1.156).

The interesting question is then the following: what happens when $x(0) \neq 0$? To answer the question, one has to find out how the optimal path looks like in general. Deriving the maximum principle condition with respect to t, we obtain that the following relationship must hold: $\dot{u} = \dot{\mu}/(2b)$. Deriving $\dot{x}$ with respect to t, and using the state equation $\dot{x} = u$, we obtain the following second-order differential equation in x:

$$\begin{aligned}\ddot{x} = \dot{u} = \frac{\dot{\mu}}{2b} &= \frac{2ax + \delta\mu}{2b} = \frac{2ax + 2b\delta u}{2b} \\ &= \frac{a}{b}x + \delta\dot{x}.\end{aligned}$$

The general solution of this equation is

$$x(t) = c_1 e^{r_1 t} + c_2 e^{r_2 t} \tag{1.159}$$

where $r_1 = (\delta - \sqrt{\delta^2 + 4a/b})/2$ and $r_2 = (\delta + \sqrt{\delta^2 + 4a/b})/2$. The constants c_1 and c_2 can be determined through the boundary conditions, i.e., the initial condition on the state $x(0) = x_0$ and the transversality condition on the costate. From the latter we get that condition (1.158) can be satisfied only if $\lim_{t\to+\infty}\mu(t)e^{-\delta t} = 0$. Thus, the transversality condition requires that

$$0 = \lim_{t\to+\infty} \mu(t)e^{-\delta t} = \lim_{t\to+\infty} 2bu(t)e^{-\delta t} = \lim_{t\to+\infty} 2b\dot{x}(t)e^{-\delta t} = c_2\infty$$

which tells us that the trasversality condition can be satisfied only for $c_2 = 0$. Thus, from (1.159) the optimal path is

$$x^*(t) = x_0 e^{t\frac{\delta - \sqrt{\delta^2 + 4a/b}}{2}} \tag{1.160}$$

with corresponding optimal control

$$u^* = \dot{x}^* = x_0 \frac{\delta - \sqrt{\delta^2 + 4a/b}}{2} e^{t\frac{\delta - \sqrt{\delta^2 + 4a/b}}{2}}. \tag{1.161}$$

Consider now the same problem with Dynamic Programming. Maximizing the right hand side of the HJB in (1.144), the optimal control must satisfy the condition

$$u^* = \arg\max\left[-ax^2 - bu^2 + J'(x)u\right]$$

from which it is $u^* = J'(x)/(2b)$, which coincides, if one recall the definition of the costate variable, with the optimal control obtained through Pontryagin's principle. The HJB equation (1.144) thus becomes

$$\begin{aligned}\delta J(x) &= -ax^2 - b\left(\frac{J'(x)}{2b}\right)^2 + \frac{[J'(x)]^2}{2b} \\ &= -ax^2 + \frac{[J'(x)]^2}{4b} \end{aligned} \tag{1.162}$$

which is a nonlinear ODE in the unknown function $J(x)$. In this case, a way to tackle the problem is to "guess" a possible value function. For instance, consider a quadratic "trial" function of the form

$$J(x) = Ax^2 ,$$

where A is a constant to be determined. With this choice of $J(x)$, the HJB equation (1.162) becomes

$$\delta Ax^2 = -ax^2 + \frac{A^2x^2}{b} ,$$

which is equivalent to

$$\left(\frac{A^2}{b} - \delta A - a\right)x^2 = 0 .$$

Thus, the trial solution works if the quantity in parenthesis is zero, i.e., if A solves the quadratic equation $A^2/b - \delta A - a = 0$. The required values of A are

$$A = \frac{b\delta \pm \sqrt{b\left(4a + b\delta^2\right)}}{2} .$$

At this point we have two possible values of A that do the trick. However, notice that the integrand function is negative, as it is the sum of two quadratic terms multiplied by -1. Thus, the value function, which gives the maximum value of this integral, can only assume nonpositive values. Thus, only the negative solution ($A < 0$) is meaningful for our problem. In this way, the following value function has been found:

$$J(x) = \frac{b\delta - \sqrt{b\left(4a + b\delta^2\right)}}{2}x^2 .$$

Now consider the optimal control, which is, as shown before,

$$u^* = \frac{J'(x)}{2b} = \frac{1}{2}\left(\delta - \sqrt{\frac{4a}{b} + \delta^2}\right)x . \tag{1.163}$$

Now from condition $\dot{x} = u$ we have that x solves the ODE

$$\dot{x} = \frac{1}{2}\left(\delta - \sqrt{\frac{4a}{b} + \delta^2}\right)x ,$$

whose solution coincides with (1.160). The optimal control as a function of time is obviously (1.161). Notice that, although both (1.161) and (1.163) represent the optimal control, this control is given in (1.161) as a function of time of the form $u^* = u(t, x_0)$ (*open-loop* control), whereas in (1.163) the optimal control is a function of the current state $u^* = u(t, x(t))$ (*closed-loop* or *feedback* control).

Following similar steps, one can study a more general linear-quadratic problem of the form:

$$\max_{u(t)} \int_0^{+\infty} -\,\mathrm{e}^{-\delta t}\left[ax^2 + bu^2\right]\mathrm{d}t \quad \text{such that:} \quad \begin{cases} \dot{x} = cx + du\,, \\ x(0) = x_0\,. \end{cases} \tag{1.164}$$

Example (A simplified Capital Accumulation Model). Consider now the optimal control problem

$$\max_{0<u(t)\le x^\alpha} \int_0^{+\infty} \mathrm{e}^{-\delta t} \log u(t)\mathrm{d}t \quad \text{such that:} \quad \begin{cases} \dot{x} = x^\alpha - \gamma x - u\,, \\ x(0) = x_0\,, \end{cases} \tag{1.165}$$

where $x = x(t) \ge 0$ can be interpreted as a physical capital that naturally grows through the technological coefficient $\alpha \in (0, 1]$, depreciates itself by an obsolescence factor $\gamma > 0$ and is reduced by current consumption u. Capital x is the state variable and consumption u is the control variable. The objective of the problem is to maximize the discounted stream of utility of consumption, assumed logarithmic. The constraint on the consumption is introduced on the one hand to impose some consumption ($u > 0$) and on the other hand to avoid that the capital is consumed ($u(t) \le x^\alpha$).

The current value Hamiltonian is

$$H^c = \log u + \mu\left[x^\alpha - \gamma x - u\right]$$

which is maximized, from condition $\frac{\partial H^c}{\partial u} = 0$, at $u = 1/\mu$. The current-value Hamiltonian (1.145) is $H^c_x = \mu\left[\alpha x^{\alpha-1} - \gamma\right]$. Through these quantities, the state-costate system of ODE that a candidate to be an optimal solution solves becomes

$$\begin{cases} \dot{x} = x^\alpha - \gamma x - \frac{1}{\mu} \\ \dot{\mu} = \mu\left[(\gamma + \delta) - \alpha x^{\alpha-1}\right] \\ x(0) = x_0 \\ \lim\limits_{t\to+\infty} \mathrm{e}^{-\delta t}\left[\log u + \mu\left(x^\alpha - \gamma x - u\right)\right] = 0 \end{cases}$$

Notice that Mangasarian's sufficient conditions are satisfied, so the necessary conditions are also sufficient.[11]

[11] This can be verified by Sylvester's criterion, being $H^c_{xx} < 0$ and $\begin{vmatrix} H^c_{xx} & H^c_{xu} \\ H^c_{ux} & H^c_{uu} \end{vmatrix} > 0$.

Instead of considering the previous system of ODE, it is instructive to translate it in a ODE system in the state-control variables. First, derive the maximum condition $u = 1/\mu$ with respect to time:

$$\dot{u} = -\frac{\dot{\mu}}{\mu^2} = \frac{\alpha x^{\alpha-1} - (\gamma + \delta)}{\mu} = u\left[\alpha x^{\alpha-1} - (\gamma + \delta)\right]$$

so that the previous system in the state-control space is translated as

$$\begin{cases} \dot{x} = x^{\alpha} - \gamma x - u \\ \dot{u} = u\left[\alpha x^{\alpha-1} - (\gamma + \delta)\right] \\ x(0) = x_0 \\ \lim\limits_{t\to+\infty} \mathrm{e}^{-\delta t}\left[\log u + \mu\left(x^{\alpha} - \gamma x - u\right)\right] = 0 \end{cases} \tag{1.166}$$

Let us study the nullclines of system (1.166). Obviously, $\dot{x} = 0 \Leftrightarrow u = f(x) = x^{\alpha} - \gamma x$. In the plane (x, u), $f(x)$ is a strictly concave function. Observe that below this curve, it is $\dot{x} > 0$. Now consider $\dot{u} = 0 \Leftrightarrow u\left[\alpha x^{\alpha-1} - (\gamma + \delta)\right] = 0$ i.e., when $u = 0$ or $x = \overline{x} = ((\gamma + \delta)/\alpha)^{\frac{1}{\alpha-1}}$: in the plane (x, u) the set of points such that $\dot{u} = 0$ are the x-axis and the vertical line $x = \overline{x}$. Moreover, on the left of $x = \overline{x}$ it is $\dot{u} > 0$. The positive quadrant of the plane (x, u) can thus be subdivided into four regions according to the signs of the regions between nullclines, as represented in Fig. 1.90, where we depicted the locus of points such that $\dot{x} > 0$ and $\dot{u} > 0$ (light yellow region), $\dot{x} > 0$ and $\dot{u} < 0$ (white region), $\dot{x} < 0$ and $\dot{u} > 0$ (gray region), $\dot{x} < 0$ and $\dot{u} < 0$ (light gray region).

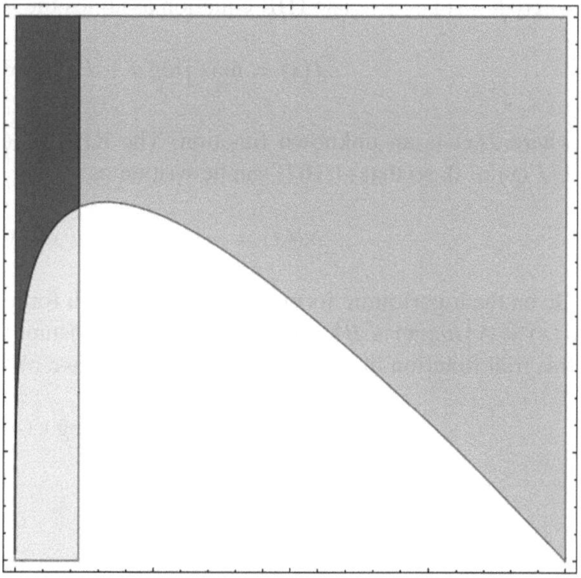

Fig. 1.90 Capital accumulation model—signs of the vector field of the optimal state-costate ODEs

System (1.166) admits three equilibrium points (x, u), obtained as points where $\dot{x} = \dot{u} = 0$: $(0, 0)$, $(\gamma^{\frac{1}{\alpha-1}}, 0)$ and $(\overline{x}, \overline{u}) = (\overline{x}, \overline{x}^{\alpha} - \gamma\overline{x})$. By the constraint on the control ($u > 0$), we disregard the first two equilibrium points, as they involve zero consumption. The analysis with the nullclines suggests that $(\overline{x}, \overline{x}^{\alpha} - \gamma\overline{x})$ is a saddle point. The same conclusions can be drawn by studying the linearization of the system about this fixed point. We perform this stability analysis in the next example in a more general setting.

The point $(\overline{x}, \overline{u})$ is the optimal equilibrium. In fact, it is a solution of (1.166), being the simplified transversality condition $\lim_{t\to+\infty} e^{-\delta t}/\overline{u} = 0$ trivially satisfied, as $\overline{u}$ is constant. When $x(0) = \overline{x}$, it is optimal to use the control $\overline{u}$ for all t to stay at $\overline{x}$.

What happens when the initial condition on the capital is out-of-equilibrium, i.e., if $x(0) \neq \overline{x}$? As $(\overline{x}, \overline{u})$ is a saddle, every trajectory starting in the gray or in the white regions departs from $(\overline{x}, \overline{u})$. For a given initial state, it can not be optimal to take a control such that the trajectory belongs to these regions, as the trajectory would not satisfy the transversality condition in (1.166).

Thus, given an initial condition $x(0) < \overline{x}$ [$x(0) > \overline{x}$], the control should be chosen in the light yellow region [light gray region] such that the path belongs to the stable manifold of the saddle point, in order to guarantee the convergence to the optimal equilibrium $(\overline{x}, \overline{u})$. If we express the optimal control as a function of the state, $u = u(x)$ such that the trajectory of the system belongs to the stable manifold of the saddle point $(\overline{x}, \overline{u})$, then $u(x)$ is the feedback optimal control for problem (1.165).

We briefly study this model with the dynamic programming approach. To simplify the problem, consider the case $\alpha = 1$, so that the state equation becomes

$$\dot{x} = \theta x - u \ ,$$

where $\theta = (1 - \gamma)$. The HJB equation assumes the form

$$\delta J(x) = \max_{u} \left[\log u + J'(x)(\theta x - u) \right] \ , \tag{1.167}$$

where $J(x)$ is an unknown function. The RHS of (1.167) is maximized for $u = 1/J'(x) > 0$, so that (1.167) can be written as

$$\delta J(x) = -\log J'(x) + \theta x J'(x) - 1 \ .$$

Given the logarithmic form of utility, we search for a solution candidate of the form $J(x) = A\left[log(x) + B\right]$, where A and B are constant to be determined. Substituting this trial function in the HJB and simplifying, we obtain the equation

$$1 - A\theta + AB\delta + \log A + \log x\,(A\delta - 1) = 0$$

which leads to the system

$$\begin{cases} A\delta - 1 = 0 \,, \\ 1 - A\theta + AB\delta + \log A = 0 \,, \end{cases}$$

so that $A = 1/\delta$ and $B = \alpha/\delta - 1 + \log\delta$. It is easy to check that function $J(x) = 1/\delta\left[\log(x+\delta) + \alpha/\delta - 1\right]$ satisfies (1.167). The optimal control is then $u^* = 1/J'(x) = \delta x$, which gives the optimal quantity to consume as a function of the current stock of capital (feedback control).

Example (A more general Capital Accumulation Model). Consider a generalization of the previous example. The production function Y depends on total capital X and on total labour force L, i.e., $Y = Y(X, L)$. Define per-capita production and capital as follows: $y = Y/L$ and $x = X/L$. Assume that $Y = Y(X, L)$ is linear homogeneous of degree 1, i.e.,

$$Y = Y(X, L) = Y(Lx, L) = LY(x, 1)$$

from which the individual production function is $y = f(x) = Y(x, 1)$. The function $f(x)$ is assumed twice differentiable with $f'(x) > 0$ and $f''(x) < 0$ and such that it satisfies the so-called *Inada* conditions:

$$\lim_{x\to 0^+} f'(x) = +\infty \quad \text{and} \quad \lim_{x\to +\infty} f'(x) = 0^+ \tag{1.168}$$

Total production is split between total consumptions C and total investments I, i.e., $Y = C + I$. The variation of the capital stock in time is

$$\dot{X} = I - \rho X = Y - C - \rho X \,,$$

where $\rho > 0$ indicates capital depreciation over time. In per-capita terms the variation of capital stock is

$$\frac{\dot{X}}{L} = y - c - \rho x \,,$$

where $c = C/L$ denotes individual consumption, i.e., the consumption of a representative agent in the economy.

On the other hand, the derivative of capital X is

$$\dot{X} = \frac{\mathrm{d}}{\mathrm{d}t}(Lx) = \dot{L}x + L\dot{x}$$

that gives $\dot{X}/L = \dot{L}/Lx + \dot{x}$. Denoting the growth rate of labour force $\dot{L}/L = n$ (assumed constant) and recalling that $y = f(x)$, we obtain the equation of capital growth:

$$\dot{x} = f(x) - c - (n + \rho)x \,.$$

The utility of a representative agent in the economy is denoted by $u(c)$, which is assumed three times differentiable with $u'(c) > 0$ and $u''(c) < 0$ and such that it satisfies the Inada conditions (1.168).

Assume that a social planner aims at maximizing the discounted value of the utility of consumers weighted by labour force over an infinite time horizon. Denoting by $\beta > 0$, the discount factor, the problem of the social planner is

$$\int_0^{+\infty} \mathrm{e}^{-\beta t} u(c)\, L \mathrm{d}t = L_0 \int_0^{+\infty} \mathrm{e}^{(n-\beta)t} u(c)\, \mathrm{d}t$$

where it is assumed that the population grows exponentially, with $L = L(t) = L_0 \mathrm{e}^{nt}$ and with initial population $L(0) = L_0$. Without loss of generality, one can normalize the initial population to 1, i.e., $L_0 = 1$. The objective of the social planner becomes

$$\max_{0 \le u \le f(x)} \int_0^{+\infty} \mathrm{e}^{-\delta t} u(c)\, \mathrm{d}t \quad \text{such that:} \quad \begin{cases} \dot{x} = f(x) - \gamma x - c \\ x(0) = x_0 \end{cases} \tag{1.169}$$

where $\delta = \beta - n > 0$ (by assumption) and $\gamma = n + \rho$. Notice that the previous example in (1.165) is a particular case of the one considered here.

Let us study the model with the maximum principle. The current-value Hamiltonian is

$$H^c = u(c) + \mu \left[f(x) - c - \gamma x \right]$$

which is maximized at $\frac{\partial H^c}{\partial u} = u'(c) - \mu = 0$, i.e., for $\mu = u'(c)$. The state-costate system of ODE that a candidate optimal solution must satisfy is

$$\begin{cases} \dot{x} = f(x) - c - \gamma x \\ \dot{\mu} = -\mu \left[f'(x) - (\gamma + \delta) \right] \\ x(0) = x_0 \\ \lim\limits_{t \to +\infty} \mathrm{e}^{-\delta t} \left[u(c) + \mu(f(x) - c - \gamma x) \right] = 0 \end{cases} \tag{1.170}$$

where also the control c appear. Notice again that Mangasarian's sufficient conditions are satisfied, so that the necessary conditions are also sufficient.[12] To get rid of the costate variable, we transform system (1.170) in an ODE system in the state-control space. Deriving the maximization condition $\mu = u'(c)$ with respect to t we get

$$\dot{\mu} = \frac{\mathrm{d}}{\mathrm{d}t} u'(c(t)) = u''(c) \dot{c} \,,$$

[12]It can be verified by Sylvester's criterion, being $H^c_{xx} = \mu f''(x) = u'(c) f''(x) < 0$ and $\begin{vmatrix} H^c_{xx} & H^c_{xu} \\ H^c_{ux} & H^c_{uu} \end{vmatrix} = u'(c) u''(c) f''(x) > 0$.

so that from the second differential equation in (1.170) we get

$$\dot{c} = -\frac{u'(c)}{u''(c)} \left[f'(x) - (\gamma + \delta)\right] \tag{1.171}$$

with the transversality condition

$$\lim_{t\to+\infty} e^{-\delta t} \left[u(c) + u'(c)(f(x) - c - \gamma x)\right] = 0. \tag{1.172}$$

Notice that by assumption it is $u''(c) \neq 0$ and $-u'(c)/u''(c) > 0$. Thus, the nullcline $\dot{c} = 0$ is the locus of points $\overline{x}$ such that

$$f'(\overline{x}) = (\gamma + \delta) \tag{1.173}$$

From the Inada conditions, for any fixed $(\gamma + \delta) > 0$ there exists a unique root $\overline{x}$ to equation (1.173). Equation (1.173) is often referred to as the *Modified Golden rule* in capital accumulation models. This nullcline is a vertical line in the (x, c) plane.

By (1.171), $\dot{c} > 0$ for $f'(x) > (\gamma + \delta)$, which occurs, being $f'(x)$ decreasing, for $x < \overline{x}$.

Now consider the nullcline $\dot{x} = 0$, which represents the set of points such that $c = f(x) - \gamma x$. Consider the function

$$c(x) = f(x) - \gamma x\ .$$

It is $c(0) = 0$ for $x = 0$ and for $c(\widetilde{x}) = 0$ such that $f(\widetilde{x}) - \gamma\widetilde{x} = 0$. Moreover, $c(x)$ has a maximum point at the point $\widehat{x}$ (*golden rule* state) such that $c'(\widehat{x}) = 0$, i.e., such that $f'(\widehat{x}) = \gamma$. In other words, the nullcline $\dot{x} = 0$ is, in the plane (x, c) an unimodal function. It is interesting to observe that $\widehat{x} > \overline{x}$, see discussion below.

Moreover, $\dot{x} > 0$ for $c < f(x) - \gamma x$ i.e., below the curve $c(x)$. Through the condition $\dot{x} = \dot{c} = 0$ we identify the equilibrium of the following system of ODE:

$$E = (\overline{x}, \overline{c}) = (\overline{x}, c(\overline{x})) = (\overline{x}, f(\overline{x}) - \gamma\overline{x})\ . \tag{1.174}$$

By graphical analysis, this equilibrium is a saddle point. The qualitative graph of the nullclines is identical to the one presented in the previous example. Notice that at the equilibrium $(\overline{x}, c(\overline{x}))$ the transversality condition is satisfied, since $c(\overline{x})$ is constant and condition $\lim_{t\to+\infty} u'(c(\overline{x}))e^{-\delta t} = 0$ implies (1.172).

The stability analysis of equilibrium (1.174) can be carried out as always considering the eigenvalues of the Jacobian matrix at equilibrium. It is

$$J(E) = \begin{bmatrix} \frac{\partial \dot{x}}{\partial x} & \frac{\partial \dot{x}}{\partial c} \\ \frac{\partial \dot{c}}{\partial x} & \frac{\partial \dot{c}}{\partial c} \end{bmatrix} = \begin{bmatrix} f'(\overline{x}) - \gamma & -1 \\ -\frac{u'(\overline{c})}{u''(\overline{c})} f''(\overline{x}) & -\frac{\left[u''(\overline{c})\right]^2 - u'''(\overline{c})u'(\overline{c})}{\left[u''(\overline{c})\right]^2} \left[f'(\overline{x}) - (\gamma + \delta)\right] \end{bmatrix}$$
$$= \begin{bmatrix} \delta & -1 \\ -\frac{u'(\overline{c})}{u''(\overline{c})} f''(\overline{x}) & 0 \end{bmatrix}$$

By (1.173), the element $J_{11} = \delta$ and $J_{22} = 0$. Clearly, Tr $(J) = \delta > 0$ and Det $(J) = -u'(\overline{c})/u''(\overline{c}) f''(\overline{x}) < 0$ so that E is indeed a saddle point. The eigenvalues are solutions of the characteristic equation in the unknown λ:

$$\lambda^2 - \delta\lambda - \frac{u'(\overline{c})}{u''(\overline{c})} f''(\overline{x}) = 0 ,$$

i.e.,

$$\lambda_{1,2} = \frac{\delta \pm \sqrt{\delta^2 + 4\frac{u'(\overline{c})}{u''(\overline{c})} f''(\overline{x})}}{2} .$$

As for the previous example, any trajectory belonging to the stable manifold of E ensures the convergence to the *optimal long-run stationary equilibrium* E in (1.174).

In the economic literature, an equilibrium such as E is often called a *turnpike* and is obtained through the modified golden rule (1.173). Why modified? The reason is the following. In general the turnpike is different from the equilibrium that maximizes the integrand function (see (1.169)). In fact, from the state equation $\dot{x} = f(x) - c - \gamma x$ we know that an equilibrium must satisfy condition $\dot{x} = 0$, i.e., $c = f(x) - \gamma x$. If we substitute this c in the integrand in (1.169) we get $u\,(f(x) - \gamma x)$, which is maximized at the $\widehat{x}$ such that $u'\,(f(\widehat{x}) - \gamma\widehat{x})\,\left(f'(\widehat{x}) - \gamma\right) = 0$, i.e., for the $\widehat{x}$ such that $f'(\widehat{x}) = \gamma$. State $\widehat{x}$ is called golden rule. Notice that the golden rule coincides with the modified golden rule in (1.173) only for $\delta = 0$.

References

1. Abraham, R.H., Gardini, L., Mira, C.: Chaos in discrete dynamical systems (A visual introduction in two dimension). Springer (1997)
2. Acemoglu, D.: Introduction to Modern Economic Growth. Princeton University Press (2008)
3. Agliari, A., Bischi, G.I., Gardini, L.: Some methods for the global analysis of dynamic games represented by noninvertible maps. In: Puu, T., Sushko, I. (Eds.), Oligopoly Dynamics: Models and Tools, pp. 31–83. Springer (2002)
4. Andronov, A.A.: Mathematical problems in the theory of self-oscillations. The first All-Union Conference on Auto-oscillations, Moscow-Leningrad GTTI, pp. 32–71 (1933) (also in Collected works, Izd. Akad. Nauk. SSSR, Moscow, (1956))
5. Bellman, R.E.: Dynamic Programming. Princeton University Press, Princeton (1957). Republished: Dover (2003)

6. Bischi, G.I., Chiarella, C., Kopel, M., Szidarovszky, F.: Nonlinear Oligopolies: Stability and Bifurcations. Springer, Berlin (2010)
7. Bischi, G.I., Kopel, M.: Equilibrium selection in a nonlinear duopoly game with adaptive expectations. J. Econ. Behav. Organ. **46**(1), 73–100 (2001)
8. Bischi, G.I., Kopel, M., Naimzada, A.: On a rent-seeking game described by a non-invertible iterated map with denominator. Nonlinear Anal. **47**(8), 5309–5324 (2001)
9. Bischi, G.I., Mammana, C., Gardini, L.: Multistability and cyclic attractors in duopoly games. Chaos Solitons Fractals **11**(4), 543–564 (2000)
10. Bischi, G.I., Naimzada, A.: Global analysis of a duopoly game with bounded rationality. In: Filar, J.A., Gaitsgory, V., Mizukami, K., (Eds.) Advances in Dynamic Games and Applications, pp. 361–385. Birkhäuser (2000)
11. Devaney, R.L.: An Introduction to Chaotic Dynamical Systems. The Benjamin/Cummings Publishing Co., Menlo Park, California (1987)
12. Gandolfo, G.: Economic Dynamics. Springer (2007)
13. Gleick, J.: Chaos: Making a New Science. Viking Penguin (1987)
14. Goodwin, R.: A growth cycle. In: Feinstein, C. (ed.) Socialism, Capitalism, and Economic Growth. Cambridge, UK: Cambridge University Press (1967)
15. Grobman, D.M.: The topological classification of the vicinity of a singular point in n-dimensional space. Math. USSR-Sb. **56**, 77–94 (1962)
16. Guckenheimer, J., Holmes, P.: Nonlinear Oscillations, Dynamical Systems and Bifurcations of Vector Fields. Springer (1983)
17. Gumowski, I., Mira, C.: Dynamique Chaotique. Cepadues Editions, Toulose (1980)
18. Hartman, P.: Ordinary Differential Equations. John Wiley, New York (1964)
19. Hommes, C.: Behavioral Rationality and Heterogeneous Expectations in Complex Economic Systems. Cambridge University Press (2013)
20. Hopf, E.: Abzweigung einer periodischen Lösung von einer stationaren Lösungeines Differentialsystems. Berichten der Mathematisch-Physischen Klasse der Sächsischen Akademie der Wissenschaften zu Leipzig XCIV, 1–22 (1942). English translation in Marsden, J.E., McCracken, M.: The Hopf Bifurcation and Its Applications. Applied Mathematical Sciences. Vol. 19. Springer-Verlag (1976)
21. Kaldor, N.: A model of the trade cycle. Econ. J. **50**(197), 78–92 (1940)
22. Kopel, M.: Simple and complex adjustment dynamics in Cournot duopoly models. Chaos Solitons Fractals **7**(12), 2031–2048
23. Laplace, P.S.: Théorie analytique des probabilitiés. V. Courcier, Paris (1820)
24. Li, T.Y., Yorke, J.A.: Period three implies chaos. Am. Math. Mon. **82**(10), 985–992 (1975)
25. Lorenz, E.: Deterministic non-periodic flow. J. Atmos. Sci. (1963)
26. Lorenz, E.N.: Predictability: does the flap of a butterfly's wings in Brazil set off a tornado in Texas? 139th Annual Meeting of the American Association for the Advancement of Science (29 Dec 1972), in Essence of Chaos (1995), Appendix 1, 181
27. Malthus, T.: An Essay on the Principle of Population. London, Printed for J. Johnson, in St. Paul's Church-Yard (1798)
28. Poincaré, H.: Science and Method. Courier Corporation (1903)
29. Puu, T.: Attractors, Bifurcations, and Chaos. Springer (2000)
30. Rosenzweig, M., MacArthur, R.: Graphical representation and stability conditions of predator-prey interaction. Am. Nat. **97**, 209–223 (1963)
31. Sethi, S.P., Thompson, G.L.: Optimal Control Theory: Applications to Management Science and Economics. Springer (2000)
32. Sydsaeter, K., Hammond, P., Seierstad, A., Strom, A.: Further Mathematics for Economic Analysis, 2 edn. Prentice Hall (2008)
33. Thom, R.: Stabilité Structurelle et Morphogénèse. Essai d'une Théorie Générale des Modèles. Benjamin, New York (1992)
34. Volterra, V.: Variazioni e fluttuazionidel numero d'individui in specie di animali conviventi. Mem. Accadem. Lincei. **6**(2), 31–113 (1926)

Chapter 2
Some Aspects on Global Analysis of Discrete Time Dynamical Systems

Anastasiia Panchuk

Abstract Dynamical systems theory distinguishes two types of bifurcations: those which can be studied in a small neighborhood of an invariant set (local) and those which cannot (global). In contrast to local bifurcations, global ones cannot be investigated by a Taylor expansion, neither they are detected by purely performing stability analysis of periodic points. Global bifurcations often occur when larger invariant sets of the system collide with each other or with other fixed points/cycles. This chapter focuses on several aspects of global bifurcation analysis of discrete time dynamical systems, covering homoclinic bifurcations as well as inner and boundary crises of attracting sets.

2.1 Introduction

Dynamical systems theory is mainly interested in asymptotic behavior of orbits depending on the initial state and how this behavior may change when varying system parameters. The important phenomenon is a bifurcation when the changes occurring in the state space cannot be obtained via a smooth transformations (the orbits before and after the bifurcation are not topologically conjugated). Two types of structural changes are distinguished: local and global ones. Local bifurcations are those which can be examined locally via an approximation of the map in a small neighborhood of some fixed point or cycle. Global bifurcations often occur when larger invariant sets of the system collide with each other or with other fixed points/cycles. Such a global bifurcation cannot be investigated by a Taylor expansion and cannot be detected by purely performing stability analysis of a periodic point. To understand what happens with orbits of the system in this case, one has to take into account global properties of the map (see [17, 27, 34] to cite a few).

A. Panchuk (✉)
Institute of Mathematics, National Academy of Sciences of Ukraine,
Tereshchenkivska str. 3, Kiev 01601, Ukraine
e-mail: anastasiia.panchuk@gmail.com

G.I. Bischi et al. (eds.), *Qualitative Theory of Dynamical Systems, Tools and Applications for Economic Modelling*, Springer Proceedings in Complexity,
DOI 10.1007/978-3-319-33276-5_2

Among common examples of global phenomena, one may list contact bifurcations, homoclinic bifurcations, and crises. Contact bifurcations are characterized by structural changes of basin and its boundary (for example, when a connected basin transforms to a nonconnected one or basin boundary becomes fractal). These bifurcations are only possible in noninvertible maps and occur due to tangencies between basin boundaries and critical curves (for more detail see, e.g., [5, 12, 25] and references therein).

Homoclinic bifurcations entail changing shape of invariant sets and are associated with appearance of homoclinic points (and, consequently, orbits) when the stable and unstable sets of a periodic point have a contact (see, for instance, [6, 8, 10]). This often allows to show strictly existence of an invariant set on which dynamics is chaotic. It can also happen, that the stable set of one periodic point intersects with the unstable set of another periodic point in which case the intersection point (and its orbit) is called heteroclinic. Homoclinic/heteroclinic orbits may also appear in a sequence of bifurcations leading to creation of closed invariant curves (see, e.g., [1, 2]).

Bifurcations called crises are related to sudden transformations of chaotic attractors and encountered in both invertible and noninvertible maps alike homoclinic bifurcations. Such swift changes occur due to a contact between a chaotic attractor and an unstable periodic orbit (or, equivalently, its stable set). Starting from [16], three types of crisis are usually distinguished: a boundary crisis at which the attractor is destroyed, an interior crisis accompanied with abrupt increase in size of the attractor, and merging crisis where several chaotic attractors collide simultaneously with an orbit on the separating basin boundary.

In the current chapter we point out several aspects related to global analysis of discrete time dynamical systems, covering homoclinic bifurcations as well as interior and boundary crises.

2.2 Preliminaries

In this section we introduce general concepts and notations used throughout the chapter.

Let us consider a smooth (or piecewise smooth) function $T : X \to X$, $X \subseteq \mathbb{R}^n$, $T = (T_1, \dots, T_n)$, which can be invertible or noninvertible. Here $n \in \mathbb{Z}^+$ with $\mathbb{Z}^+$ denoting the set of all positive integer numbers. Recall that:

- For an arbitrary point $\mathbf{x} \in X$ the value $T(\mathbf{x})$ is called a *rank-1 image* of $\mathbf{x}$ or simply an *image* of $\mathbf{x}$.
- From definition of T it follows that $T(\mathbf{x}) \in X$. Hence, we may define $T(T(\mathbf{x}))$, $T(T(T(\mathbf{x})))$, and so on. For shortness, the composition of t consecutive applications of the function T to the vector $\mathbf{x}$ is abbreviated as $T^t(\mathbf{x})$, i.e.,

$$\underbrace{T(T(\dots T}_{t}(\mathbf{x})\dots)) \equiv \underbrace{T \circ T \dots \circ T}_{t}(\mathbf{x}) \stackrel{def}{=} T^t(\mathbf{x}).$$

- For any $\mathbf{x} \in X$ the value $T^t(\mathbf{x})$, $t \in \mathbb{Z}^+$, is called a *rank-t image* of $\mathbf{x}$.
- For an arbitrary $\mathbf{x} \in X$ the value $\mathbf{y} \in X$ such that $T^t(\mathbf{y}) = \mathbf{x}$, $t \in \mathbb{Z}^+$, is called a *rank-t preimage* of $\mathbf{x}$. The set of all rank-t preimages is denoted as $T^{-t}(\mathbf{x})$. For shortness, we refer to a rank-1 preimage of $\mathbf{x}$ simply as a *preimage* of $\mathbf{x}$.

It is important to mention, that for an arbitrary $\mathbf{x} \in X$ its rank-t image, $t \in \mathbb{Z}^+$, exists and is unique. This is not necessarily true for preimages of $\mathbf{x}$ that may be multiple ($T^{-t}(\mathbf{x})$ has more than one element) or missing ($T^{-t}(\mathbf{x}) = \varnothing$), in case where T is noninvertible.

For avoiding confusion we emphasize that whenever the symbol $T^{-1}(\mathbf{x})$ is used, it is meant to denote the set of *all preimages* of a certain point $\mathbf{x}$. However, writing simply T^{-1} (without $\mathbf{x}$) we have in mind an *inverse function* of T. In case where T is a bijection (a one-to-one and onto), each $\mathbf{x} \in X$ has exactly one (rank-1) preimage (the set $T^{-1}(\mathbf{x})$ consists of a single point). Then, the function T is called *invertible* and the inverse function T^{-1} can be uniquely defined. On the other hand, if T is not bijective, then it is *noninvertible* and the set $T^{-1}(\mathbf{x})$ may consist of more than one value or be empty depending on $\mathbf{x}$ (hence, the inverse function T^{-1} is either not defined uniquely or not defined for some $\mathbf{x} \in X$).

Example 2.1 Consider the function $T_\mu = \mu\sqrt{x}$, $0 < \mu \leq 1$, with $X = [0, 1]$. It is clear that $T_\mu(X) = [0, \mu]$. Consequently, for $\mu < 1$ the points $\bar{x} > \mu$ do not have preimages, that is, $T_\mu^{-1}(\bar{x}) = \varnothing$ and the inverse function T_μ^{-1} is only defined for $x \in [0, \mu]$. Nevertheless, the dynamical system related to T_μ do not produce complex behavior since T_μ is monotonically increasing.

Example 2.2 Consider the function $T = 4x(1 - x)$ with $X = [0, 1]$. The image of X is $T(X) = [0, 1] = X$, but every $x \in X$, except for $\bar{x} = 1$, has two preimages $y_{\mathscr{L}} < 0.5$ and $y_{\mathscr{R}} > 0.5$. The inverse function T^{-1} can not be uniquely defined, though one may define two distinct inverse functions of T. That is, $T_{\mathscr{L}}^{-1} : [0, 1] \to [0, 0.5]$ and $T_{\mathscr{R}}^{-1} : [0, 1] \to [0.5, 1]$ with obvious equality $T_{\mathscr{L}}^{-1}(1) = T_{\mathscr{R}}^{-1}(1) = 0.5$.

Let us consider now a *discrete* (or discrete time) dynamical system (DDS for short) represented by the iterative relation

$$\mathbf{x}_{t+1} = T(\mathbf{x}_t) \quad t \in \mathbb{Z}^*, \tag{2.1}$$

where $\mathbb{Z}^* = \mathbb{Z}^+ \cup \{0\}$ is a set of all non-negative integers, or, in equivalent notation,

$$\mathbf{x}' = T(\mathbf{x}), \tag{2.1$'$}$$

where $\mathbf{x}'$ denotes the next iterate under T (image of x). Having in mind the evolutionary process (2.1) the function T is often referred to as a *map* or a *mapping*. The set X then serves as the *state space*, while an n-dimensional real vector $\mathbf{x} = (x_1, \ldots, x_n) \in X$ is the *state variable*.

For every initial state $\mathbf{x}_0 \in X$ of the system (2.1) or (2.1′) at the time $t = 0$, the sequence of successive images of $\mathbf{x}_0$ constitutes a *discrete trajectory* or an *orbit*:

$$\mathscr{O}(\mathbf{x}_0) = \left\{T^t(\mathbf{x}_0)\right\}_{t=0}^{\infty} = \{\mathbf{x}_0, \mathbf{x}_1, \ldots, \mathbf{x}_t, \ldots\}. \tag{2.2}$$

Here T^0 denotes the identity function (map), namely, $T^0(\mathbf{x}) = \mathbf{x}$ for any $\mathbf{x} \in X$.

The main task in studying a dynamical system (2.1) is to understand the asymptotic behavior of its orbits depending on the initial condition $\mathbf{x}_0$ and how this develops with changing parameters. For instance, an orbit can "stuck" at some invariant set or diverge to infinity, and this behavior may be different depending on $\mathbf{x}_0$. It may also happen that an orbit diverges from some invariant set, but then comes back to it again due to existence of a homoclinic orbit.

Recall that a set $A \subset X$ is called *invariant* under T or T*-invariant* if it is mapped onto itself $T(A) = A$. The two simplest kinds of invariant sets (and, hence, the simplest asymptotic behavior of orbits of (2.1)) are

- a *fixed point* $\mathbf{x}^*$ such that $T(\mathbf{x}^*) = \mathbf{x}^*$, and
- a *k-cycle*, $\mathscr{C}_k = \{\mathbf{x}_1^*, \ldots, \mathbf{x}_k^*\}$ such that $\mathbf{x}_{i+1}^* = T(\mathbf{x}_i^*)$, $\mathbf{x}_1^* = T(\mathbf{x}_k^*)$.

However, the structure of an invariant set A may be much more complex than just a point or a finite set of points, and this fact led to introducing such terms as *strange attractor* [15, 29] and *chaotic attractor* [9, 21]. There exist many possible definitions of chaos in dynamical systems theory, some of them being stronger than the others. Here we follow the definition given by Devaney [7].

Definition 2.1 Consider a map $T : X \to X$ and let a set $A \subset X$ be invariant under T. The restriction $T|_A : A \to A$ is called *chaotic* on A if

(i) there exist infinitely many periodic orbits dense in A,
(ii) $T|_A$ is *topologically transitive*, that is, for any pair of open sets $U, V \subset A$ there exists $t \in \mathbb{Z}^+$ such that $T^t(U) \cap V \neq \varnothing$.

The set A is also often said to be *chaotic*.

To be precise, in his definition Devaney also includes the third property that $T|_A$ has sensitive dependence on initial conditions,[1] although this property can be derived from the other two (see, e.g., [4]). The chaos of the described type is also called topological chaos, that is having positive topological entropy.

It is worth to mention that condition (ii) is often replaced by

(ii′) there exists an aperiodic orbit dense in A.

Conditions (ii′) and (ii) are not connected in general, though under additional requirements on A they become equivalent. Namely, if A has no isolated points then (ii′)

[1]The map $T : X \to X$ is said to have *sensitive dependence on initial conditions* if there exists $\delta > 0$ such that for any $\mathbf{x} \in X$ and any neighborhood $U(\mathbf{x})$ there exist $\mathbf{y} \in U(\mathbf{x})$ and $t \geq 0$ such that $|T^t(\mathbf{x}) - T^t(\mathbf{y})| > \delta$.

implies (ii). The opposite is true if A contains a countable dense subset and is of second category[2] (see, e.g., [32]).

One might expect that chaotic behavior is more natural for DDS with dimension $n \geq 2$, however, already one-dimensional *noninvertible* maps can show non-regular dynamics of this kind. One of the most known examples is a family of quadratic maps such as logistic map [31] or conjugated to it Myrberg map [26].

In the next sections we describe a couple of phenomena due to which a chaotic attractor may appear or be reshaped. For that we first introduce such important concepts as stable and unstable sets of a fixed point or a cycle.

2.3 Stable and Unstable Sets

The stable and the unstable set of a fixed point or a cycle are important concepts for studying behavior of a dynamical system globally. They are related to boundaries of basins of attraction, saddle-connections and birth of closed invariant curves, homoclinic tangles, appearance and modification of chaotic attractors. Classically, these notions are defined for maps of dimension greater than one, which are characterized by possibility of both expansion and contraction in the same invariant set. However, an extension of the idea to a class of one-dimensional noninvertible maps was given already in 1969 by Sharkovsky [30] and then developed in [22, 25] among the others.

2.3.1 Stable Manifold Theorem

Let consider a DDS (2.1) and denote

$$DT(\mathbf{x}) = \begin{pmatrix} \dfrac{\partial T_1(\mathbf{x})}{\partial x_1} & \cdots & \dfrac{\partial T_1(\mathbf{x})}{\partial x_n} \\ \cdots & \cdots & \cdots \\ \dfrac{\partial T_n(\mathbf{x})}{\partial x_1} & \cdots & \dfrac{\partial T_n(\mathbf{x})}{\partial x_n} \end{pmatrix}$$

the Jacobian matrix of the map T at the point $\mathbf{x}$. And let $\lambda_i = \lambda_i(\mathbf{x})$, $1 \leq i \leq n$, be the eigenvalues of $DT(\mathbf{x})$.

Definition 2.2 A fixed point $\mathbf{x}^*$ of the map T is called *hyperbolic* if $DT(\mathbf{x}^*)$ has no eigenvalues on the unit circle, that is, $|\lambda_i(\mathbf{x}^*)| \neq 1$, $1 \leq i \leq n$.

For sake of shortness, we drop the symbol $(\mathbf{x}^*)$ in notation of eigenvalues whenever it is clear which point is meant. If all λ_i, $1 \leq i \leq n$, of $DT(\mathbf{x}^*)$ are inside the unit

[2] A subset A of a topological space X is said to be of *second category* in X if A cannot be written as the countable union of subsets which are nowhere dense in X.

circle ($|\lambda_i| < 1$), then $\mathbf{x}^*$ is *attracting* (more precisely, asymptotically stable), while in case when all $|\lambda_i| > 1$, the point $\mathbf{x}^*$ is called *expanding*. In the intermediate case, when some of eigenvalues of $\mathbf{x}^*$ are inside the unit circle and others are outside it, the point $\mathbf{x}^*$ is referred to as *saddle* or *unstable*.[3] Clearly, for a one-dimensional map the concepts of expanding and unstable fixed points are equivalent.

Let rearrange the eigenvalues of $DT(\mathbf{x}^*)$ so that $|\lambda_i| < 1$ for $1 \leq i \leq i_0$ and $|\lambda_i| > 1$ for $i_0 < i \leq n$ and consider the sets of related eigenvectors. Let $E^s(\mathbf{x}^*)$ denote a subspace of dimension i_0 spanned on the eigenvectors $\{\mathbf{v}_1, \ldots, \mathbf{v}_{i_0}\}$ corresponding to eigenvalues being inside the unit circle. Then $E^s(\mathbf{x}^*)$ is called a *stable eigenspace* of $\mathbf{x}^*$. Similarly, the eigenvectors $\{\mathbf{v}_{i_0+1}, \ldots, \mathbf{v}_n\}$ corresponding to eigenvalues outside the unit circle define an *unstable eigenspace* $E^u(\mathbf{x}^*)$ (whose dimension is clearly $n - i_0$).

Example 2.3 Let consider a linear mapping $T_{\text{lin}} : \mathbb{R}^2 \to \mathbb{R}^2$,

$$T_{\text{lin}} : \begin{cases} x_1' = 2x_1 + x_2, \\ x_2' = x_1 + x_2. \end{cases}$$

It obviously has a fixed point $\mathbf{x}^* = (0, 0)$ with Jacobian

$$DT_{\text{lin}}(0, 0) = \begin{pmatrix} 2 & 1 \\ 1 & 1 \end{pmatrix}. \tag{2.3}$$

The eigenvalues of $DT_{\text{lin}}(0, 0)$ are $\lambda_1 = (3 - \sqrt{5})/2$, $\lambda_2 = (3 + \sqrt{5})/2$ with the corresponding eigenvectors $\mathbf{v}_i = (1/(\lambda_i - 2), 1)$, $i = 1, 2$. Since $\lambda_1 < 1$, the stable eigenspace is

$$E^s(0, 0) = \{\alpha \mathbf{v}_1 \mid \alpha \in \mathbb{R}\}, \tag{2.4}$$

that is, the line defined by the vector $\mathbf{v}_1$. Similarly, the vector $\mathbf{v}_2$ gives the unstable eigenspace

$$E^u(0, 0) = \{\alpha \mathbf{v}_2 \mid \alpha \in \mathbb{R}\}, \tag{2.5}$$

being the line orthogonal to $E^s(0, 0)$ (see Fig. 2.1).

Now, we directly come to introducing the notions of stable and unstable sets. First, the classical case is considered where the mapping T is a diffeomorphism, that is, a smooth bijective function. Then, modified versions of the definitions are given so that to cover the case of noninvertible and/or piecewise smooth maps.

[3] Some authors use also the term *repelling* in this case, though it might be confusing since there is more strict definition of a repelling set.

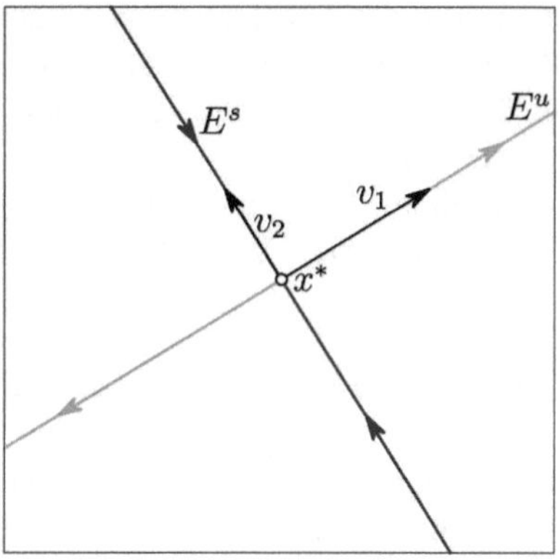

Fig. 2.1 Stable E^s and unstable E^u eigenspaces of the saddle fixed point $\mathbf{x}^*$

2.3.1.1 Diffeomorphisms: Stable and Unstable Manifolds

Definition 2.3 Let the mapping T be a diffeomorphism, and let $\mathbf{x}^*$ be a hyperbolic fixed point of the system (2.1). The sets

$$W^s(\mathbf{x}^*) = \left\{\mathbf{y} \in X \ : \ \lim_{t\to\infty} T^t(\mathbf{y}) = \mathbf{x}^*\right\}, \tag{2.6}$$

$$W^u(\mathbf{x}^*) = \left\{\mathbf{y} \in X \ : \ \lim_{t\to\infty} T^{-t}(\mathbf{y}) = \mathbf{x}^*\right\} \tag{2.7}$$

are called the *stable* and the *unstable set* of $\mathbf{x}^*$, respectively.

Note that if $\mathbf{x}^*$ is an expanding fixed point, then its stable set is empty, while its unstable set is some n-dimensional area (the shape of this area depends on presence of other invariant sets of the system). On the contrary, if $\mathbf{x}^*$ is an asymptotically stable fixed point, then its unstable set is empty, while its stable set is just its basin of attraction.

The Stable Manifold Theorem (see, e.g., [3, 34]) guarantees that $W^s(\mathbf{x}^*)$ and $W^u(\mathbf{x}^*)$ exist and are manifolds of the same smoothness as T.

Theorem 2.1 (Stable Manifold Theorem) *Let T be a diffeomorphism and $\mathbf{x}^*$ be a hyperbolic fixed point of the DDS (2.1). Then $W^s(\mathbf{x}^*)$ is as smooth as T manifold and the stable eigenspace $E^s(\mathbf{x}^*)$ is tangent to $W^s(\mathbf{x}^*)$ at the point $\mathbf{x}^*$.*

Applying Theorem 2.1 to the inverse T^{-1}, the same can be stated about the unstable set $W^u(\mathbf{x}^*)$, that is, $W^u(\mathbf{x}^*)$ is as smooth as T manifold and $E^u(\mathbf{x}^*)$ is its tangent space at $\mathbf{x}^*$. For sake of shortness, we drop the symbol $(\mathbf{x}^*)$ in notation of eigenspaces and stable/unstable manifolds whenever it is clear which point is meant.

Example 2.4 In case of linear mapping T (such as the one in Example 2.3), the stable/unstable manifold W^s/W^u of the fixed point $\mathbf{x}^* = (0, 0)$ coincides with the stable/unstable eigenspace E^s/E^u. We demonstrate this for the stable manifold, and the unstable one is treated likewise. First, we show that E^s is invariant under

T. Let $\mathbf{x} \in E^s$ which means that $\mathbf{x} = \alpha \mathbf{v}_1$ with a certain $\alpha \in \mathbb{R}$. Then $T(\mathbf{x}) = T(\alpha \mathbf{v}_1) = \alpha \lambda_1 \mathbf{v}_1$ and, hence, $T(\mathbf{x}) \in E^s$. And vice verse: consider $T(\mathbf{x}) \in E^s$, that is, $T(\mathbf{x}) = \alpha \mathbf{v}_1$, which is equivalent to $T(\lambda_1 \mathbf{x}/\alpha) = \lambda_1 \mathbf{v}_1$. The latter clearly implies $\mathbf{x} = \alpha \mathbf{v}_1/\lambda_1 \in E^s$.

It remains to verify that for any $\mathbf{y} \in E^s$ there holds $\lim_{t\to\infty} T^t(\mathbf{y}) = \mathbf{x}^*$. Indeed, $T^t(\mathbf{y}) = \alpha \lambda_1^t \mathbf{v}_1$, $\alpha \in \mathbb{R}$. Obviously, $\lim_{t\to\infty} \|T^t(\mathbf{y})\| = 0$ since $|\lambda_1| < 1$.

2.3.1.2 Noninvertible Maps: Stable and Unstable Sets

Now, we consider the more general case where T is not a diffeomorphism, namely, T may be noninvertible and/or piecewise smooth. The definition (2.6) for the stable set $W^s(\mathbf{x}^*)$ remains unchanged, but the expression (2.7) for the unstable set $W^u(\mathbf{x}^*)$ has to be modified with taking into account that the points can have more than one preimage.

Definition 2.4 Let T be a continuous piecewise smooth map and consider a hyperbolic fixed point $\mathbf{x}^*$ of the DDS (2.1). The locus of points having a sequence of preimages that tends towards $\mathbf{x}^*$, that is,

$$W^u(\mathbf{x}^*) = \left\{ \mathbf{y} \in X \ : \ \exists \, \{\mathbf{z}_t\}_{t=0}^{\infty}, \ \mathbf{z}_0 = \mathbf{y}, \ T(\mathbf{z}_{t+1}) = \mathbf{z}_t \ \text{ such that } \ \lim_{t\to\infty} \mathbf{z}_t = \mathbf{x}^* \right\}, \tag{2.8}$$

is called the *unstable set* of $\mathbf{x}^*$.

Note that in this definition *not all* preimages of $\mathbf{y}$ tend towards $\mathbf{x}^*$, but *there exists* an infinite sequence of preimages of $\mathbf{y}$ having $\mathbf{x}^*$ as a limit point. Moreover, in case of a noninvertible T, the stable set of an expanding fixed point may be nonempty (in contrast to invertible maps). More precisely, it consists of all preimages of this fixed point. Similarly, the unstable set of an asymptotically stable fixed point $\mathbf{x}^*$ may be nonempty.

Clearly, in case of noninvertible or non-smooth T, the Stable Manifold Theorem (Theorem 2.1) cannot be applied directly. We additionally require that (i) the point $\mathbf{x}^*$ is not *degenerate*, that is, $\det DT(\mathbf{x}^*) \neq 0$ (the Jacobian have no zero eigenvalues), and (ii) T is smooth at $\mathbf{x}^*$. If these two conditions are satisfied there exists a neighborhood $U(\mathbf{x}^*)$ such that $T|_{U(\mathbf{x}^*)}$ is a diffeomorphism. Hence, Theorem 2.1 guarantees existence of *local stable* $W^s_{\text{loc}}(\mathbf{x}^*) \subset U(\mathbf{x}^*)$ and *local unstable* $W^u_{\text{loc}}(\mathbf{x}^*) \subset U(\mathbf{x}^*)$ manifolds, which are tangent at $\mathbf{x}^*$ to the stable and unstable eigenspaces $E^s(\mathbf{x}^*)$ and $E^u(\mathbf{x}^*)$, respectively. Then the *global stable* and *global unstable sets* can be alternatively defined as

$$W^s(\mathbf{x}^*) = \cup_{t=0}^{\infty} T^{-t}(W^s_{\text{loc}}(\mathbf{x}^*)), \tag{2.9}$$

$$W^u(\mathbf{x}^*) = \cup_{t=0}^{\infty} T^{t}(W^u_{\text{loc}}(\mathbf{x}^*)). \tag{2.10}$$

Example 2.5 Let consider a nonlinear function $T_{\text{nlin}} : \mathbb{R}^2 \to \mathbb{R}^2$,

$$T_{\text{nlin}} : \begin{cases} x_1' = 2x_1 + x_2, \\ x_2' = x_1 - x_1^2 + x_2, \end{cases}$$

which is clearly noninvertible. It has a fixed point $\mathbf{x}^* = (0, 0)$ with the same Jacobian (2.3) as in Example 2.3. Consequently, the stable E^s and unstable E^u eigenspaces of $\mathbf{x}^*$ are given by (2.4) and (2.5), respectively. The stable W^s and unstable W^u sets of $\mathbf{x}^*$ cannot be derived in analytic form, but one may approximate them numerically. In Fig. 2.2 the sets W^s and W^u together with E^s and E^u are shown.

In general, stable and unstable sets of a fixed point $\mathbf{x}^*$ have the following properties (see, e.g., [25]):

1. W^s is backward invariant, but not necessarily (forward) invariant (mapped only *into* itself), that is,

$$T^{-1}(W^s) = W^s,$$
$$T(W^s) \subseteq W^s.$$

2. W^u is invariant, but not necessarily backward invariant:

$$T(W^u) = W^u,$$
$$T^{-1}(W^u) \supseteq W^u.$$

3. Both W^s and W^u are not necessarily manifolds and may have self intersections.
4. When T is continuous, the set W^u is connected, while W^s may consist of disjoint connected components.

The notions of stable and unstable sets can be generalized for a k-cycle $\mathcal{C}_k = \{\mathbf{x}_1^*, \ldots, \mathbf{x}_k^*\}$ by considering T^k, for which every point $\mathbf{x}_i^*$ is a fixed point. Then under

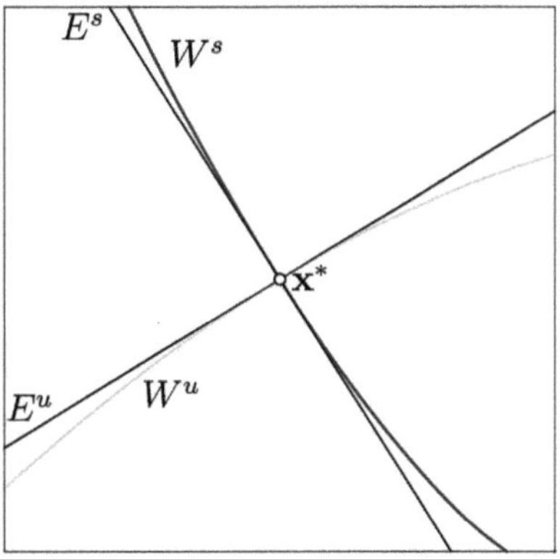

Fig. 2.2 Stable W^s and unstable W^u sets of the saddle fixed point $\mathbf{x}^*$ together with related eigenspaces

the stable (unstable) set of $\mathscr{C}_k$ one mean the union of all stable (unstable) sets $W^s(\mathbf{x}_i^*)$ ($W^u(\mathbf{x}_i^*)$).

As one can see, the stable and unstable sets are not defined only in a neighborhood of a fixed point (or a cycle), but in the larger area of the state space and may even extend to infinity. Due to this reason, they often play an important role when studying global properties of the map T. In particular,

1. stable sets of saddle points or cycles may define boundaries of basins of attraction;
2. stable sets of saddle cycles may establish saddle-connections (homoclinic/heteroclinic loops) inducing creation of closed invariant curves.
3. stable and unstable sets may attain a very complex configuration generating a homoclinic tangle, which causes then appearance of an invariant chaotic set;
4. contact of the stable set of a saddle periodic point with another invariant set may lead to a crisis bifurcation.

The situation (1) is briefly explained immediately below by using a simple example. The phenomenon (3) is described in detail in Sect. 2.4. Interior and boundary crises (4) are considered in Sect. 2.5. The item (2) is beyond the scopes of the current chapter, and the reader is referred to [1, 2] and references therein.

2.3.2 Stable Sets of Saddle Points and Basins

Let us consider a two-parametric family $H_{a,b} : \mathbb{R}^2 \to \mathbb{R}^2$ such that

$$H_{a,b} : \begin{pmatrix} x \\ y \end{pmatrix} \to \begin{pmatrix} a - by - x^2 \\ x \end{pmatrix}. \tag{2.11}$$

The map (2.11) is called Hénon map[4] (see, e.g., [7, 18]). Note that $H_{a,b}$ has only one nonlinear term, so that, it is indeed one of the simplest nonlinear maps in higher dimensions.

It is easy to calculate the Jacobi matrix of $H_{a,b}$:

$$DH_{a,b} = \begin{pmatrix} -2x & -b \\ 1 & 0 \end{pmatrix},$$

whose determinant is $\det DH_{a,b} = b$. Clearly, if $b \neq 0$, then the mapping (2.11) is invertible, and its inverse is given as

$$H_{a,b}^{-1} : \begin{pmatrix} x \\ y \end{pmatrix} \to \begin{pmatrix} x \\ -\dfrac{a}{b} - \dfrac{y}{b} + \dfrac{x^2}{b} \end{pmatrix}. \tag{2.12}$$

[4]In the original paper of M. Hénon this map is written in slightly different form, namely, $\widetilde{H}_{a,b}$: $(x, y) \to (1 + y - ax^2, bx)$, but topological conjugacy between $\widetilde{H}_{a,b}$ and $H_{a,b}$ can be easily shown.

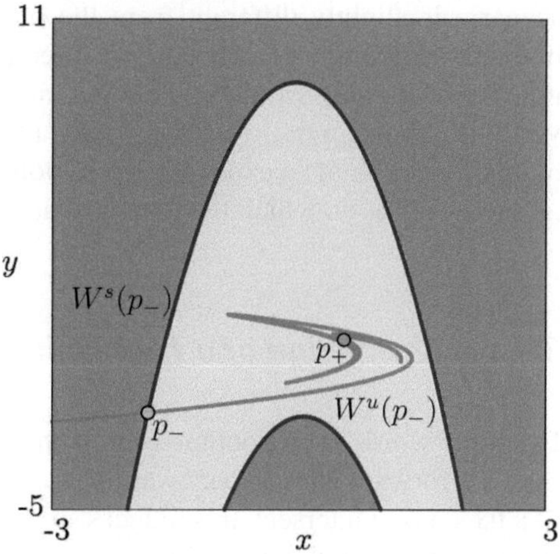

Fig. 2.3 Phase space of the Hénon map $H_{a,b}$ with $a = 1$, $b = 0.3$. The basin of the attracting fixed point $\mathbf{p}_+$ is confined by the stable set of the saddle $\mathbf{p}_+$ (namely, it is the *area* between the *two parabola-shaped curves*). Orbits starting in the *dark-gray region* diverge to infinity

It can be shown that for any $a \in \mathbb{R}, b \in \mathbb{R}\backslash\{0\}$ the map $H_{a,b}$ is topologically conjugate to $H_{A,B}^{-1}$ with $A = -a/b^2$, $B = -1/b$. Hence, it is enough to consider $|b| \leq 1$.

Note that in case $b = 0$, the function $H_{a,0}$ is noninvertible and maps the entire plane onto a parabola $P = \{(x, y) : x = a - y^2\}$, which implies topological conjugacy between the restriction $H_{a,0}|_P$ and the Myrberg map [26].

We set $a = 1$, $b = 0.3$, for which $H_{1,0.3}$ has two fixed points: a saddle $\mathbf{p}_- = (x_-, x_-)$ and a sink $\mathbf{p}_+ = (x_+, x_+)$. In Fig. 2.3 a part of the phase space is shown. The basin of attraction $\mathscr{B}(\mathbf{p}_+)$ of the sink $\mathbf{p}_+$ is confined by the stable set $W^s(\mathbf{p}_-)$ (a part of this stable set appears in Fig. 2.3 as two parabola-shaped curves). The orbits starting 'outside' $\mathscr{B}(\mathbf{p}_+)$ (in the dark-gray region) diverge to infinity. Since $\mathbf{p}_+$ is attracting, its stable set is simply its basin $\mathscr{B}(\mathbf{p}_+)$, and one branch of $W^u(\mathbf{p}_-)$ clearly tends to $\mathbf{p}_+$ with $t \to \infty$ (the other branch diverges to infinity). The unstable set is $W^u(\mathbf{p}_+) = \varnothing$, because $H_{a,b}$ is invertible.

2.4 Homoclinic Bifurcations

In this section we explain such notions as a homoclinic orbit and a homoclinic bifurcation. A homoclinic orbit is often a basic tool for rigorously showing existence of chaotic dynamics [10, 33]. In spaces with dimension greater than one, a homoclinic bifurcation is related to a homoclinic tangle. The latter represents a structure where stable and unstable sets of a saddle fixed point twist and interlace in a very complex manner. Whenever this intersection is transverse, there exists an invariant set on which the restriction of the map is chaotic.

As for a one-dimensional DDS, there is no concept of a saddle point. However, if the map is noninvertible, the stable set of a repelling fixed point may be nonempty (as was mentioned above). Then, homoclinic bifurcations are also possible but the scenario is slightly different from the one involving a saddle point. Namely, the homoclinic orbit has to be mapped directly to a repelling fixed point after a finite number of iterates. Moreover, homoclinic orbits of this kind also appear in noninvertible higher-dimensional maps. As was proved by Marotto [22], a non-degenerate homoclinic orbit of an expanding fixed point (a snap-back repeller) is associated with an invariant set on which the map is chaotic.

2.4.1 The Notion of a Homoclinic Orbit

In simple words, a homoclinic orbit is the one that tends to the same invariant set in both the forward and the backward processes. Whenever the stable and unstable sets of a fixed point intersect, this induces such a homoclinic orbit. We give all definitions for fixed points and remark that to get similar definitions for k-cycles it is enough to consider the k-th iterate of the map T^k.

Definition 2.5 Let $\mathbf{x}^*$ be a hyperbolic fixed point of a map T. Any point $\mathbf{y} \in W^s(\mathbf{x}^*) \cap W^u(\mathbf{x}^*)\backslash\{\mathbf{x}^*\}$ is called *homoclinic*.

If a homoclinic point exists, then infinitely many homoclinic points must also exist, accumulating in a neighborhood of $\mathbf{x}^*$. Intuitively, this can be understood by observing that the images of $\mathbf{y}$ and its suitable preimages are also homoclinic points, which converge to $\mathbf{x}^*$.

Definition 2.6 A sequence of images and suitable preimages of a homoclinic point $\mathbf{y}$, which converge to $\mathbf{x}^*$, is called a *homoclinic orbit* of $\mathbf{x}^*$.

In other words, a homoclinic orbit is an infinite set of points

$$\mathscr{O}_h(\mathbf{x}^*) = \{\ldots, \mathbf{y}_{-t}, \ldots, \mathbf{y}_{-1}, \mathbf{y}_0, \mathbf{y}_1, \ldots, \mathbf{y}_t, \ldots\} \tag{2.13}$$

such that $T^t(\mathbf{y}_{-t}) = \mathbf{y}_0$ and $\mathbf{y}_t = T^t(\mathbf{y}_0)$ with

$$\lim_{t\to\infty} \mathbf{y}_{-t} = \mathbf{x}^*, \quad \lim_{t\to\infty} \mathbf{y}_t = \mathbf{x}^*.$$

It is sometimes important to distinguish between critical and noncritical homoclinic orbits. A homoclinic orbit is called *critical* if it contains a critical point, that is, a point at which the map T is not locally invertible. Otherwise, a homoclinic orbit is *noncritical* (see, e.g., [13]). For instance, in case of a unimodal map a homoclinic orbit including the local extremum point is critical.

2.4.2 Homoclinic Orbits for One-Dimensional Maps

By analyzing a simple example we explain how homoclinic orbits appear in scalar maps and what is a homoclinic bifurcation in this case. Consider a famous logistic map $T_\mu(x) = \mu x(1-x)$, $T_\mu : [0,1] \to [0,1]$. In Fig. 2.4a the bifurcation diagram of T_μ is plotted, where the dashed curve indicates the fixed point $x^* = 1 - 1/\mu$ when it becomes unstable, while the black vertical line corresponds to the parameter value $\mu = \mu^* \approx 3.6786$, at which a homoclinic bifurcation occurs. The graphs of T_μ with $\mu < \mu^*$, $\mu = \mu^*$, and $\mu > \mu^*$ appear in Fig. 2.4b, c and d, respectively. A few steps of the forward orbit of the extremum point $x_e = 1/2$ are also plotted.

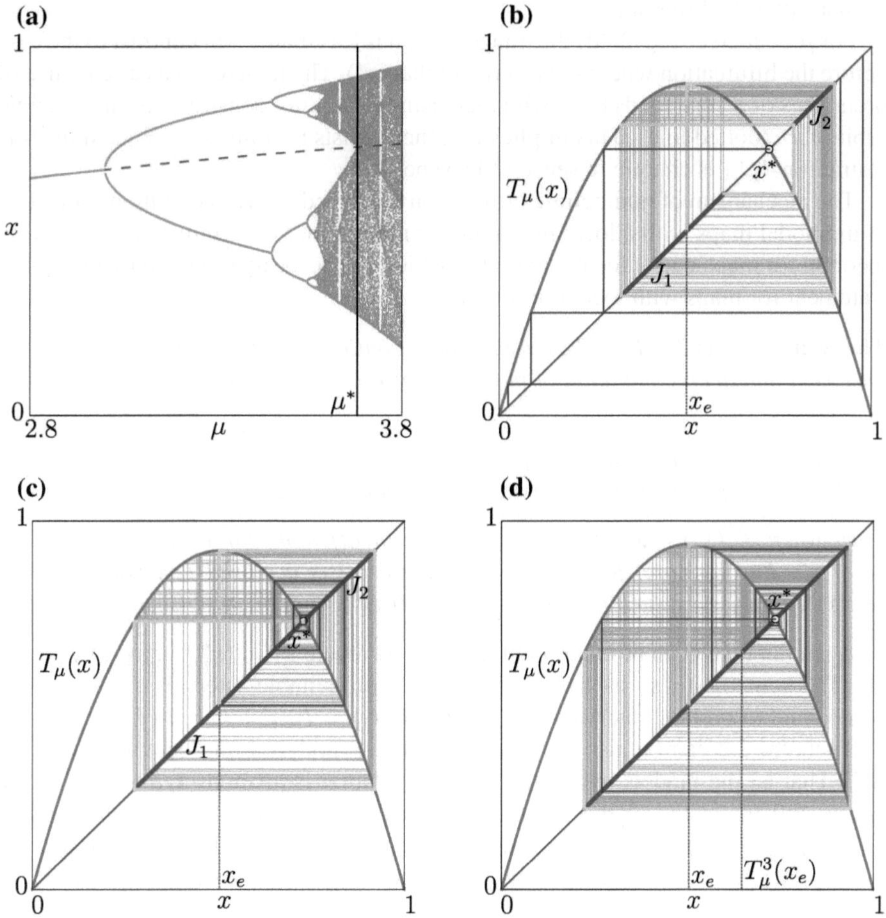

Fig. 2.4 Logistic map: A bifurcation diagram in (**a**); and the map dynamics (**b**) before, **c** at, and **d** after the first homoclinic bifurcation of the fixed point $\mathbf{x}^*$

As can be noticed in Fig. 2.4b, typical asymptotic orbits of T_μ (see the shown staircase diagram) never exit intervals $J_1 \cup J_2$ which are bounded by the first four images of the extremum point x_e. Note that $T_\mu^3(x_e)$ is greater than x^*. As for the fixed point x^*, the set of all its preimages can be divided into two sequences. The first (left) sequence approaches zero with increasing rank. The second (right) sequence consists of points which do not have preimages. This implies that x^* cannot have homoclinic orbits.

In Fig. 2.4c the point x_e is clearly homoclinic. On one hand, in three steps x_e is mapped into x^*, that is, $T^3(x_e) = x^*$. On the other hand, there exists an appropriate backward orbit of x_e that tends to x^*, since the fixed point is repelling. This moment corresponds to the first homoclinic bifurcation of x^*. Due to the equality $T^3(x_e) = x^*$, the two intervals J_1 and J_2 merge at x^* creating a single invariant interval J, which persists after the bifurcation.

For $\mu > \mu^*$ (see Fig. 2.4d) the image $T_\mu^3(x_e)$ is less than x^* (in contrast to the case before the bifurcation where it was greater than x^*). Then, there exists a sequence of preimages of x^* that tends to x^* with increasing rank, thus, constituting a homoclinic orbit of x^* (dotted line). This implies that there exists a subset $\Lambda \subset J$ such that the restriction $T_\mu|_\Lambda$ is chaotic in sense of Devaney.

The mechanism of homoclinic bifurcation described above is common for a class of unimodal maps with a local maximum or minimum. We formulate the following theorem for maps with a local maximum, which can be easily modified to get similar statement for maps with local minimum.

Theorem 2.2 *Let $T : I \to I$ be a unimodal continuous map of the interval $I \subset \mathbb{R}$ into itself and denote the point of maximum of T as x_e. Suppose that*

1. *T is smooth in $I \backslash \{x_e\}$;*
2. *T has a unique unstable fixed point x^*;*
3. *there exists a sequence of preimages of x_e approaching x^*.*

Then when $T^3(x_e) = x^*$, there occurs a (critical) homoclinic orbit of the fixed point x^*. Furthermore, for $T^3(x_e) < x^*$ there exists a closed invariant set $\Lambda \subseteq [T^2(x_e), T(x_e)] \subseteq I$ such that the restriction $T|_\Lambda$ is chaotic in sense of Devaney.

A similar result can be clearly obtained for a k-cycle $\mathscr{C}_k = \{\mathbf{x}_1^*, \ldots, \mathbf{x}_k^*\}$ by applying Theorem 2.2 to $\mathbf{x}_j^*$ as fixed points of T^k on suitable intervals I_j. The latter correspond to cyclical intervals of T, that is $T(I_j) = I_{j+1}$, $j < k$, $T(I_k) = I_1$.

Theorem 2.2 can be also used when studying models with "backward dynamics". That is, the models where the iterative relation is given as $x_t = T(x_{t+1})$, but one is still interested in the behavior of the forward values of the state variable $(x_t, x_{t+1}, x_{t+2}, \ldots)$ [11]. Among economic examples there may be mentioned the overlapping generations (OLG) model [14] and the cash-in-advance model [23].

2.4.3 Homoclinic Tangles

In this section we turn to saddle points in dynamical systems of dimension greater than one and their homoclinic orbits. A homoclinic orbit causes the stable and unstable sets of a saddle point $\mathbf{x}^*$ fold, twist and intersect in a rather complex way. Due to its complexity, this structure is referred to as a *homoclinic tangle* (see, e.g., [1, 17]) and often the mechanism of its appearance is as follows. First, a homoclinic tangency between stable $W^s(\mathbf{x}^*)$ and unstable $W^u(\mathbf{x}^*)$ sets occurs. Then tangency turns to transverse crossing that persists for a certain parameter range. Finally, the homoclinic tangle is destroyed through the second tangency between the same branches of $W^s(\mathbf{x}^*)$ and $W^u(\mathbf{x}^*)$ (see Fig. 2.5a–c). Moreover, whenever $W^s(\mathbf{x}^*)$ and $W^u(\mathbf{x}^*)$ intersect transversely, there exists an invariant set Λ such that the restriction of the map to Λ is chaotic (see, e.g., [17, 34]).

Note that asymptotic behavior of the related branches of $W^s(\mathbf{x}^*)$ and $W^u(\mathbf{x}^*)$ differ before and after the homoclinic tangle. Namely, if before the first tangency the suitable branch of $W^s(\mathbf{x}^*)$ tends towards some invariant set A_1, then after the second tangency this branch must tend towards another invariant set A_2. The same is true for $W^u(\mathbf{x}^*)$, i.e., it must come from different invariant sets B_1 and B_2 before and after the tangle. Knowing this, one can discover existence of a homoclinic tangle studying asymptotic behavior of stable and unstable sets of a fixed point.

The phenomena described above can be clearly generalized to a saddle k-cycle $\mathscr{C}_k = \{\mathbf{x}_1^*, \ldots, \mathbf{x}_k^*\}$ by considering k-th iterate T^k. However, since there are k saddle fixed points for T^k, there are two situations possible. The first situation is complete analogue of what happens for a single fixed point when the stable and unstable sets of *the same* point of the cycle intersect. Namely, $W^s(\mathbf{x}_i^*) \cap W^u(\mathbf{x}_i^*) \neq \varnothing$, $1 \leq i \leq k$. The second situation is when the stable set of each $\mathbf{x}_i^*$ intersects the unstable set of *the next* point $\mathbf{x}_{i+1}^*$ (we assume $i + 1 = 1$ for $i = k$), that is, $W^s(\mathbf{x}_i^*) \cap W^u(\mathbf{x}_{i+1}^*) \neq \varnothing$. In this case, it is sometimes said that there exist *heteroclinic orbits* for points $\mathbf{x}_i^*$, $1 \leq i \leq k$, and the structure of their stable and unstable sets is referred to as a *heteroclinic tangle*.

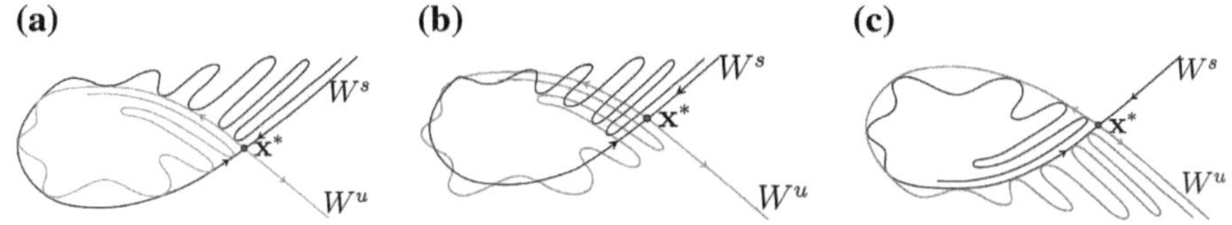

Fig. 2.5 A homoclinic tangle: **a** the first homoclinic tangency; **b** transverse intersection of the stable and unstable manifolds; **c** the second homoclinic tangency

2.4.4 Homoclinic Orbits in 2D: Smale Horseshoe

In two-dimensional systems, a prototypical example for chaotic behavior related to homoclinic orbits of saddle points is a *Smale horseshoe*. Thorough understanding of asymptotic behavior related to this structure is essential for understanding certain aspects of global dynamics of specific real systems. The map possessing the Smale horseshoe was introduced by Smale [33] while studying dynamics of orbits of the van der Pohl oscillator. The action of the map is defined geometrically by squeezing the square, then stretching the result into a long strip, and, finally, folding the strip into the shape of a horseshoe as shown in Fig. 2.6 (see also [3, 7, 34]).

More formally, let us consider a region Δ consisting of three components: a central square Q with side length 1 and two semidisks D_1 and D_2 at either end (that is, Δ is shaped like a "stadium"). The horseshoe map F takes Δ inside itself according to the following prescription. First, F linearly contracts Q in the vertical direction by a factor $\delta < 1/2$, while the regions D_1 and D_2 are contracted so as to remain semidisks attached to the resulting rectangle. Then, F expands Q in the horizontal direction by a factor $1/\delta$ so that it is long and thin (D_1 and D_2 remain unchanged). Finally, F puts Q back inside Δ in a horseshoe-shaped figure. The regions D_1 and D_2 are then mapped into D_1. We remark that $F(\Delta) \subset \Delta$ and that F is one-to-one, however, since F is not onto, the inverse F^{-1} is not globally defined. In particular, only the points belonging to the two horizontal stripes shaded in Fig. 2.6 has rank-1 preimages in the square Q.

Figure 2.7 serve to clarify this point, where only the region Q is shown for compactness. The two stripes which has rank-1 preimages are denoted H_0 and H_1 and both has height δ. The preimages of H_0 and H_1 constitute two vertical stripes V_0 and V_1 (both of the width δ) also shown in Fig. 2.7a. For sake of simplicity, we assume that V_0 and V_1 are mapped linearly onto H_0 and H_1. This assumption implies that F preserves horizontal and vertical lines in Q. For later use we note the following property of F.

LP Let $\ell_h \subset Q$ be a horizontal line segment of length a, such that $F(\ell_h) \subset Q$. Then, $F(\ell_h)$ is also a horizontal line segment whose length is a/δ. Similarly, let $\ell_v \subset Q$, $F(\ell_v) \subset Q$, be a vertical line segment of length b, then, $F(\ell_v)$ is also a vertical line segment whose length is $b\delta$.

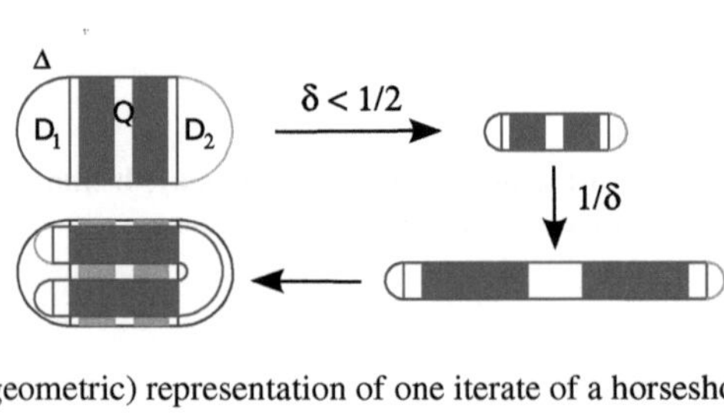

Fig. 2.6 Schematic (geometric) representation of one iterate of a horseshoe map

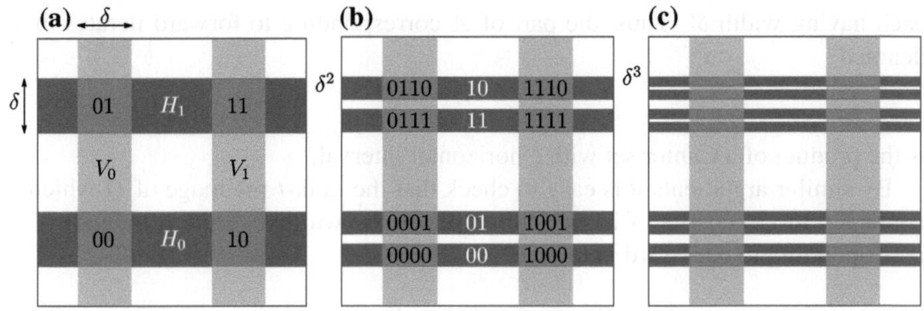

Fig. 2.7 Consecutive construction of images of the square Q under action of the Smale horseshoe map F. **a** The first image $F(Q) \cap Q$ consists of *two horizontal stripes* H_0 and H_1 of height δ. **b** The second image $F^2(Q) \cap Q$ is represented by *four stripes* of height δ^2. **c** The third image $F^3(Q) \cap Q$ is *eight* δ^3-height *stripes*

In other words, F preserves vertical and horizontal line segments belonging to Q if they are mapped into Q.

Now, we turn to describing asymptotic dynamics of orbits of F. First, note that F is a contraction in D_1, hence, there is a unique attracting fixed point $\mathbf{p} \in D_1$, that is, $\lim_{t\to\infty} F^t(\mathbf{y}) = \mathbf{p}$ for any $\mathbf{y} \in D_1$. Since $F(D_2) \subset D_1$, all forward orbits in D_2 behave likewise. Similarly, if $\mathbf{y} \in Q$ but $F^t(\mathbf{y}) \notin Q$ for some $t > 0$, then we must have $F^t(\mathbf{y}) \in D_1 \cup D_2$, so that the orbit $\mathcal{O}(\mathbf{y})$ also converges to $\mathbf{p}$. Consequently, to understand the forward orbits of F, it suffices to consider the set of points whose forward orbits lie for all time in Q, that is,

$$\Lambda = \{\mathbf{y} \in Q \; : \; F^t(\mathbf{y}) \in Q, \; t \geq 0\} = \cap_{t=-\infty}^{\infty} F^t(Q). \tag{2.14}$$

It can be shown, that most points eventually leave the square Q under the action of F, and the remaining invariant set Λ is a Cantor set. Recall that a *Cantor set* is an uncountable set with no connected subsets (except individual points). In a space of dimension greater than one such a set is sometimes referred to as a *Cantor dust*.

Construction of the invariant set Λ is an inductive process, and it is convenient to construct separately those parts corresponding to forward iterates and backward iterates, taking further their intersection to obtain Λ. As it has been already said, $Q_1 = Q \cap F(Q) = H_0 \cup H_1$. Since only points belonging to $V_0 \cup V_1$ stay in Q under F, only the points belonging to the intersection of H_0, H_1 and V_0, V_1 are mapped into Q under next iteration of F. In Fig. 2.7a these four squares are marked by symbols 00, 01, 10, and 11. The image $Q_2 = F(Q_1) \cap Q = F^2(Q) \cap F(Q) \cap Q$ (of the squares 00, 01, 10, and 11) is made up of *four* horizontal stripes as in Fig. 2.7b (each stripe is of height δ^2). Similarly, only the points belonging to the intersection of Q_2 and V_0, V_1 are mapped into Q under F generating $Q_3 = F(Q_2) \cap Q = F^3(Q) \cap F^2(Q) \cap F(Q) \cap Q$, which is constituted by *eight* horizontal stripes of height δ^3 (see Fig. 2.7c). Inductively, the set $Q^t = \cap_{j=0}^{t} F^j(Q)$, $F^0(Q) = Q$, consists of 2^t horizontal stripes each having width δ^t. Thus, the part of Λ corresponding to forward iterates of F denoted

$$\Lambda_+ = \cap_{t=0}^{\infty} F^t(Q)$$

is the product of a Cantor set with a horizontal interval.

By similar arguments it is easy to check that the rank-t preimage of Q which is contained in Q consists of 2^t vertical stripes of the width δ^t. Hence, the part of Λ corresponding to backward iterates

$$\Lambda_- = \cap_{t=-\infty}^{0} F^t(Q)$$

is the product of a Cantor set with a vertical interval. This implies that the invariant set $\Lambda = \Lambda_+ \cap \Lambda_-$ is also a Cantor set (a two-dimensional Cantor dust). In Fig. 2.8 the first three approximations of Λ are shown.

It is possible to show further, that this invariant Cantor set Λ is in one-to-one correspondence with the set Σ_2 of bi-infinite binary sequences. Recall that Σ_2 is defined as

$$\Sigma_2 = \{(s) = (\dots s_{-2}s_{-1} \cdot s_0 s_1 s_2 \dots) \ : \ s_j = 0 \text{ or } 1\}$$

with the metric $d((s),(t)) = \sum_{i=-\infty}^{\infty} 2^{-|i|}|s_i - t_i|$. Introduce also the *shift* map σ as

$$\sigma(\dots s_{-2}s_{-1} \cdot s_0 s_1 s_2 \dots) = (\dots s_{-2}s_{-1}s_0 \cdot s_1 s_2 \dots),$$

which is known to be chaotic. It can be proved that the restriction $F|_\Lambda$ is topologically conjugate to σ, which proves that $F|_\Lambda : \Lambda \to \Lambda$ is chaotic as well (see, e.g., [7]).

What has that got to do with homoclinic orbits? one may ask. Indeed, above we have mentioned only one fixed point $\mathbf{p} \in D_1$ which is attracting, and hence any point in Q which eventually leaves Q belongs to the stable set of $\mathbf{p}$. However, the map F has two more fixed points belonging to Λ which are saddles (their existence can be proved by using symbolic sequences of zeros and ones, for more detail see

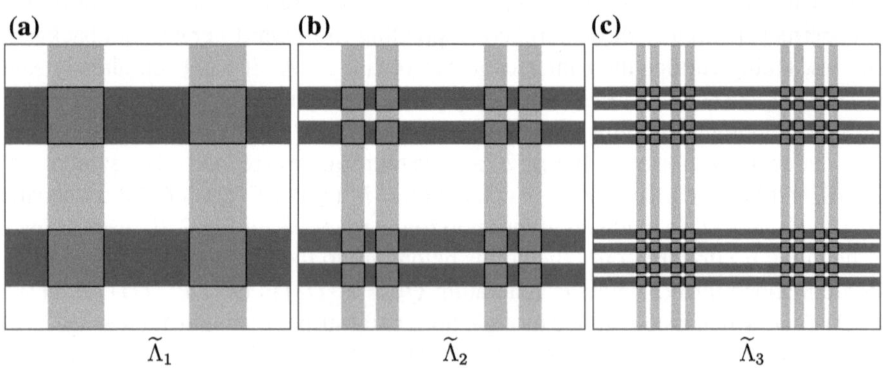

Fig. 2.8 First three approximations of the invariant Cantor set Λ

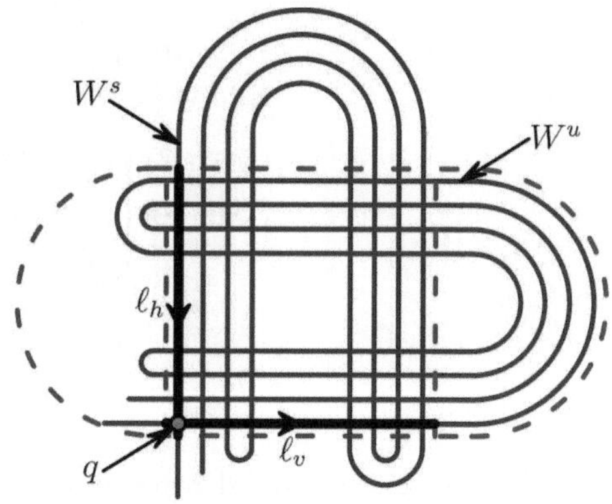

Fig. 2.9 Transverse homoclinic intersection of the stable W^s and unstable W^u sets of a saddle fixed point q for a horseshoe map

[7, 34]). The stable and unstable sets of these points have much more complex structure. For example, consider a fixed point $\mathbf{q} \in V_0 \cap H_0 \subset Q$. Figure 2.9 shows $\mathbf{q}$ together with $W^s(\mathbf{q})$ and $W^u(\mathbf{q})$, which clearly mimic the "fold-and-pleat" behavior of F. Any point that lies on the vertical segment ℓ_v through $\mathbf{q}$ belongs to $W^s(\mathbf{q})$. Indeed, since at each iteration of F the segment ℓ_v is squeezed by factor δ, the points from ℓ_v constantly approach $\mathbf{q}$. Clearly, all preimages $F^{-t}(\ell_v)$, $t \geq 1$, belong to $W^s(\mathbf{q})$ as well. Due to property LP these preimages constitute a set of vertical segments of length 1. More precisely, $F^{-t}(\ell_v)$ consists of 2^t such segments. Similarly, the horizontal segment ℓ_h through $\mathbf{q}$ belongs to $W^u(\mathbf{q})$, as well as all its forward images $F^t(\ell_h)$. It can be checked, that $F^t(\ell_h) \subset \Delta$ is a "snake-like" curve that cuts across Q exactly 2^t times. This inevitably implies that $W^s(\mathbf{q}) \cap W^u(\mathbf{q}) \neq \varnothing$, and, hence, there exists infinite number of points which are homoclinic to $\mathbf{q}$.

Let us show that as soon as for an invertible map T there exists a transverse homoclinic orbit (that is, the stable and the unstable sets intersect transversally), then there also exists a related horseshoe-like structure. We start with a rectangle R containing a fixed point $\mathbf{x}^*$ of the map T. Successive iterates of T stretch R out along the unstable set $W^u(\mathbf{x}^*)$. Under iterates of T^{-1}, it is stretched out along the stable set $W^s(\mathbf{x}^*)$. In particular, there exist numbers l and k such that $T^{-l}(R)$ extends along the stable set to include some homoclinic point $\mathbf{y}$, and $T^k(R)$ extends along the unstable set to include $\mathbf{y}$ as well (see Fig. 2.10). Thus T^{k+l} is a horseshoe map with the "square" domain $Q = T^{-l}(R)$ and its image $T^k(R)$. This can be summarized as

Theorem 2.3 [33] *Let T be a diffeomorphism of the plane, and let $\mathbf{x}^*$ be a saddle fixed point. If the stable and unstable manifolds of $\mathbf{x}^*$ cross transversally, then there is a hyperbolic horseshoe for some iterate of T.*

Note that in order to have exactly the dynamics of the ideal horseshoe, there must be uniform stretching and contraction at points in the invariant set, which is unlikely to happen in a real system. Thus, the domains which are squares and rectangles for the pure horseshoe map in real systems are somewhat deformed (though being still topologically conjugate to "ideal" shapes).

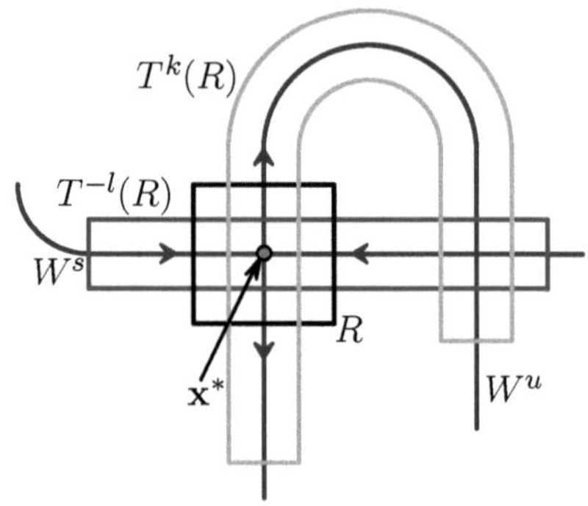

Fig. 2.10 A horseshoe generated from a homoclinic orbit. The domain $T^{-l}(R)$ corresponds to the square Q, while the domain $T^k(R)$ is its *horseshoe-shape* image. The iterate T^{k+l} is a horseshoe map

Example 2.6 Theorem 2.3 guarantees that presence of a transverse homoclinic point implies chaotic orbits (although we cannot say anything about whether they are attracting or not). Let us illustrate this by an example. We consider the Hénon map $H_{a,b}$ defined in (2.11) with $a = 2.1, b = 0.3$. Figure 2.11a (cf. Fig. 2.10) shows the stable and the unstable sets of the saddle point $\mathbf{p}_+ = (x_+, x_+)$ which is marked by a circle. The initial area $\widetilde{R}$ is not a rectangle but a parallelepiped snapped onto the related eigenvectors. The rank-4 image $R_4 = H^4_{a,b}(\widetilde{R})$ of $\widetilde{R}$ and its rank-2 preimage $R_{-2} = H^{-2}_{a,b}(\widetilde{R})$ are also shown. Figure 2.11b is a zoomed window marked by letter 'b' in Fig. 2.11a, in which one can clearly see that the intersection $R_4 \cap R_{-2} \neq \varnothing$. Hence, $H^6_{a,b}$ is topologically conjugate to the Smale horseshoe.

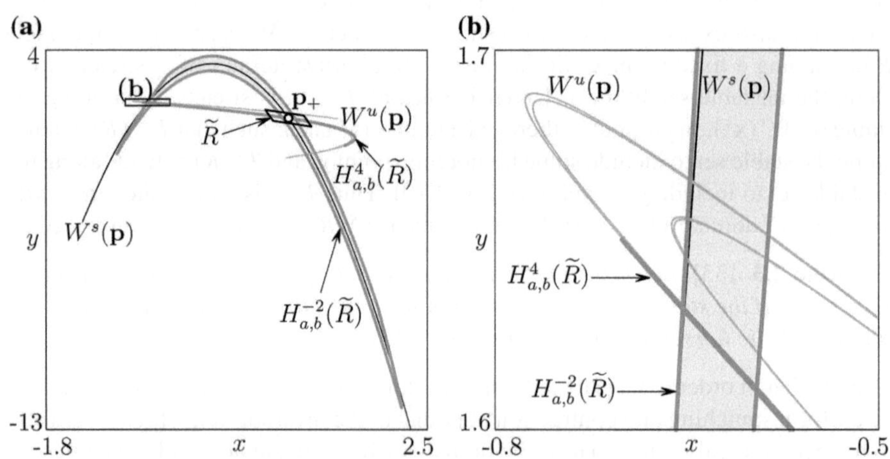

Fig. 2.11 Horseshoe structure in the Hénon map. Plot **b** is a zoom of the *rectangle* marked '*b*' in **a**

2.5 Crises

In this section we describe bifurcations which concern transformation or disappearance of chaotic attractors. Chaotic attractors are unlike fixed points or cycles not only because they are infinite sets. A periodic point can not abruptly change its periodicity or location while existent. With varying parameters it can only change its stability characteristics or disappear through the related bifurcation. On the contrary, a chaotic attractor may undergo sudden dramatic transformation under small variations of the map parameters. Such sharp changes are called *crises* and can include sudden appearance or disappearance of the attractor or a discontinuous variation in its size or shape [3, 16, 27]. Very commonly, these transformations occur when a chaotic attractor has a contact with some unstable invariant set (e.g., the stable set of a saddle periodic point).

2.5.1 *Interior Crisis*

Interior crisis is a bifurcation due to which a chaotic attractor $\mathcal{Q}$ experiences a sudden change in size and shape. In simple words, this phenomenon can be explained as follows. Let $\mathbf{x}^*$ be an unstable periodic point with nonempty stable $W^s(\mathbf{x}^*)$ and unstable $W^u(\mathbf{x}^*)$ sets, which belongs to the interior of the basin of $\mathcal{Q}$. At some critical (bifurcation) parameter value the attractor $\mathcal{Q}$ has a contact with $\mathbf{x}^*$ (or, equivalently, with $W^s(\mathbf{x}^*)$). After the bifurcation the point $\mathbf{x}^*$ joins $\mathcal{Q}$ which implies the whole unstable set $W^u(\mathbf{x}^*)$ to be "swallowed up" by $\mathcal{Q}$ as well. This clearly leads to swift transformation of $\mathcal{Q}$.

To explain this in more detail, we analyze a simple example. Let consider Ikeda map family $g : \mathbb{R}^2 \to \mathbb{R}^2$ given by

$$g : \begin{pmatrix} x \\ y \end{pmatrix} \mapsto \begin{pmatrix} A + bx\cos(m) - by\sin(m) \\ by\cos(m) + bx\sin(m) \end{pmatrix} \tag{2.15}$$

with

$$m = \phi - \frac{q}{1 + x^2 + y^2}. \tag{2.16}$$

The system having more general form was proposed first by Ikeda [19] as a model of light going around across a nonlinear optical resonator. In [20] the original map was reduced to the simplified form given above.

We fix all the parameters except q. Figure 2.12a, b show the chaotic attractor $\mathcal{Q}_q$ of g for two different parameter values: before ($q = 7.1$) and after ($q = 7.3$) the crisis. There are two different attractors here: (i) the sink fixed point $\mathbf{p}$ whose basin is the shaded area and (ii) the chaotic attractor $\mathcal{Q}_q$ whose basin is shown white. The two basins are separated by the stable set of the saddle fixed point $\mathbf{q}$.

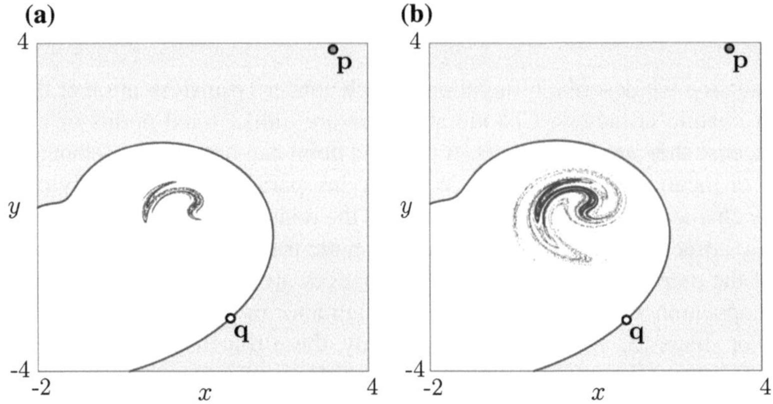

Fig. 2.12 State space of the Ikeda map with $A = 0.84, b = 0.9, \phi = 0.4$ and **a** $q = 7.1 < q_c$, **b** $q = 7.3 > q_c$. The basins of the stable fixed point **p** and the chaotic attractor $\mathcal{Q}_q$ are separated by the stable set of the saddle point **q**

Simple numerical experiments for intermediate values of q (between 7.1 and 7.3) show that the shape of $\mathcal{Q}_q$ varies only slightly from that shown in Fig. 2.12a for $7.1 < q \leq q_c \approx 7.24$. For $q > q_c$, however, the attractor in this region of the plane suddenly becomes much larger. So that, the attractor does not continuously increase in size as q passes q_c, as it happens in local bifurcations for periodic points.

Let us discover what happens at a crisis value q_c. It appears that for all q being close to q_c there exist a saddle 5-cycle $\mathcal{C}_5$ in the vicinity of $\mathcal{Q}_q$ (in fact, there exist several 5-cycles, but only one is relevant to the phenomenon investigated). For $q < q_c$ the cycle $\mathcal{C}_5$ does not belong to the attractor $\mathcal{Q}_q$. As q approaches q_c from below, the distance between $\mathcal{Q}_q$ and $\mathcal{C}_5$ (equivalently, $W^s(\mathcal{C}_5)$) goes to zero. At $q = q_c$ the attractor $\mathcal{Q}_q$ and the cycle $\mathcal{C}_5$ ($W^s(\mathcal{C}_5)$) collide. Then, after the crisis the chaotic attractor $\mathcal{Q}_q$ absorbs $\mathcal{C}_5$ together with its unstable set $W^u(\mathcal{C}_5)$. The result of such crossing between chaotic attractor and the stable set of some periodic point can be formalized as follows (see, e.g., [3]).

Lemma 2.1 (Lambda or Inclination Lemma) *Let T be a diffeomorphism of the plane, and let $\mathbf{x}^*$ be a hyperbolic saddle fixed point of T. Suppose that a curve L crosses the stable manifold $W^s(\mathbf{x}^*)$ transversally. Then each point in the unstable manifold $W^u(\mathbf{x}^*)$ is a limit point of $\cup_{t>0}T^t(L)$.*

The proof of Lemma 2.1 can be found, e.g., in [28].

In other words, if a curve L crosses the stable manifold $W^s(\mathbf{x}^*)$ transversally, then forward iterates of L limit on the entire unstable manifold $W^u(\mathbf{x}^*)$ (see Fig. 2.13). Specifically, it means that for each point $\mathbf{y} \in W^u(\mathbf{x}^*)$ and for any ε-neighborhood $U_\varepsilon(\mathbf{y})$, there exists $t > 0$ such that $T^t(L) \cap U_\varepsilon(\mathbf{y}) \neq \varnothing$. Similar property is known to be true also for noninvertible maps, but there is no rigorous proof for general case.

Using Lemma 2.1, we can interpret the crisis in the Ikeda example above. Figure 2.14a shows that as a parameter q comes close to a critical value q_c, the

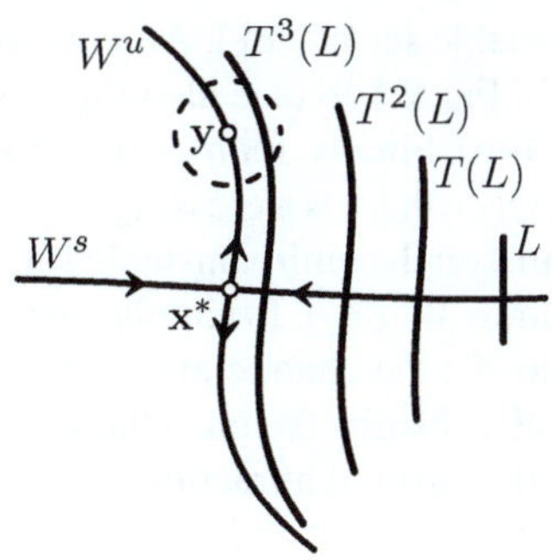

Fig. 2.13 Whenever a curve L crosses the stable set of a saddle $\mathbf{x}^*$ transversally, forward iterates of L approach the unstable set of $\mathbf{x}^*$

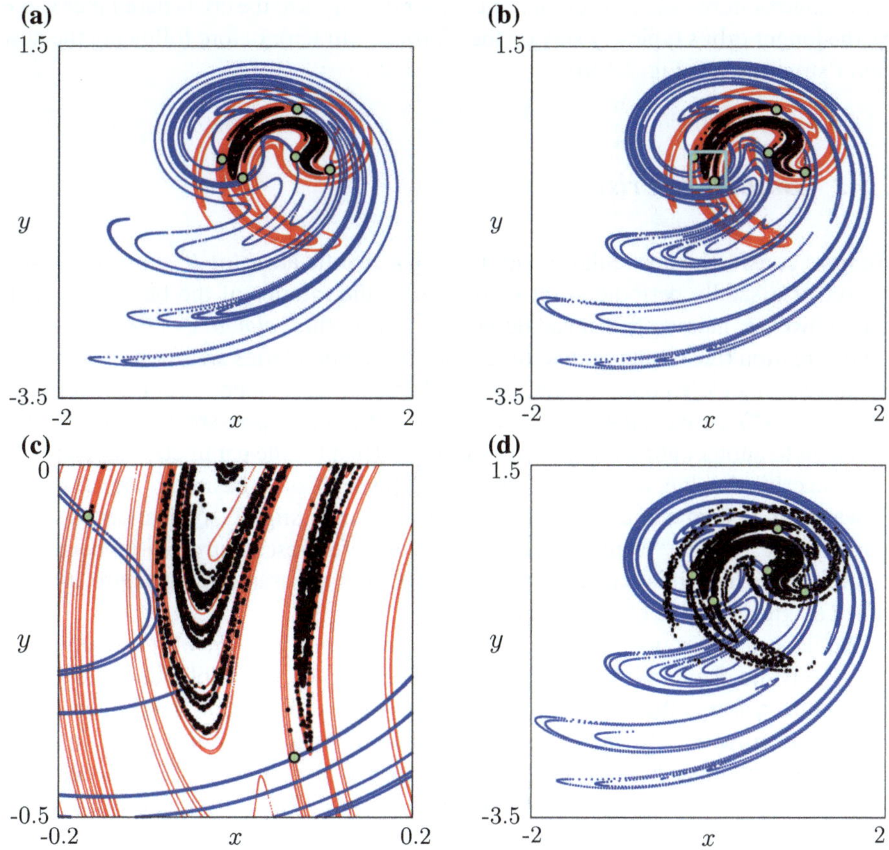

Fig. 2.14 Interior crisis for the chaotic attractor $\mathscr{Q}_q$ of the Ikeda map. **a** Before the crisis, $q = 7.1$; **b** at the crisis value $q = q_c \approx 7.24$; **c** zoom of box outlined *light-blue* in (**b**); **d** after the crisis $q = 7.3$. *Green points* mark the saddle 5-cycle $\mathscr{C}_5$, whose stable and unstable sets are shown by *blue* and *red lines*, respectively. The other parameters are $A = 0.84$, $b = 0.9$, $\phi = 0.4$

attractor $\mathcal{Q}_q$ approaches the stable set $W^s(\mathcal{C}_5)$. At q_c the outer edge of $\mathcal{Q}_{q_c}$ is tangent to this stable set, as shown in Fig. 2.14b (see also Fig. 2.14c which is a zoom of the rectangle indicated in Fig. 2.14b). Finally, for $q > q_c$ the attractor has crossed the stable set $W^s(\mathcal{C}_5)$ (Fig. 2.14d). Once there is a crossing, Lemma 2.1 tells us that forward iterates of portions of $\mathcal{Q}_q$ limit on the entire unstable set $W^u(\mathcal{C}_5)$. Consequently, for $q > q_c$ the attractor $\mathcal{Q}_q$ contains $W^u(\mathcal{C}_5)$. In simple words, having a contact with the stable set of the saddle cycle $\mathcal{C}_5$, the chaotic attractor $\mathcal{Q}_q$ absorbs this cycle, but it inevitably follows that $\mathcal{Q}_q$ also absorbs the unstable set of this cycle. It is important that $W^u(\mathcal{C}_5)$ is contained in the basin of attraction of $\mathcal{Q}_q$ for q near q_c. In this case it is said that there is an *interior crisis* at $q = q_c$, and sudden increase of the attractor size is a specific feature of such crises.

Note that the structure of the "smaller" attractor (which is relevant for $q < q_c$) is still apparent in Fig. 2.14d for $q > q_c$. It appears darker since orbits spend a larger percentage of iterates in this region. The closer $q > q_c$ is to the crisis parameter value q_c, the longer orbits typically stay on the "former" attractor before following the new larger structure (cf. Fig. 2.12b).

2.5.2 Boundary Crises

We have considered above the situation when a chaotic attractor $\mathcal{Q}$ contacts the stable set of a saddle periodic point belonging to the interior of the basin $\mathcal{B}(\mathcal{Q})$. It may, however, happen that a saddle point $\mathbf{x}^*$ is on the *boundary* of $\mathcal{B}(\mathcal{Q})$ before the bifurcation (recall that in invertible maps basin boundaries are often consisted of stable sets of saddle points). Then, it is said that at the moment of contact between $\mathcal{Q}$ and $W^s(\mathbf{x}^*)$ there occurs a *boundary crisis*. In this case there are points in $W^u(\mathbf{x}^*)$ which go to another attractor (perhaps infinity). Then for the parameter value greater than the critical value, the chaotic attractor $\mathcal{Q}$ (as well as its basin) no longer exists. However, if a parameter exceeds the critical value only slightly, typical orbits spend many iterates on the "former" chaotic attractor before escaping from its neighborhood. This behavior is called *transient chaos*, and the transient structure itself is called a "ghost" of the chaotic attractor $\mathcal{Q}$.

We illustrate this again by using the Ikeda map family (2.15) with $b = 0.9$, $\phi = 0.4$, $q = 6$ and changing A through $A_c \approx 1$. For $A < A_c$, the stable set of the saddle fixed point $\mathbf{q}$ forms the boundary between the basins of the sink fixed point $\mathbf{p}$ and the chaotic attractor $\mathcal{Q}_A$. In Fig. 2.15a it can be seen that one branch of $W^u(\mathbf{q})$ goes to $\mathcal{Q}_A$, and the other branch goes to $\mathbf{p}$. For $A = A_c$ the attractor $\mathcal{Q}_{A_c}$ collides with the boundary of its basin (that is, $W^s(\mathbf{q})$), and for $A > A_c$ it no longer exists (and all points from the former basin of $\mathcal{Q}_A$ approach $\mathbf{p}$). However, as shown in Fig. 2.15b, some orbits spend many iterates on what was the structure of $\mathcal{Q}_A$ before crossing the stable set $W^s(\mathbf{q})$ and converging to $\mathbf{p}$.

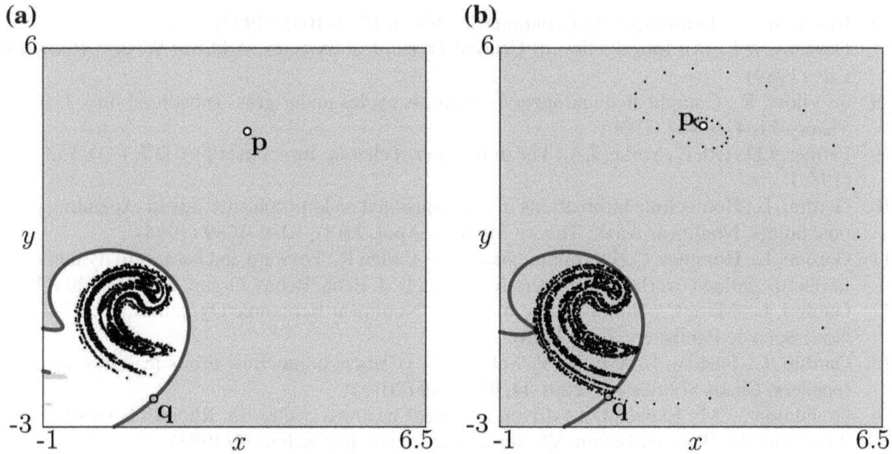

Fig. 2.15 Boundary crisis for the chaotic attractor $\mathscr{Q}_A$ of the Ikeda map. **a** Before the crisis, $A = 0.95$, the stable set of the saddle fixed point **q** separates the two basins, $\mathscr{B}(\mathscr{Q}_A)$ and $\mathscr{B}(\mathbf{p})$. **b** After the crisis, $A = 1.003$, there is a "ghost" attractor. Other parameters are $b = 0.9$, $\phi = 0.4$, $q = 6$

2.6 Conclusions

Dynamical systems theory distinguishes two types of bifurcations: those which can be studied in a small neighborhood of an invariant set (local) and those which cannot (global). In this chapter we focused on several aspects of global bifurcation analysis of discrete time dynamical systems, such as homoclinic bifurcations and crises. There are also other global phenomena being important when investigating a dynamical system. Among them there are bifurcations related to appearance of closed invariant curves and those which cause qualitative changes in the basin structure (for example, when a connected basin becomes nonconnected). For further reading and deeper understanding of all these aspects, we may suggest [1, 24, 25] and references therein.

References

1. Agliari, A., Bischi, G.I., Dieci, R., Gardini, L.: Global bifurcations of closed invariant curves in two-dimensional maps: a computer assisted study. Int. J. Bifurcat. Chaos **15**(4), 1285–1328 (2005)
2. Agliari, A., Gardini, L., Puu, T.: Some global bifurcations related to the appearance of closed invariant curves. Math. Comput. Simul. **68**, 201–219 (2005)
3. Alligood, K.T., Sauer, T.D., Yorke, J.A.: Chaos: An Introduction to Dynamical Systems. Springer, New York (1996)
4. Banks, J., Brooks, J., Cairns, G., Davis, G., Stacey, P.: On Devaney's definition of chaos. Am. Math. Mon. **99**(4), 332–334 (1992)
5. Bischi, G.I., Gardini, L., Kopel, M.: Analysis of global bifurcations in a market share attraction model. J. Econ. Dyn. Control **24**, 855–879 (2000)

6. Brock, W.A., Hommes, C.H.: Econometrica **65**(5), 1059–1095 (1997)
7. Devaney, R.L.: An Introduction to Chaotic Dynamical Systems. Addison-Wesley, Redwood City (1989)
8. de Vilder, R.: Complicated endogenous business cycles under gross substitutability. J. Econ. Theory **71**, 416–442 (1996)
9. Farmer, J.D., Ott, E., Yorke, J.A.: The dimension of chaotic attractors. Phys. D **7**(1–3), 153–180 (1983)
10. Gardini, L.: Homoclinic bifurcations in n-dimensional endomorphisms, due to expanding periodic points. Nonlinear Anal.: Theory Methods Appl. **23**(8), 1039–1089 (1994)
11. Gardini, L., Hommes, C.H., Tramontana, F., de Vilder, R.: Forward and backward dynamics in implicitly defined overlapping generations models. J. Econ. Behav. Organ. **71**, 110–129 (2009)
12. Gardini, L., Mira, C.: Noninvertible maps. In: Nonlinear Economic Dynamics, pp. 239–254. Nova Science Publishers, Inc. (2010)
13. Gardini, L., Sushko, I., Avrutin, V., Schanz, M.: Critical homoclinic orbits lead to snap-back repellers. Chaos Solitions Fractals **44**, 433–449 (2011)
14. Grandmont, J.M.: Expectations driven nonlinear business cycles. In: Rheinisch-Westfälische Akademie der Wissenschaften. VS Verlag für Sozialwissenschaften (1993)
15. Grebogi, C., Ott, E., Pelikan, S., Yorke, J.A.: Strange attractors that are not chaotic. Phys. D **13**, 261–268 (1984)
16. Grebogi, C., Ott, E., Yorke, J.A.: Crises, sudden changes in chaotic attractors, and transient chaos. Phys. D **7**(1–3), 181–200 (1983)
17. Guckenheimer, J., Holmes, P.: Nonlinear oscillations, dynamical systems and bifurcations of vector fields. Applied Mathematical Sciences, vol. 42. Springer (1983)
18. Hénon, M.: A two-dimensional mapping with a strange attractor. Commun. Math. Phys. **50**(1), 69–77 (1976)
19. Ikeda, K.: Multiple-valued stationary state and its instability of the transmitted light by a ring cavity system. Opt. Commun. **30**, 257–261 (1979)
20. Ikeda, K., Daido, H., Akimoto, O.: Optical turbulence: chaotic behavior of transmitted light from a ring cavity. Phys. Rev. Lett. **45**(9), 709–712 (1980)
21. Li, T.Y., Yorke, J.A.: Period three implies chaos. Am. Math. Mon. **82**(10), 985–992 (1975)
22. Marotto, F.R.: Snap-back repellers imply chaos in r^n. J. Math. Anal. Appl. **63**(1), 199–223 (1978)
23. Michener, R.W., Ravikumar, B.: Chaotic dynamics in a cash-in-advance economy. J. Econ. Dyn. Control **22**, 1117–1137 (1998)
24. Mira, C., Gardini, L.: From the box-within-a-box bifurcations organization to the julia set. Part I. Int. J. Bifurcat. Chaos **19**(1), 281–327 (2009)
25. Mira, C., Gardini, L., Barugola, A., Cathala, J.C.: Chaotic Dynamics in Two-Dimensional Noninvertible Maps. Nonlinear Science. World Scientific, Singapore (1996)
26. Myrberg, P.J.: Iteration der reellen polynome zweiten grades iii. Annales AcademiæScientiarum Fennicæ **336**, 3–18 (1963)
27. Ott, E.: Chaos in Dynamical Systems. Cambridge University Press (1993)
28. Palis, J., de Melo, W.: Geometric Theory of Dynamical Systems: An Introduction. Springer, New York (1982)
29. Ruelle, D.P., Takens, F.: On the nature of turbulence. Commun. Math. Phys. **20**(3), 167–192 (1971)
30. Sharkovskii, A.N.: Problem of isomorphism of dynamical systems. In: Proceedings of 5th International Conference on Nonlinear Oscillations, vol. 2, pp. 541–544 (1969)
31. Sharkovskii, A.N., Maistrenko, Y.L., Romanenko, E.Y.: Difference Equations and Their Applications. Kluwer Academic Publisher, Dordrecht (1993)
32. Silverman, S.: On maps with dense orbits and the definition of chaos. Rocky Mt. J. Math. **22**(1), 353–375 (1992)
33. Smale, S.: Differentiable dynamical systems. Bull. Am. Math. Soc. **73**, 747–817 (1967)
34. Wiggins, S.: Global Bifurcations and Chaos: Analytical Methods, Applied Mathematical Sciences, vol. 73. Springer, New York (1988)

Chapter 3
Dynamical Analysis of Cournot Oligopoly Models: Neimark-Sacker Bifurcation and Related Mechanisms

Anna Agliari, Nicolò Pecora and Alina Szuz

Abstract This chapter describes some properties of the nonlinear dynamics emerging from two oligopoly models in discrete time. The target of this chapter is the investigation of some local and global bifurcations which are responsible for the changes in the qualitative behaviors of the trajectories of discrete dynamical systems. Two different kinds of oligopoly models are considered: the first one deals with the presence of differentiated goods and gradient adjustment mechanism, while the second considers the demand function of the producers to be dependent on advertising expenditures and adaptive adjustment of the moves. In both models the standard local stability analysis of the Cournot-Nash equilibrium points is performed, as well as the global bifurcations of both attractors and (their) basins of attraction are investigated.

3.1 Introduction

The object of the present chapter is to describe some properties of nonlinear dynamics emerging from oligopoly models in discrete time. The target of our analysis is the investigation of some bifurcations which are responsible for the changes in the qualitative behaviors of the trajectories of the iterative process.

We consider two different kinds of oligopoly models: the first one deals with the presence of differentiated goods and gradient adjustment mechanism, while the second considers the demand function of the producers to be dependent on advertising expenditures and adaptive adjustment of the moves. In both models, we perform the

A. Agliari (✉) · N. Pecora
Department of Economics and Social Sciences, Catholic University,
Via Emilia Parmense 84, 29100 Piacenza, Italy
e-mail: anna.agliari@unicatt.it

N. Pecora
e-mail: nicolo.pecora@unicatt.it

A. Szuz
Independent Researcher, Cluj-Napoca, Romania
e-mail: alina.ghirvu@ubbcluj.ro

G.I. Bischi et al. (eds.), *Qualitative Theory of Dynamical Systems, Tools and Applications for Economic Modelling*, Springer Proceedings in Complexity,
DOI 10.1007/978-3-319-33276-5_3

local stability analysis of the Cournot-Nash equilibrium point as well as a global analysis of dynamics to study bifurcations of both attractors and basins of attraction.

Being the dynamics of such models described by maps of the family $T : X \to X$, $X \subset \mathbb{R}^2$, we will see that a particular kind of bifurcation may occur, related to a pair of complex conjugated eigenvalues which crosses the unit circle, namely the Neimark-Sacker bifurcation (NS henceforth). The NS bifurcation is associated with the existence of closed invariant curves around the bifurcating fixed point.

In the nonlinear map describing the Cournotian competition with differentiated products and gradient adjustment mechanism the steady state may be destabilized via *supercritical* NS bifurcation. Such a bifurcation gives rise to an attracting closed curve around the unstable equilibrium and, through global analysis, we shall also show that different multistability situations (i.e., coexistence of attractors) may arise.

On the other hand, the oligopoly model with advertising costs allows us to analyze the effects of the occurrence of a *subcritical* NS bifurcation in which the destabilization of the equilibrium point is due to its merging with a repelling closed curve existing when the point is still stable. The occurrence of a subcritical NS bifurcation has important implications in economic models since it can be associated with corridor stability, due to the bounded basin of attraction of the stable equilibrium (its boundary being the repelling closed curve), and to catastrophic effects, since after the bifurcation the trajectories may either jump to a different attractor far from the equilibrium or diverge. Moreover, the map describing this second oligopoly setting is piecewise smooth (PW henceforth), and we shall show that even border collision bifurcations (BCB henceforth) may cause multistability phenomena as well. We recall that BCB are typical occurrences in PW map and are related to invariant sets, such as attractors or manifolds, having a contact with the border of a region where the map changes its definition. Such a contact can cause abrupt changes either in the structure or in the stability property of the colliding invariant set. Seminal papers on such topic are by Nusse et al. [16] and Nusse and Yorke [17, 18].

3.2 Some Remarks on Neimark-Sacker Bifurcations

The aim of this section is to briefly illustrate some theoretical aspects associated with the occurrence of NS bifurcation and the appearance/disappearance of closed invariant curves, that will be the objects of the analysis developed in the following parts.

To this extent, we first recall that generally a steady state loses stability through a NS bifurcation when its Jacobian matrix has two complex eigenvalues lying on the unit circle. Two kind of NS bifurcations can be distinguished:

- *supercritical* when, immediately after the bifurcation, the unstable steady state is surrounded by an *attracting closed curve* corresponding to periodic or quasi-periodic dynamics;

- *subcritical* when, immediately before the bifurcation, the stable equilibrium is surrounded by a *repelling closed curve* which shrinks and at the bifurcation merges with the fixed point leaving a repelling focus.

A typical way of investigating the occurrence of a NS bifurcation is to start from the analysis of a two-dimensional bifurcation diagram.[1] In so doing we can detect the kind of the occurring NS bifurcation and, sometimes, to find the so-called *Chenciner* points. These points correspond to the degeneracy of the NS bifurcation, they belong to the bifurcation curve and separate cases in which either a subcritical or a supercritical bifurcation occurs. Moreover, a typical and well-known structure of the bifurcation diagram, in a two-dimensional parameter plane, is given by the so-called *Arnold's tongues* issuing from a NS bifurcation curve (on this we refer to some classical texts, e.g. [10, 15], and other works like [1, 6, 8], to cite a few). Inside any tongue at least 2 cycles exist, one stable and a saddle, and a closed invariant curve exists, made up by the unstable set of the saddle cycle that connects the periodic points of the stable cycle. These periodicity regions follow the Farey summation rule, which is also known in the literature as "adding rule" [9]. This implies that the tongues are organized so that between any two periodicity regions related to the rotation numbers, say m_1/n_1 and m_2/n_2, there exists a periodicity region related to the rotation number $(m_1 + m_2)/(n_1 + n_2)$. The rational rotation is not generic only for parameter values taken exactly on the NS bifurcation curve, while soon after the bifurcation the rational rotation becomes generic: infinitely many periodicity regions fill the parameter plane densely [9]. Generally, inside any tongue we have an attracting set formed by a saddle-node connection, that is, the unstable set of the saddle n-cycle reaches the node n-cycle thus forming a closed attracting curve. The boundaries of a m/n tongue are saddle-node bifurcation curves in the case of smooth maps and BCB curve if we are dealing with PW maps.

Let us recall that the *stable* and *unstable* sets of a saddle S^* are defined as

$$W^s(S^*) = \left\{ x : \lim_{n \to +\infty} T^n(x) = S^* \right\},$$

$$W^u(S^*) = \left\{ x : \lim_{n \to +\infty} T_{jn}^{-n}(x) = S^* \right\},$$

respectively, where T_{jn}^{-n} means a suitable sequence of inverses.

In the subcritical case, the periodicity regions exist when the fixed point is still stable, implying multistability situations, but generally immediately after their appearance no saddle-node connection exists. The repelling closed curve involved with the subcritical NS bifurcation appears at the crossing of a (global) bifurcation curve γ together with an attracting one (*Chenciner bifurcation*). The curve γ originates from a Chenciner point and enters the region in which the fixed point is stable, crossing the periodicity regions. The global bifurcations occurring at the crossing of γ are an interesting and challenging field of research.

---

[1]Obviously the map has to depend on at least two parameters.

In particular, in the case of smooth maps, the appearance/disappearance of closed invariant curves is associated with a saddle-connection. This particular configuration is defined as a closed invariant curve formed by the merging of a branch of the stable set of a periodic point of a saddle cycle with the unstable branch of another periodic point of the same saddle, thus forming a closed connection among the periodic points of the saddle. We shall call such a situation *homoclinic loop* that can also involve a saddle cycle of period k, being related to the forward iterate map T^k, but in this case we can also obtain a *heteroclinic loop*: indeed, the map T^k exhibits k saddles points and a branch of the stable set of a saddle may merge with a branch of another periodic point of the saddle cycle. Stated in other words, if S_i, $i = 1, \ldots, k$, are the periodic points of the saddle cycle and $\alpha_{1,1} \cup \alpha_{2,i}$ $(\omega_{1,i} \cup \omega_{2,i})$ are the unstable (stable) sets of S_i, then a heteroclinic loop is given by the merging, for example, of the unstable branch $\alpha_{1,i}$ of S_i with the stable branch $\omega_{1,j}$ of a different periodic point S_j (see Fig. 3.1). Then each periodic point of the saddle cycle is connected with another one, and an invariant closed curve is so created connecting the periodic points of the saddle cycle.

Dealing with discrete maps homoclinic and heteroclinic loops are frequently replaced by homoclinic tangles. This means that a tangency between an unstable branch $W_1^u(S^*) = \cup\alpha_{1,i}$ with a stable one $W_1^s(S) = \cup\omega_{1,i}$ occurs, followed by transverse crossings of the two sets, followed by another tangency of the same sets, but on opposite side. For major details see [1, 7, 12].

For PW maps only a few works devoted their attention to the investigation on how invariant curves appear/disappear in these peculiar contexts. Below we shall show that not only homoclinic bifurcations are involved, but also border collision bifurcations may occur at the crossing of the curve γ.

3.3 A Cournot Duopoly Model with Differentiated Products: Supercritical NS Bifurcation

In the first oligopoly model we analyze, we consider a Cournotian game with differentiated goods in which boundedly rational firms apply a gradient adjustment mechanism to update the quantity produced in each period (see [2] for a complete investigation).

The demand functions of the two players are derived from an underlying CES utility function

$$U(q_1, q_2) = q_1^\alpha + q_2^\alpha, \quad 0 < \alpha \leq 1, \tag{3.1}$$

which is maximized subject to the budget constraint

$$p_1 q_1 + p_2 q_2 = 1, \tag{3.2}$$

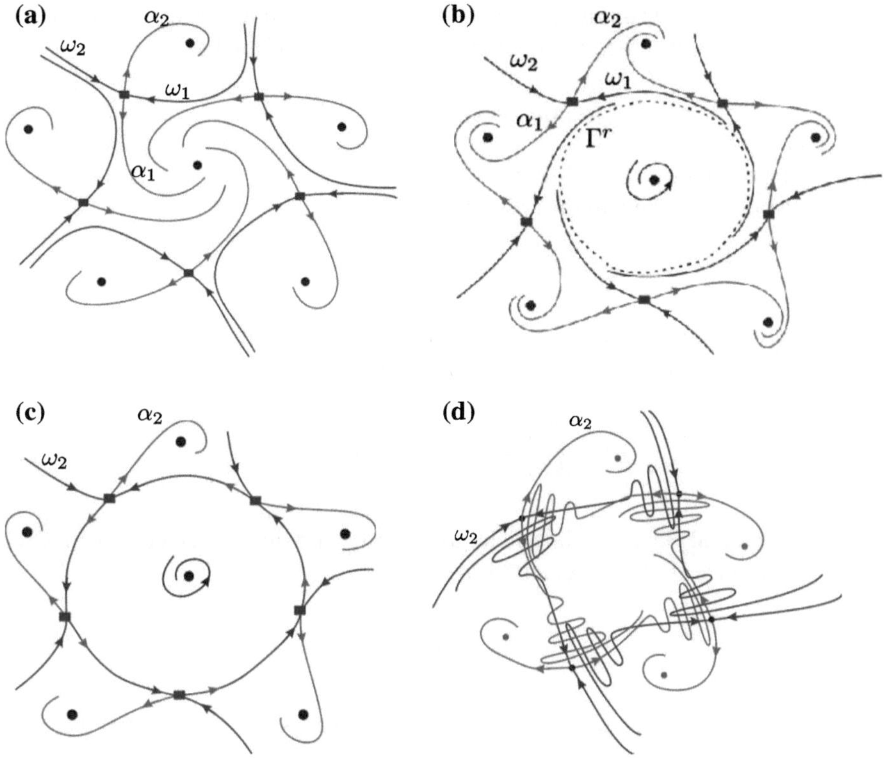

Fig. 3.1 The bifurcation mechanism associated with appearance of two closed invariant curves. **a** Stable/unstable sets before the bifurcation. **b** Two closed curves appear. **c** Heteroclinic loop. **d** Homoclinic tangle

where α gives the degree of substitutability/differentiation among the commodities, p_1 and p_2 are the prices of good 1 and 2 respectively and we assume the consumer's exogenous income equal to 1.

Maximizing (3.1) subject to (3.2) results in the inverse demand functions

$$p_1 = \frac{q_1^{\alpha-1}}{q_1^\alpha + q_2^\alpha}, \quad p_2 = \frac{q_2^{\alpha-1}}{q_1^\alpha + q_2^\alpha} \tag{3.3}$$

for goods 1 and 2 respectively.[2] From the inverse demand functions, we observe that if $\alpha = 1$ the commodities are indistinguishable and, accordingly, the consumers regard them as identical; lower values of α makes the commodities conceived as interchangeable but not quite identical. Decreasing the exponent parameter α makes

[2]We refer to [5] (Appendix A) for the mathematical computations that lead to the demand functions represented by (3.3).

the goods less close substitutes and as $\alpha \to 0$ the commodities become independent (see also [14]). Further, we assume linear cost functions given by

$$c_i(q_i) = c_i q_i, \quad i = 1, 2, \tag{3.4}$$

where c_i are constant marginal costs.

Then the profit of the i-th firm becomes

$$\Pi_i(q_i, q_j) = p_i(q_i, q_j) q_i - c_i q_i, \quad i, j = 1, 2, \quad i \neq j. \tag{3.5}$$

From the profit maximization we are able to compute the Nash equilibrium, which is unique and it is given by

$$E^* = \left(\frac{\alpha c_1^{\alpha-1} c_2^{\alpha}}{(c_1^{\alpha} + c_2^{\alpha})^2}, \frac{\alpha c_1^{\alpha} c_2^{\alpha-1}}{(c_1^{\alpha} + c_2^{\alpha})^2} \right).$$

Boundedly rational players update their quantities by an adjustment mechanism based on a local estimate of the marginal profit

$$\Phi_i(q_i, q_j) = \frac{\partial \Pi_i}{\partial q_i}.$$

A firm increases (decreases) its quantity if it perceives positive (negative) marginal profit, according to

$$q_i(t+1) = q_i(t) + k_i \Phi_i(q_i, q_j), \tag{3.6}$$

where $k_i > 0$, $i = 1, 2$, is a coefficient that "tunes" the speed of adjustment of firm i's quantity at time $t+1$ with respect to a marginal change in profits when q_i varies at time t.

Therefore, under the above assumptions, the two-dimensional system that characterizes the dynamics of the differentiated Cournot duopoly can be written as follows:

$$T = \begin{cases} q_1' = q_1 + k_1 \left(\frac{\alpha q_1^{\alpha-1} q_2^{\alpha} - c_1 (q_1^{\alpha} + q_2^{\alpha})^2}{(q_1^{\alpha} + q_2^{\alpha})^2} \right), \\ q_2' = q_2 + k_2 \left(\frac{\alpha q_2^{\alpha-1} q_1^{\alpha} - c_2 (q_1^{\alpha} + q_2^{\alpha})^2}{(q_1^{\alpha} + q_2^{\alpha})^2} \right), \end{cases} \tag{3.7}$$

where $'$ denotes the unit-time advancement operator, i.e., if q_i is quantity produced at time t then q_i' is production at time $t+1$.

Due to the presence of the denominator, it is obvious that T is defined only at points such that $(q_1, q_2) \neq (0, 0)$; furthermore from an economic point of view we are only interested in the study of positive trajectories, i.e. with points belonging to the positive quadrant of the plane $\mathbb{R}^2$. Indeed we will consider the feasible region as the set of points in the plane defined by

$$F = \{(q_1, q_2) : q_1 > 0, q_2 > 0\}. \tag{3.8}$$

3.3.1 Local Stability Analysis

In order to study the local stability of the unique Nash equilibrium we localize the eigenvalues of the Jacobian matrix of T evaluated at E^*. Making use of the Jury's conditions, we can state that the Nash equilibrium is locally asymptotically stable if

$$\begin{cases} -A^3 k_1 k_2 + D_1 B k_2 + D_2 C k_1 > 0, \\ A^4 k_1 k_2 - 2D_1 B k_2 - 2D_2 C k_1 + D_1 D_2 > 0, \end{cases} \tag{3.9}$$

where $A = c_1^\alpha + c_2^\alpha$, $B = [c_2^\alpha(\alpha - 1) - c_1^\alpha(\alpha + 1)]$, $C = [c_2^\alpha(1 + \alpha) + c_1^\alpha(1 - \alpha)]$, $D_1 = \alpha c_1^{\alpha-2} c_2^\alpha$ and $D_2 = \alpha c_2^{\alpha-2} c_1^\alpha$. The two conditions (3.9) define a region in the plane of the speeds of adjustment (k_1, k_2) whose shape is like the shaded area in Fig. 3.2.[3] This region is bounded by the two branches of hyperbola, whose equation is given by the vanishing of the left hand side of

$$-A^3 k_1 k_2 + D_1 B k_2 + D_2 C k_1 > 0,$$

and the curve represented by the vanishing of the left hand side of

$$A^4 k_1 k_2 - 2D_1 B k_2 - 2D_2 C k_1 + D_1 D_2 > 0.$$

For values of (k_1, k_2) inside the stability region the Nash equilibrium E^* is a stable steady state. The boundaries given by two branches of hyperbola on the left and on the right represent bifurcation curves at which E^* loses its stability through a period doubling (or *flip*) bifurcation. The hyperbola in the central portion represents the bifurcation curve at which the Nash equilibrium is destabilized via NS bifurcation. The lines OI_1 and OI_2 of Fig. 3.2 represent pairs of (k_1, k_2) for which $Tr^2 J - 4det J = 0$ and separate real and complex eigenvalues regions.[4] From the stability conditions we can obtain direct information on the effects of the speed of adjustment, k_1 and k_2, on the local stability of E^*. In particular, an increase of the parameters k_i, with the other parameters fixed, may turn the Nash equilibrium unstable through a flip or a NS bifurcation.

[3] We computed the stability region for $c_1 = 0.1$ and $c_2 = 0.9$ to better visualize the NS bifurcation curve. With different values of marginal costs, the structure of the stability region does not change and only the stable focus region becomes smaller and smaller.

[4] When $c_1 = c_2 = c$, such a condition reduces to $(16c^4\,(k_2 - k_1)^2)/\alpha^2 < 0$ which is never satisfied. Hence no NS bifurcation can occur in the case of symmetric game.

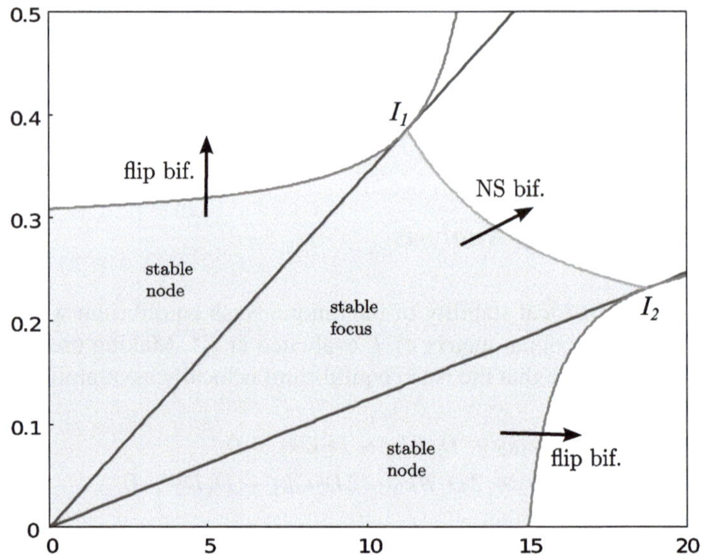

Fig. 3.2 Stability region of the Nash equilibrium. The *shaded gray area* represents the region of local asymptotic stability of the Nash equilibrium in the parameter plane of speeds of adjustment (k_1, k_2). It is noteworthy that if $c_1 = c_2$, the two intersection points I_1 and I_2 coincides

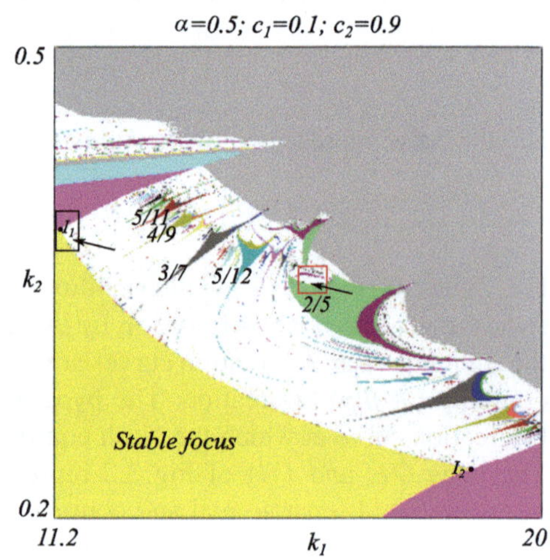

Fig. 3.3 2D bifurcation diagram of the map T in the (k_1, k_2) parameter plane. The different colors are associated with cycles of different period. *White points* correspond to cycles of large period, quasi-periodic trajectories (when close to the NS bifurcation curve) or to complex dynamics. *Gray points* correspond to unfeasible trajectories

Numerical simulations allow us to find out that the NS bifurcation is of supercritical type and it gives rise to an attractive closed invariant curve around the unstable equilibrium, which is a focus. In the 2D bifurcation diagram of Fig. 3.3 the so-called Arnold's tongues issue from the NS bifurcation curve, as we said in Sect. 3.2.

Fig. 3.4 Enlargement of the bifurcation diagram around the co-dimension 2 bifurcation point $R1:2$. The *dots* on the *arrow* correspond to the sequence of Fig. 3.5

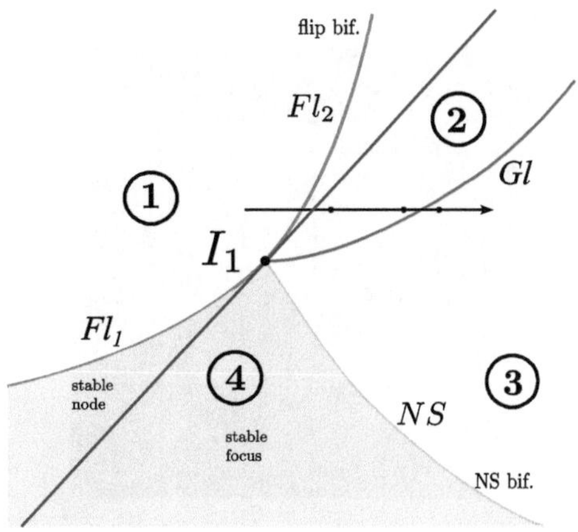

3.3.2 Co-dimension 2 Bifurcation

As the 2D bifurcation diagram of Fig. 3.3 shows, many periodicity regions exist, which are organized following the Farey structure. In order to show some dynamic features that take place in this Cournot setting with differentiated goods, we analyze two portions of the 2D bifurcation diagram, highlighted through squares in Fig. 3.3. In particular we first analyze the dynamics around the intersection point I_1 between the NS and the flip bifurcation curves (see Fig. 3.2). In so doing, we study the global bifurcations occurring around the co-dimension 2 bifurcation, the strong resonance $R1:2$ (see [13]), where the eigenvalues are $\lambda_{1,2} = -1$ (an enlargement is reported in Fig. 3.4).

In region 1, the Nash equilibrium E^* is a saddle and coexists with a stable cycle C of period 2 (created at the crossing of the curve Fl_1). Its stable set separates the basins of attraction of the two stable fixed points of the second iterate of the map. Following the path indicated by the arrow in Fig. 3.4, the crossing of the curve $Fl2$ causes a second flip bifurcation of the equilibrium point E^*, that becomes an unstable node, and a saddle cycle $\widetilde{C}$ of period 2 appears. In Fig. 3.5a, the Nash equilibrium E^* is turned into an unstable focus. The attractor of the map is still the 2-cycle C, and the two fixed points of T^2 (second iterate of T) have basins of attraction separated by the stable set of $\widetilde{C}$. The unstable set of $\widetilde{C}$, which is depicted in gray, gives rise to a saddle-node connection: in particular, in Fig. 3.5a, the branch $\alpha_1(\widetilde{C_2})$ converges to C_1 while the branch $\alpha_2(\widetilde{C_2})$ converges to C_2 and the two branches of $\alpha(\widetilde{C_1})$ behave analogously. We also observe that a branch of the unstable set of the saddle 2-cycle issuing from the point $\widetilde{C_2}$, i.e. $\alpha_1(\widetilde{C_2})$, approaches the point $\widetilde{C_1}$ before converging to C_1, signaling that a global bifurcation is likely to occur. In fact, as the parameter k_2 slightly increases, we observe the coexistence of the stable 2-cycle C with an attractive closed invariant curve (see Fig. 3.5b). The appearance of such a curve may be due to a saddle-connection: the unstable set issuing from the point $\widetilde{C_1}$ (previously converging to C_2) reaches the periodic point $\widetilde{C_2}$ (becoming one of its stable branches)

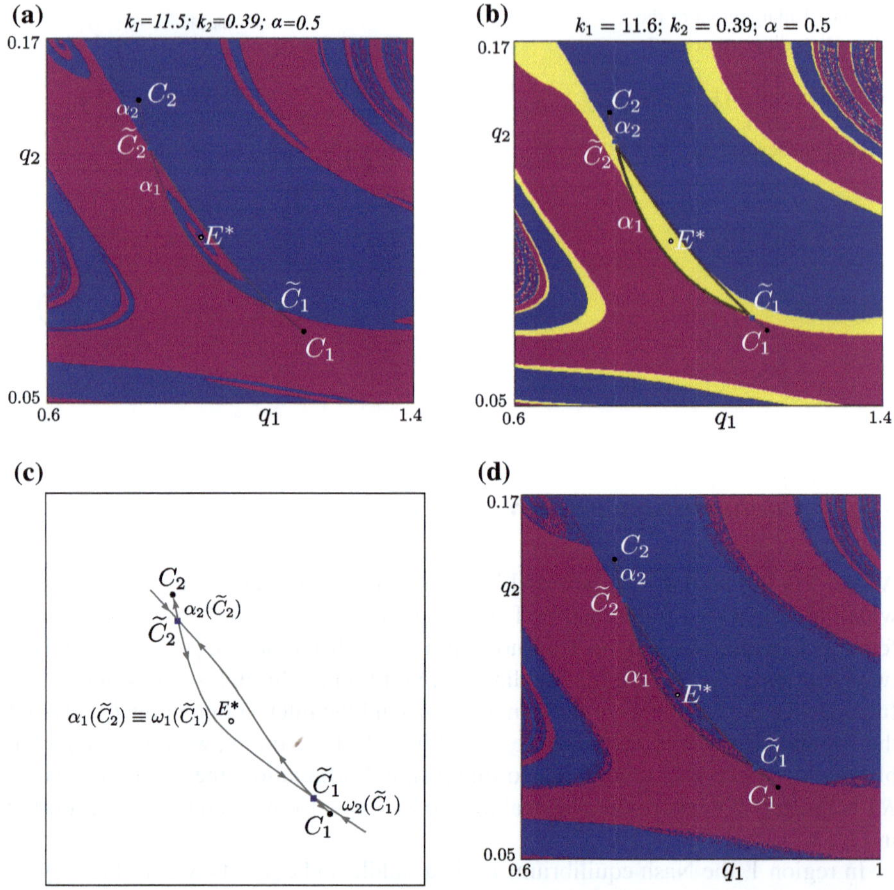

Fig. 3.5 Global bifurcation leading to the appearance of an attracting closed curve. The basins of attraction of the second iterate of the map are represented in order to show the invariant sets of the saddles. **a** Saddle node connection made up by the unstable set of $\widetilde{C}$. **b** An attracting closed curve coexists with the stable 2-cycle C. **c** A qualitative sketch of the saddle connection associated with the appearance of the curve. **d** New fractal portions of the basin of attraction suggesting the possible occurrence of a homoclinic tangle

and, vice versa, the unstable set issuing from the point $\widetilde{C_2}$ (previously converging to C_1) reaches the periodic point $\widetilde{C_1}$ (becoming one of its stable branches) leading to a closed curve (this is shown qualitatively in Fig. 3.5c). As a confirm, one can notice that before the bifurcation the basin of attraction (for the map T^2) of the point C_1 (and similarly for C_2) includes the stable sets of both $\widetilde{C_1}$ and $\widetilde{C_2}$ while, after the bifurcation, for the second iterate of the map the basin of each attracting node is bounded by the stable set of only one saddle fixed point. Moving towards the NS bifurcation curve denoted by NS in Fig. 3.4, the cycles C and $\widetilde{C}$ merge and disappear in a saddle-node bifurcation. The closed invariant curve remains the unique attractor and becomes smaller and smaller. Finally, at the crossing of the NS curve, a reverse supercritical NS bifurcations occurs and it leaves the Nash equilibrium E^* as the unique attractor.

Before concluding, we observe that the bifurcation mechanism sketched in Fig. 3.5c is simply a schematic representation. Indeed we are dealing with a discrete model and thus it is possible that a homoclinic tangle occurs, that is in a certain parameter range the contact between the stable and unstable set is opened by their quadratic tangency, at which homoclinic orbits appear (and related complex dynamics), followed by transverse intersection and closed by a second quadratic tangency at the opposite side which destroys all the homoclinic orbits. Figure 3.5d seems to suggest this occurrence: the basins of attraction may have a fractal structure close to

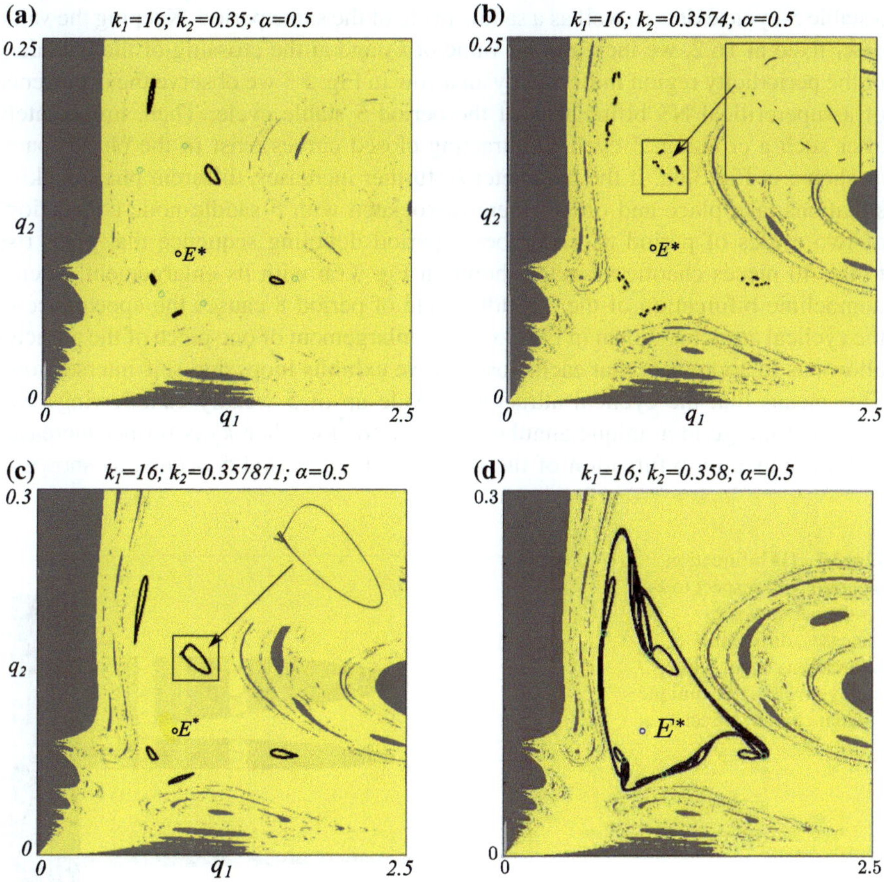

Fig. 3.6 A route to chaotic dynamics. The parameter values are chosen close to the periodicity region of the 5-cycle. The *green circles* denote the period 5 saddle cycle. **a** 5 cyclical attracting closed curves due to NS bifurcation. **b** 40-piece chaotic attractor due to a period doubling sequence. **c** Cyclical attractor made up of 5 weakly chaotic rings. **d** A unique annular chaotic attractor

the saddles and to their preimages. These particular basins' regions are similar to the ones that can be already observed in Fig. 3.5a, but they appear only in the portions of the phase-space that will become the basin of attraction of the closed curve (compare Fig. 3.5b, d).

3.3.3 A Route to Chaos

We now analyze the dynamics of the model when the parameters are chosen close to the tongue associated with a 2/5 attracting cycle, as highlighted through the red square in Fig. 3.3. This allows us to show that besides the period doubling sequence, the model exhibits also a different route to chaotic dynamics. In such a periodicity region, a stable 5-cycle exists as well as a saddle cycle of the same period. Keeping the value of k_1 fixed at 16.2, we increase the value of k_2 and at the crossing of the boundary of the periodicity region indicated by an arrow in Fig. 3.3 we observe the occurrence of a supercritical NS bifurcation of the period 5 stable cycle. Then, immediately after such a crossing, 5 cyclical attracting closed curves exist in the phase space, as shown in Fig. 3.6a. If the parameter k_2 further increases, different phase locking situations take place and one of them (associated with 5 saddle-node connections of two cycles of period 8) undergoes a period doubling sequence that gives rise to the 40 pieces chaotic attractor shown in Fig. 3.6b with its enlargement. Then a homoclinic bifurcation of the 5 saddle cycle of period 8 causes the appearance of the cyclical attractor shown in Fig. 3.6c; the enlargement of one piece of the attractor allows us to appreciate that each closed curve exhibits loops and self-intersections. This means that the cyclical attractor is made up of 5 weakly chaotic rings (see [15]) that merge in a unique annular chaotic attractor when k_2 is further increased and a homoclinic bifurcation of the saddle cycle of period 5 occurs, as shown in

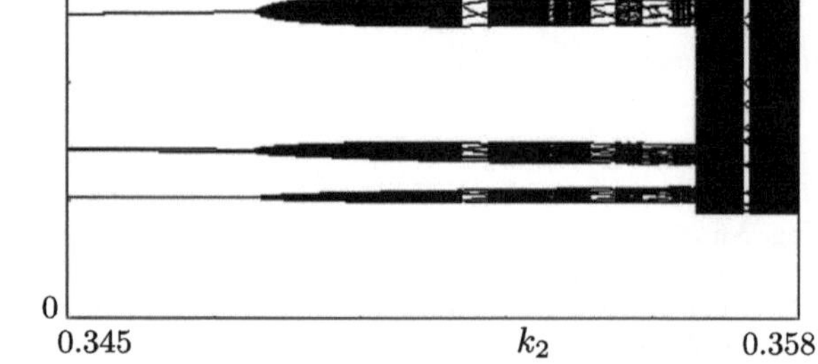

Fig. 3.7 1D bifurcation diagram with respect to k_2. As the parameter k_2 increases, different bifurcations occur which finally give rise to a unique annular chaotic attractor

Fig. 3.6d. The bifurcation diagram in Fig. 3.7 shows the latter bifurcation path and summarizes the sequence of bifurcations leading to the appearance of the annular chaotic attractor.

3.4 An Advertising Cournot Model: Subcritical NS Bifurcation

The second example we propose is a particular case of a Cournot triopoly where firms face markets in several countries. The basic features of this model are the unimodal reaction functions, obtained with the assumptions of isoelastic demand function, constant marginal costs, and an adaptive adjustment of the strategic variable.

The model is based on the following assumptions:

(i) there are three firms on the market that produce perfect substitute goods; they produce the commodity quantities q_i and distribute their products in several countries making use of x_i quantities of advertising ($i = 1, 2, 3$);
(ii) x_i is the strategic variable, to focus on the marketing issue; we disregard the production costs and c_i is the cost per unit of advertisement;
(iii) two of the competing firms produce the same commodity, equally behave in production and marketing policy, (i.e., $q_1 = q_3$ and $x_1 = x_3$) and have equal advertisement cost.

Further, we assume that the exponents of the Cobb-Douglas utility functions depend on advertising expenditures by the competitors, more precisely that the exponents are the shares of each firm in total advertising expenditures of all competitors. Then the consumers' demand is:

$$U = q_1^{\frac{x_1}{x_1+X_1}} q_2^{\frac{x_2}{x_2+X_2}} q_3^{\frac{x_3}{x_3+X_3}}, \tag{3.10}$$

where X_i, $i = 1, 2, 3$, denotes the advertising expenditure of competitors of firm i.

Therefore, utility maximizing consumer disposing of one monetary unit in the budget spends

$$p_i q_i = \frac{x_i}{x_i + X_i} \tag{3.11}$$

on each commodity. From the producer point of view, (3.11) represents the revenue of the firm. In this way, we can consider both the situation when the producers set prices or quantities, as this will not affect the results.

Finally, the optimization of profit of producer i leads to the reaction functions depending on the expected productions of competitors ($X_i^{(e)}$):

$$r_i(X_i^{(e)}) = \sqrt{\frac{X_i^{(e)}}{c_i}} - X_i^{(e)}. \tag{3.12}$$

From the economic point of view (3.12) only makes sense as long as $0 \leq X_i^{(e)} \leq 1/c_i$, otherwise reaction and profit become negative and the producer can decide either to withdraw or to modify his strategy.

To close the model, we assume that competitors move making use of the "adaptive expectation" mechanism. They give a weight $\theta \in [0, 1]$ to the best calculated reply for their competitors (3.12) and $(1-\theta)$ to their own previous move x_i. Thus any competitor moves according

$$x_i' = \begin{cases} \theta\left(\sqrt{\frac{X_i}{c_i}} - X_i\right) + (1-\theta)\,x_i, & 0 \leq X_i \leq \frac{1}{c_i}, \\ (1-\theta)\,x_i, & X_i > \frac{1}{c_i}, \end{cases} \tag{3.13}$$

with $i = 1, 2, 3$.

Now, considering assumption (iii) the model (3.13) becomes a 2D map with $X_1 = x_1 + x_2$, $X_2 = 2x_1$. Then the object of our study is the 2D nonlinear map:

$$T : \begin{cases} x' = T_1(x, y), \\ y' = T_2(x, y), \end{cases} \tag{3.14}$$

where

$$T_1(x, y) = \begin{cases} \theta\left(\sqrt{\frac{x+y}{c_1}} - (x+y)\right) + (1-\theta)x, & 0 \leq x + y \leq \frac{1}{c_1}, \\ (1-\theta)x, & x + y > \frac{1}{c_1}, \end{cases} \tag{3.15}$$

and

$$T_2(x, y) = \begin{cases} \theta\left(\sqrt{\frac{2x}{c2}} - 2x\right) + (1-\theta)y, & 0 \leq x \leq \frac{1}{2c_2}, \\ (1-\theta)y, & x > \frac{1}{2c_2}. \end{cases} \tag{3.16}$$

For a sake of simplicity in (3.14) we have denoted with x and y the phase variables x_1 and x_2.

To have meaningful trajectories from the economic point of view, we restrict the analysis of T to the positive orthant of R_2 (feasible region).

The map T in (3.14) depends on three parameters, the two marginal costs, $c_1 > 0$ and $c_2 > 0$, and the "adaptive" coefficient $\theta \in [0; 1]$. However a simple change of coordinates[5] allows us to show that only two parameters, $(c_2/c_1; \theta)$, are essential to study the dynamics of T. Then in the following, without loss of generality, we shall consider the map T with $c_1 = 1$ and the parameter c_2 has to be interpreted as the ratio of the marginal costs.

[5]The coordinate change is $\Psi(x, y) = (c_1 x, c_1 y)$.

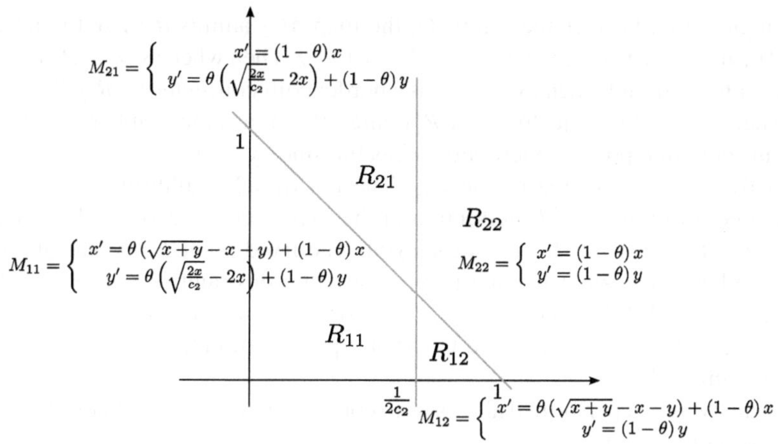

Fig. 3.8 The PW smooth map T. The different branches of the map T and related definition regions

The map T is a continuous piecewise map, then there exist four regions in which T assume a different definition. They are

$$R_{11} = \left\{(x_1, x_2) : 0 \le x_1 \le \frac{1}{2c_2}, 0 \le x_1 + x_2 \le 1\right\}, \tag{3.17}$$

$$R_{12} = \left\{(x_1, x_2) : x_1 > \frac{1}{2c_2}, 0 \le x_1 + x_2 \le 1\right\}, \tag{3.18}$$

$$R_{21} = \left\{(x_1, x_2) : 0 \le x_1 \le \frac{1}{2c_2}, x_1 + x_2 > 1\right\}, \tag{3.19}$$

$$R_{22} = \left\{(x_1, x_2) : x_1 > \frac{1}{2c_2}, x_1 + x_2 > 1\right\} \tag{3.20}$$

and are illustrated in Fig. 3.8.

3.4.1 Local Stability Analysis

To look at the fixed points of the map M in (3.13) we consider separately the four regions R_{ij} $(i, j = 1, 2)$ defined in (3.20). We shall denote with M_{ij} the map defined in R_{ij}.

The maps M_{22} and M_{21} admit a unique fixed point $O = (0, 0) \in R_{11}$, then they have a "virtual" steady state; moreover the eigenvalues of the Jacobian matrix are $\lambda_1 = \lambda_2 = 1 - \theta$. This means that any trajectory either in R_{22} or R_{21} leaves the regions in a finite number of steps.

In region R_{12}, besides the origin O, the map M_{12} admits the fixed point $P^* = (1/4, 0)$, always with eigenvalues $\lambda_1 = \lambda_2 = 1 - \theta$. Then when $c_2 > 2$, P^* is a stable equilibrium of M and when $c_2 = 2$ it has a border collision, entering R_{11}. This means that when $c_2 < 2$ the trajectories in R_{12} enter R_{11} in a finite number of steps and, from an economic point of view, are unfeasible since $y < 0$.

Finally, we consider the region R_{11}. The not trivial equilibrium of M_{11} is the Cournot equilibrium point (intersection of the two reaction curves (3.12)) given by $E^* = \left(2c_2/(2+c_2)^2, 2\,(2-c_2)/(2+c_2)^2\right)$. The fixed point E^* is feasible only if $c_2 \leq 2$ and, when feasible, belongs to R_{11}, since the constraint $x_1 + x_2 < 1$ is always satisfied by E^*. While, when $c_2 = 2$, E^* belongs to the constraint $x_1 = 1/2c_2$ and to $x_2 = 0$, and if c_2 increases the Cournot equilibrium enters the region R_{12} and becomes unfeasible.

The analysis just performed allows us to conclude that the map M admits a unique not trivial equilibrium:

1. The Cournot equilibrium E^* if $c_2 < 2$;
2. P^* if $c_2 > 2$, which is always stable.

When $c_2 = 2$ the two fixed points merge and belong to the border line separating regions R_{11} and R_{12}.

Henceforth, we restrict our analysis to the case $c_2 < 2$ to only consider the dynamic behaviors associated with the Cournot equilibrium point. The localization of the eigenvalues of the Jacobian matrix $J(E^*)$ of T evaluated at E^* allows us to state the following

Proposition 3.1 *Let* $(\theta, c_2) \in \Omega = [0, 1] \times (0, 2)$. *The fixed point*

$$E^* = \left(\frac{2c_2}{(c_2+2)^2}, \frac{2(2-c_2)}{(c_2+2)^2}\right)$$

is locally stable if

$$\theta < \theta_{ns} := \frac{2c_2(10-c_2)}{(c_2+2)^2}.$$

Proof See [3]. □

Moreover, it is possible to obtain that the eigenvalues of $J(E^*)$ are complex conjugated in $\widetilde{\Omega} = \left\{(\theta, c_2) : 0 < \theta < 1 \wedge 0 < c_2 < 13 - 3\sqrt{17}\right\}$. Following [11] (Theorem 3.5.2) it is possible to prove the following

Proposition 3.2 *If*

1. $(\theta, c_2) \in \widetilde{\Omega}$ *with* $c_2 \notin \left\{10 - 4\sqrt{6}, \bar{c}_2\right\}$, *where*

$$\bar{c}_2 = \frac{16}{3} + \frac{16\sqrt{7}}{3}\cos\frac{\arctan(\psi) + 2\pi}{3} \quad \textit{with} \quad \psi = \frac{27\sqrt{47}}{563}$$

2. *and* $P_6(c_2) < 0$, *where* $P_6(c_2) = 5c_2^6 + 120c_2^5 - 1372c_2^4 + 4224c_2^3 - 4752c_2^2 + 2176c_2 - 320$,

then at

$$\theta = \theta_{ns} = \frac{2c_2(10 - c_2)}{(c_2 + 2)^2} \tag{3.21}$$

the fixed point E^* *undergoes a subcritical Neimark-Sacker bifurcation.*

Proof Proposition 3.1 states that if $\theta < \theta_{ns}$ the fixed point is a stable focus (the two eigenvalues forming a complex conjugated pair) and it becomes an unstable focus when θ exceeds θ_{ns}. Then the complex eigenvalues of $J(E^*)$ have modulus one when (3.21) holds. Moreover, it is possible to verify that if $c_2 \notin \left\{10 - 4\sqrt{6},\ \bar{c}_2\right\}$ then $\lambda^n \neq 1$, $n = 1, 2, 3, 4$. Thus strong resonance cases are excluded. Finally, computing the coefficients d and a of Theorem 3.5.2 in [11, p. 162], we obtain $d = (10 - c_2)/8 > 0$ and $a > 0$ if $P_6(c_2) < 0$.[6] This proves that a subcritical NS bifurcation takes place at $\theta = \theta_{ns}$. □

From Proposition 3.2, we also obtain that the parameter value $(\theta_{ns}(\bar{c}_2), \bar{c}_2) \in \widetilde{\Omega}$ corresponds to a $1:3$ resonant case and $\left(\theta_{ns}\left(10 - 4\sqrt{6}\right), 10 - 4\sqrt{6}\right) \in \widetilde{\Omega}$ to a $1:4$ resonant case. This means that at these parameter values the closed invariant curve might appear in a very peculiar way, or there might be several invariant curves bifurcating from the fixed point.

Numerical investigation allows us to find out that condition (b) holds if $c_2 < \hat{c}_{ch}$, with $\hat{c}_{ch} \approx 0.2769$. The parameter values $\hat{C} = \left(\theta_{ns}\left(\hat{c}_{ch}\right), \hat{c}_{ch}\right)$ correspond to a Chenciner point. This means that in the parameter space $\widetilde{\Omega}$ a (global) bifurcation curve γ originates from $\hat{C}$ and enters the region in which E^* is stable. The crossing of such a curve causes the appearance of two invariant closed curves, one attracting and one repelling, the latter being involved in the occurring NS bifurcation. The study of the global bifurcations occurring along the curve γ is our present aim.

As in Sect. 3.3, we start our analysis looking at a 2D-bifurcation diagram, shown in Fig. 3.9. In such a figure we can observe that the Arnold's tongues have a particular "sausage" shape, typical of PW maps. Indeed, the boundaries of these periodicity regions are given by BCB's and their "narrowed" portions correspond to the crossing of one periodic point through a boundary of a region R_{ij}, associated with a border collision which preserves the existing attractor.

Comparing Fig. 3.9 with Fig. 3.3 a second peculiarity emerges: in the former figure the Arnold tongues appear when the fixed point is still stable and intersect the θ_{ns} bifurcation curve. As we said in Sect. 3.2, this is due to the Neimark-Sacker bifurcation (NS) of subcritical type, as we proved, and indicates that a pair of cycles must appear in the phase-space and coexists with the stable fixed point E^* (see also [4]).

---

[6] For major details see [3].

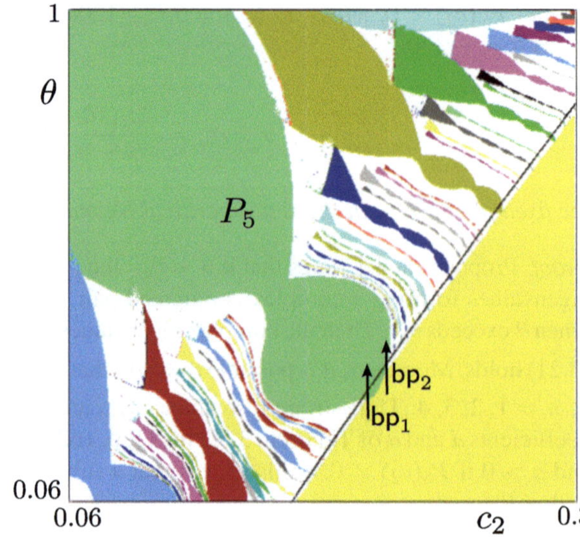

Fig. 3.9 2D bifurcation diagram. The Arnold's tongues of the PW map T have the "sausage" structure

To investigate the bifurcation mechanisms leading to the appearance of the invariant closed curves, at the crossing of the curve γ, we follow the bifurcation paths indicated with an arrow in Fig. 3.9, since simple cycles of period 5 will be involved.

3.4.2 Appearance of Curves Due to Homoclinic Bifurcation

We start considering the bifurcation path bp_1 in which $c_2 = 0.1306$. At the crossing of the boundary of the region P_5 a pair of cycles of period 5 appears, a saddle cycle S and an attracting one C. As shown in Fig. 3.10a, at their appearance the two cycles are very close to each other and close to the line separating region R_{11} from region R_{21}. This suggests that the appearance of the cycles may be due to a "saddle-node" BCB.

The basins of attraction of the two coexisting attractors are separated by the stable set $W^s(S) = \omega_1 \cup \omega_2$ of the saddle cycle S and no invariant curve exists immediately after the occurrence of the BCB. Indeed, the branch α_1 of the unstable set $W^u(S) = \alpha_1 \cup \alpha_2$ of S goes to E^* while α_2 converges to the cycle C. Increasing the parameter θ the stable branch ω_1 approaches the unstable branch α_1, as we can observe in Fig. 3.10b. More precisely, if we consider separately the periodic points S_s, with $s = 1, \ldots, 5$, of the saddle cycle S and $\omega_i = \bigcup_{s=1}^{5} \omega_{i,s}$, $\alpha_i = \bigcup_{s=1}^{5} \alpha_{i,s}$ with $i = 1, 2$, we have that $\omega_{1,s}$ is closer and closer to $\alpha_{1,s+1}$ ($s = 1, \ldots, 5$ and $\alpha_{1,6}$=$\alpha_{1,1}$). This means that a heteroclinic connection (or "heteroclinic loop") between the periodic points of the saddle cycle may be near to occur. This seems exactly the same situation described in Sect. 3.2. Indeed, if we slightly increase θ as in Fig. 3.10c (passing from 0.5635 to 0.5636), we observe that two invariant closed curves appear,

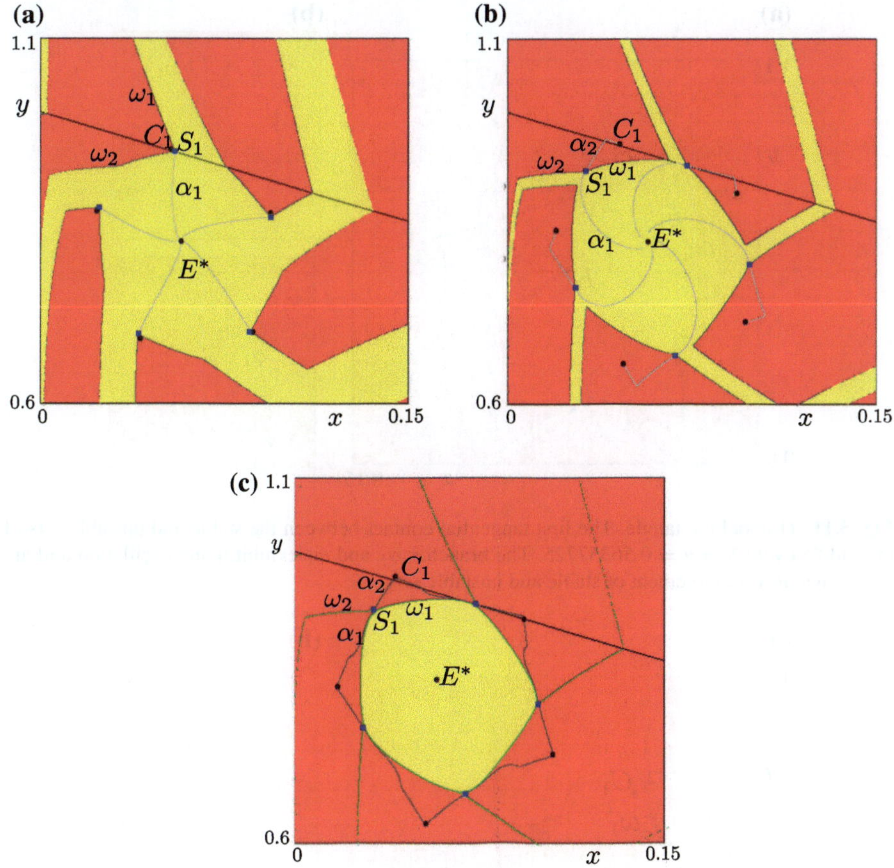

Fig. 3.10 Appearance of two invariant closed curves. Along the bifurcation path bp_1 a homoclinic bifurcation occurs. **a** $\theta = 0.562683$. The cycle of period 5 appears via BCB. **b** $\theta = 0.5635$. No closed curves exist. **c** $\theta = 0.5636$. Two invariant closed curves appear

one stable made up by the unstable set $W^u(S)$ which connects the periodic points of C and one unstable which bounds the basin of attraction of E^*. This means that the curve γ associated with the Chenciner point has been crossed.

To confirm that a homoclinic bifurcation occurs, we propose two further figures. Figure 3.11 represents the phase-plane immediately before the bifurcation: no invariant closed curve exists but we can observe that both stable and unstable set are wandering and quite tangent (see the enlargements in Fig. 3.11b). Then a homoclinic tangle start to develop and transverse crossing between the two invariant sets arise in a small parameter range. At the closure of the homoclinic tangle, the branches α_1 and ω_1 are again tangent, but at the opposite side, as Fig. 3.12 shows, and two invariant curves appear. To sum up, we remark that along the bifurcation curve bp_1

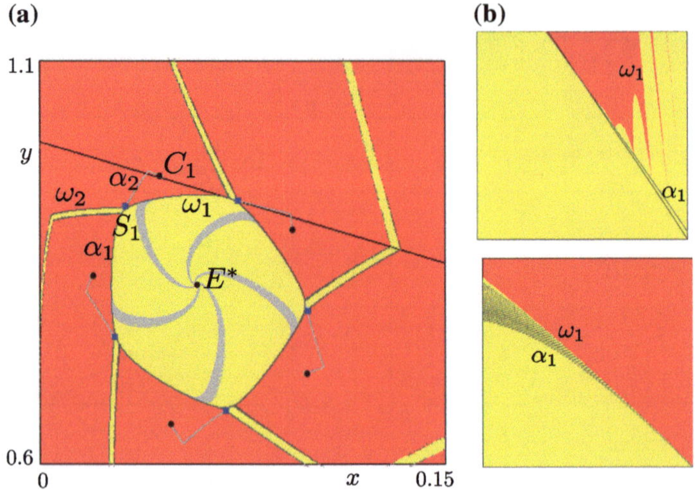

Fig. 3.11 Homoclinic tangle. The first tangential contact between the stable and unstable sets of the saddle cycle S. **a** $\theta = 0.56357725$. The branches α_1 and ω_1 exhibit many oscillation and are quite tangent. **b** Enlargement of stable and unstable sets

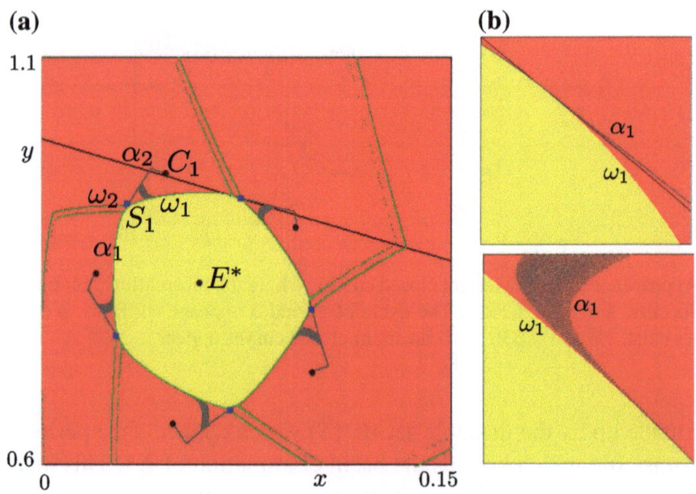

Fig. 3.12 Homoclinic tangle. The second tangential contact between the stable and unstable sets of the saddle cycle S. **a** $\theta = 0.56357729$. The branches α_1 and ω_1 have exchanged their mutual position and two invariant closed curves exist, very close each other. **b** Enlargement of the stable and unstable sets

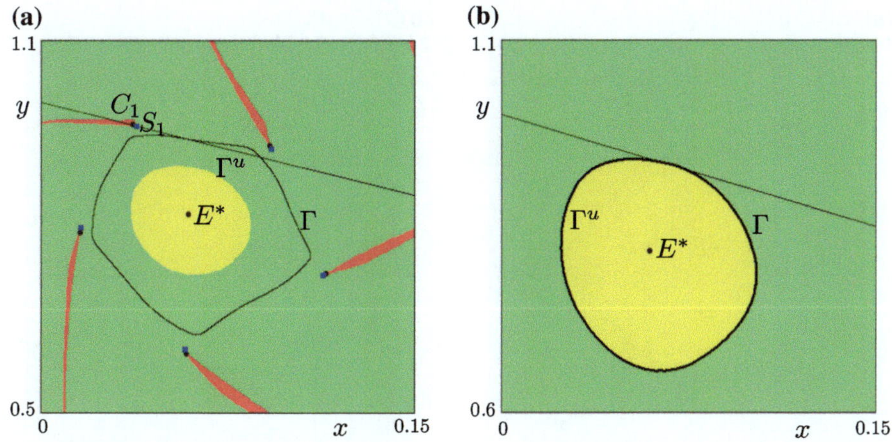

Fig. 3.13 Appearance of two invariant closed curves. Along the bifurcation path bp_2 a BCB causes the appearance of two invariant closed curves. **a** $\theta = 0.5825$. Three coexisting attractors. **b** $\theta = 0.5795$. Appearance of the closed invariant curves

the crossing of the curve γ (that is, the appearance of two invariant closed curves) occurs when a homoclinic bifurcation takes place.

3.4.3 Appearance of Curves Due to BCB

A different bifurcation mechanism arises along the bifurcation path bp_2 where $c_2 = 0.135$. Indeed, as shown in Fig. 3.13a, when the two cycles of period 5 appear, at $\theta = 0.5825$, the two curves Γ and Γ^u already exist. The bifurcation leading to the appearance of C and S seems again a "saddle-node" BCB and after its occurrence we have the coexistence of three attractors: the period 5 cycle C whose basin of attraction is bounded by the stable set of the saddle cycle S, the attracting closed curve Γ and the Cournot equilibrium E^*. A repelling closed curve Γ^u separates the basins of attraction of these two last attractors. This means that the bifurcation curve γ do not intersect the periodicity region P_5 along bp_2.

To investigate when the two invariant closed curves appear we decrease the parameter θ and we find that at $\theta = 0.5795$ the two curves are quite indistinguishable (see Fig. 3.13b) and the attracting one appears quite tangent to the border separating region R_{11} from region R_{21}. This suggests that a BCB can be the cause of the appearance of Γ and Γ^u.

To show that this is really the case we consider the map M_{11}, since in the parameter plane under scrutiny the Cournot equilibrium point is a fixed point of it.

The global analysis of the map M_{11} allows us to show that when being stable the fixed point E^* has a basin of attraction which contains all the feasible trajectories. If we consider $\theta = 0.579$, very closed to the value previously identified, the basin of attraction of E^* is bounded by a repelling closed curve Γ^u (see Fig. 3.14a) an its very small portion (denoted with a white arrow) belongs to region R_{21}. This implies

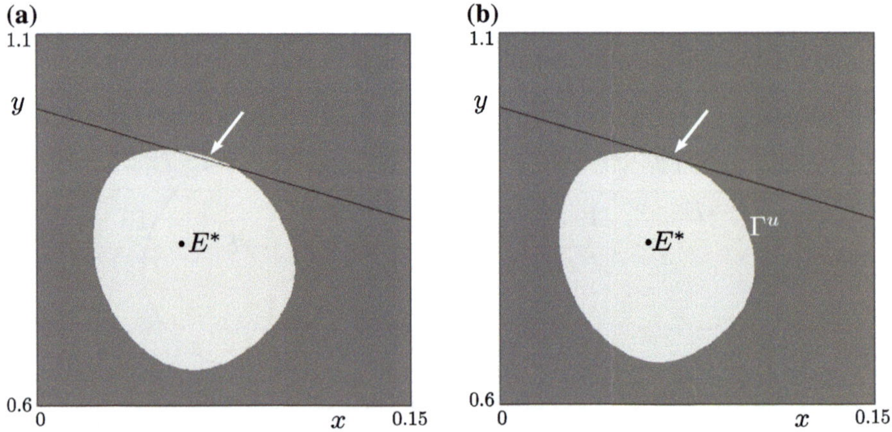

Fig. 3.14 The map M_{11}. A repelling closed curve bounds the set of feasible trajectories. **a** $\theta = 0.579$. The basin of attraction of E^*. **b** $\theta = 0.5795$. Border collision of an invariant closed curve

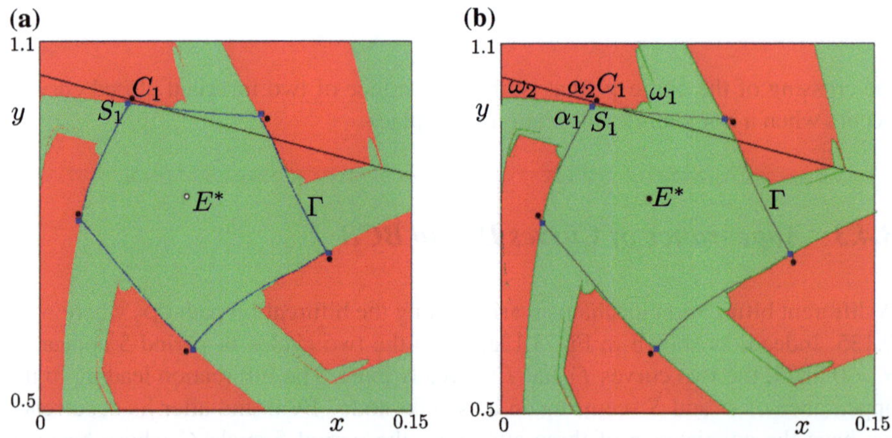

Fig. 3.15 After the occurrence of the Neimark-Sacker bifurcation. At $\theta = 0.5895$ only two attractors survive. Their basins of attraction are separated by the stable set of S. **a** The attracting closed curve coexists with the period 5 cycle C. **b** The unstable and stable sets of the saddle cycle S

that when we consider the map M such portion will be iterated with the map M_{21} and consequently all the unfeasible points disappear from the phase space of map M.

Coming back to the map M_{11}, when θ increases this portion becomes smaller and smaller and finally, at $\theta = 0.5795$ it disappears (see Fig. 3.14b). When the portion disappears, the boundary of the set of feasible trajectories of M_{11} becomes tangent to the border line $x_1 + x_2 = 1$. This means that Γ^u is an invariant set of M_{11} belonging to the region R_{11} and then is an invariant set also of the map M. Moreover the points of region R_{21}, all unfeasible for M_{11}, have a different behavior when iterated by M, that is a further attractor has to be appear. This is exactly what we have observed in Fig. 3.13b. Then we can conclude that along the bifurcation path bp_2 a different bifurcation mechanism occurs when the curve γ is crossed, since a BCB has caused the appearance of two invariant closed curves.

Nevertheless, a homoclinic bifurcation occurs even along bp_2. Increasing the parameter θ, firstly we observe the occurrence of the subcritical NS bifurcation. Then E^* becomes an unstable focus and only two attractors survive, the period 5 cycle C and the closed curve Γ, as in Fig. 3.15a. The basins of attraction are still separated by the stable set $W^s(S)$ of the saddle S. But, as we can observe in Fig. 3.15b, the branch ω_1 of $W^s(S)$ exhibits some fluctuations before converging to S and it is very close to the branch α_1 of $W^u(S)$. As we have seen above, this is exactly the prelude of the occurrence of a homoclinic bifurcation. Indeed, if we slightly increase the parameter θ the branches ω_1 and α_1 change their reciprocal position and, in particular, ω_1 now come from E^* (see Fig. 3.16). Then a homoclinic bifurcation has occurred and it has

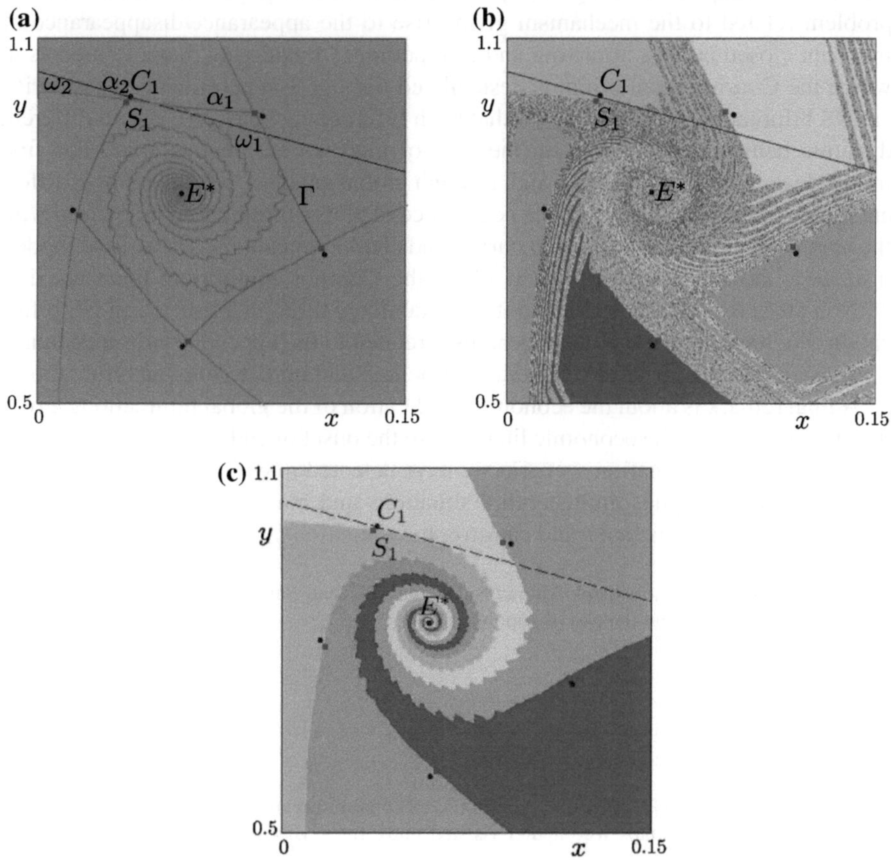

Fig. 3.16 The 5-cycle C is the unique attractor. A homoclinic bifurcation has caused the disappearance of the attracting closed curve Γ. **a** $\theta = 58995$. The closed curve Γ has disappeared. **b** $\theta = 0.5898$. The basins of attraction of the 5 fixed points of the map M^5 are strongly mixed. **c** $\theta = 58995$. The complex structure of the basins of attraction does not exist

caused the disappearance of the attracting closed curve Γ. To illustrate the effect of the occurring homoclinic bifurcation we consider the 5th iterate of the map M. The map M^5 undergoes five simultaneous homoclinic bifurcations, as proved by the fractal structure of the basins of attraction of its stable fixed points C_s, $s = 1, \ldots, 5$, due to the presence of a chaotic repellor (see Fig. 3.16b). This particular structure of the basins immediately disappears when the second tangential contact between ω_1 and α_1 occurs, as shown in Fig. 3.16.

3.5 Conclusions

Several studies have been devoted to the NS bifurcations of fixed points in two-dimensional maps describing oligopoly models. In this paper we focused on the problem related to the mechanism giving rise to the appearance/disappearance of invariant closed curves, attracting and/or repelling. Considering duopoly models in which the Cournot equilibrium is destabilized through a supercritical or a subcritical NS bifurcation, we have shown that such bifurcations may give rise to different dynamic behavior, depending on the type of map we are dealing with. The first model deals with a smooth map and, through global analysis, we have shown different multistability situations and the occurrence of global bifurcations associated with the appearance of invariant closed curves and chaotic dynamics. The second model, instead, is described by a PW map where the Cournot equilibrium point coexists with an attracting closed curve and it is destabilized through a subcritical NS bifurcation. We have shown that the mechanism related to the appearance/disappearance of invariant closed curves is related to homoclinic and border collision bifurcations.

A final remark is about the economic implication of the global bifurcations: given the attention paid in the economic literature to the onset of endogenous and long-run fluctuations, the bifurcation scenario we have detected may find important applications. Indeed, it implies multistability situations and may deserve to explain phenomena like hysteresis loops and catastrophic transitions.

Acknowledgments Authors thank Ahmad Naimzada and Tönu Puu for their many valuable suggestions and remarks about the two oligopoly settings.

References

1. Agliari, A., Bischi, G.I., Dieci, R., Gardini, L.: Global bifurcations of closed invariant curves in two-dimensional maps: a computer assisted study. Int. J. Bifurcat. Chaos **15**(4), 1285–1328 (2005)
2. Agliari, A., Naimzada, A., Pecora, N.: Nonlinear dynamics of a Cournot duopoly game with differentiated products. Appl. Math. Comput. (2016)
3. Agliari, A., Puu, T., Szuz, A.: The dynamics of a particular triopoly game when two competitors become identical. Mimeo

4. Agliari, A., Gardini, L., Puu, T.: Global bifurcation in duopoly when the Cournot point is destabilized via a subcritical neimark bifurcation. Int. Game Theory Rev. **8**(1), 1–20 (2006)
5. Ahmed, E., Elsadany, A.A., Puu, T.: On bertrand duopoly game with differentiated goods. Appl. Math. Comput. **251**, 169–179 (2015)
6. Boyland, P.L.: Bifurcations of circle maps: Arnol'd tongues, bistability and rotation intervals. Commun. Math. Phys. **106**, 353–381 (1986)
7. Chenciner, A.: Bifurcations de points fixes elliptiques : III—orbites pèriodiques de "petites" pèeriodes et èlimination rèsonnante des couples de courbes invariantes. Publications de l'I.H.E.S. **66**(1), 5–91 (1987)
8. Gardini, L., Puu, T., Sushko, I.: The hicksian floor-roof model for two regions linked by interregional trade. Chaos Solitons Fractals **18**, 593–612 (2003)
9. Gardini, L., Sushko, I.: Doubling bifurcation of a closed invariant curve in 3d maps. ESAIM: Proc. **36**, 180–188 (2012)
10. Guckenheimer, J., Holmes, P.: Nonlinear Oscillations, Dynamical Systems, and Bifurcations of Vector Fields, vol. 42. Springer (1983)
11. Guckenheimer, J., Holmes, P.: Nonlinear Oscillations and Dynamical Systems and and Bifurcations of Vector Fields. Springer, New York (1985)
12. Gumovsky, I., Mira, C.: Dynamique Chaotique: Transformations Ponctuelles. Transition Ordre - Désordre. Collection Nabla. Cépaduès Édition, Toulouse (1980)
13. Kuznetsov, Y.A.: Elements of Applied Bifurcation Theory. Springer (2013)
14. Martin, S.: Advanced Industrial Economics, 2 edn. Wiley-Blackwell (2002)
15. Mira, C., Gardini, L., Barugola, A., Cathala, J.C.: Chaotic Dynamics in Two-dimensional Noninvertible Maps. Nonlinear Science. World Scientific, Singapore (1996)
16. Nusse, H., Ott, E., Yorke, J.: Border-collision bifurcations: an explanation for observed bifurcation phenomena. Phys. Rev. E **49**, 1073–1076 (1994)
17. Nusse, H.E., Yorke, J.A.: Border-collision bifurcations including 'period two to period three' for piecewise smooth systems. Phys. D **57**, 39–57 (1992)
18. Nusse, H.E., Yorke, J.A.: Border-collision bifurcations for piecewise smooth one-dimensional maps. Int. J. Bifurcat. Chaos **5**(1), 189–207 (1995)

Chapter 4
Some Dynamical Models in Regional Economics: Economic Structure and Analytic Tools

Ingrid Kubin, Pasquale Commendatore and Iryna Sushko

Abstract This chapter aquaints the reader with the use of dynamic models in regional economics. The focus is on the New Economic Geography (NEG) approach. A brief comparison is provided between NEG and other economic approaches to investigate regional inequalities. The analytic structure of a general multi-regional model is described, and some simple examples are presented, where the number of regions is assumed to be small to obtain more easily analytic and numerical results. Tools from the mathematical theory of dynamical systems are drawn to study the qualitative properties of such multi-regional model.

4.1 Introduction

In the following, we introduce the reader to the use of dynamic models in regional economics. The focus will be on the New Economic Geography (NEG) approach initiated by Krugman in his seminal contribution, [23].

Section 4.1 sketches the basic mechanism inherent in NEG models and briefly compares it to other economic approaches to regional inequality; Sect. 4.2 develops the analytic structure of a general multi-regional NEG model and presents some

I. Kubin
Department of Economics, Institute for International Economics
and Development (WU Vienna University of Economics and Business),
1 Welthandelsplatz, 1020 Vienna, Austria
e-mail: ingrid.kubin@wu.ac.at

P. Commendatore
Department of Law, University of Naples Federico II,
16 Via Mezzocannone, 80134 Naples, NA, Italy
e-mail: commenda@unina.it

I. Sushko (✉)
Institute of Mathematics, National Academy of Sciences of Ukraine,
Tereshchenkivska str. 3, Kiev 01601, Ukraine
e-mail: sushko@imath.kiev.ua

G.I. Bischi et al. (eds.), *Qualitative Theory of Dynamical Systems, Tools and Applications for Economic Modelling*, Springer Proceedings in Complexity,
DOI 10.1007/978-3-319-33276-5_4

examples with a small number of regions (3 or 4); finally, Sect. 4.3 applies methods and tools from the dynamic systems theory to this model.

The central question in NEG is how the spatial distribution of economic activity looks like in the long-run. Will economic activity be equally distributed among regions, agglomerated in a few regions, or unevenly distributed over many regions. For illustrative purposes, Fig. 4.1 shows the gross domestic product per inhabitant

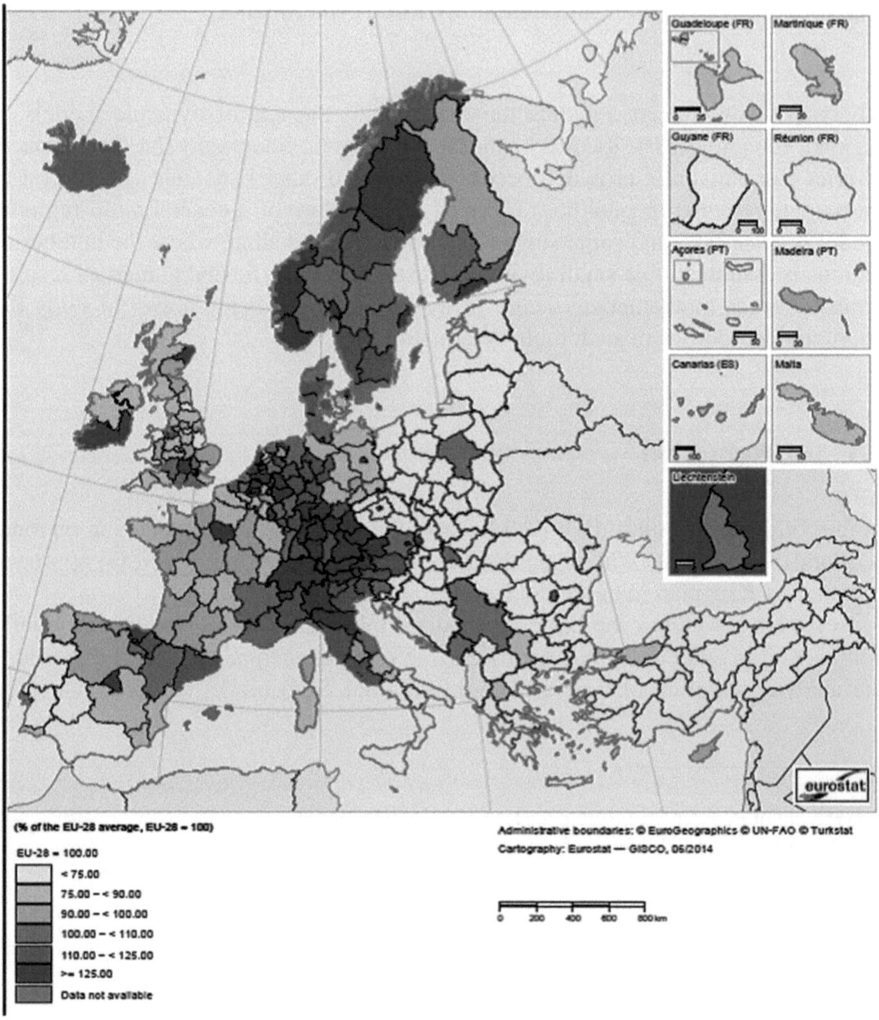

Fig. 4.1 Gross domestic product per inhabitant, in purchasing power standards by NUTS 2 regions 2011 (*Source* Eurostat)

for the European NUTS[1] 2 regions in 2011. It reveals quite distinct patterns with a high income level in the central regions and a low income level in the eastern and southern periphery. The NEG aims at providing explanations for the emergence of such regional core-periphery patterns.

Before going specifically to the NEG models, we sketch two other macroeconomic explanation patters also widely used in regional economics, namely a Solow growth model adapted for a multi-region setting and the Heckscher-Ohlin model applied to interregional trade.

The Solow growth model assumes that regions are endowed with a growing amount of productive factors (e.g., capital, labor and technical knowledge or human capital) that are used to produce one homogenous commodity. Markets are assumed to be perfectly competitive. Typically, even in a multi-regions setting, region are not much interrelated: Sometimes technological spillovers are accounted for; but in most cases the models do not allow for commodity trade or factor mobility. Regions may be different, because of specific growth rates of population and technical knowledge, because of specific "accumulation or saving rates" for physical and human capital, and because of specific distances to the respective steady state. Thus, the regional GDPs may grow at different rates and questions of regional convergence or divergence can be studied; many empirical studies are based on this theoretical framework (for further reading see, e.g., [1, 6, 25]).

In contrast, neoclassical models (in particular the Ricardian and Heckscher-Ohlin model) of international and interregional trade assume that regions are endowed with a given amount of productive factors (e.g., labor, capital, technical knowledge), but differentiate between several sectors (e.g., agriculture and manufacturing) that vary wrt productivity (Ricardian model) or factor intensity in production (Heckscher-Ohlin model). Again, all markets are assumed to be perfectly competitive. Regions are connected by commodity trade but not by factor mobility and regions differ wrt the technical knowledge (Ricardian model) or wrt the (relative) factor endowment (Heckscher-Ohlin model). In consequence of deeper trade integration, the model predicts that regions will specialize their production in a subset of the possible sectors—thus, regionally differentiated patterns of economic activity emerge (for further reading see, e.g., [21]).

Finally, most models of the New Economic Geography also assume that regions are endowed with a given amount of productive factors (e.g., labor of various skill levels) and differentiate between several sectors (e.g., agriculture and manufacturing). However, in contrast to the approaches presented above and crucial for the NEG models, not all markets are perfectly competitive, but some of them are monopolistically competitive (as a rationale for this market structure, a technology with fixed inputs is assumed for the sectors under consideration). In a NEG perspective, regions are not only connected by commodity trade, but also by factor mobility. Regions are assumed to be initially identical, in particular productive factors are equally distributed between regions. The main achievement of the New Economic Geography is to

[1] The Nomenclature of Territorial Units for Statistics.

show that even with initially identical regions deeper trade integration may trigger a self reinforcing process, which leads to agglomeration of economic activities in some of the regions (in most of the cases this process works via factor mobility). This approach, thus, provides an endogenous explanation for an uneven distribution of economic activity over space (while the approaches presented above have to recur to exogenously given differences).

Since the seminal contribution by Krugman [23] an entire family of models were developed, mainly differentiated according to which factor is assumed to mobile between the regions (for an overview see [5]). In the core-periphery model (originally developed in [23]), workers (and their consumption expenditures) are assumed to be mobile; in the footloose capital model (originally developed in [26]), firms and capital move simultaneously, but the capital owners (and their expenditures) do not move; finally, in the footloose entrepreneur model (originally developed in [22]) firms, capital and capital owners (the entrepreneurs and their expenditures) move between regions.

All NEG models share a similar mechanism for explaining agglomeration processes; in the following we provide a more detailed account of this process. We try to be fairly general; however, in some instances we have to base our exposition on one particular approach. We choose the footloose entrepreneur (FE) model, in which the productive factors are low and high qualified labor, the latter being the entrepreneurs who are mobile between regions. This choice is primarily driven by the fact that the FE model allows analytic solutions to a larger extent than other NEG models. Note in passing, that the assumed mobility pattern—higher mobility for firms, entrepreneurs and high qualified workers, and lower mobility for less qualified workers—also corresponds to stylized facts in the European Union.

What are the main ingredients in Krugman's cookbook for explaining endogenous agglomeration processes?

First, most NEG models assume that the manufacturing sector is characterized by Dixit Stiglitz monopolistic competition with iso-elastic demand functions. With this market form, profits are higher in the bigger market, or—to be more precise—in the market with higher expenditures for commodities; and prices are set as constant mark-ups on marginal costs (including transport cost). Typically, marginal costs are assumed to be constant. Second, NEG models explicitly specify several (at least two) locations and shipping commodities to another region involves transport costs (typically assumed to be of the iceberg type). Thus, the location of a firm matters: Although firms are selling to all markets, factor remunerations (profits) are higher if the local market (served without transport costs) is bigger. Third, NEG models assume that factors are mobile and move to a location where the remuneration is higher, or more precisely where the indirect utility is higher. The dynamic equation governing the factor mobility is at the core of every NEG model.

How can these three elements explain self reinforcing agglomeration processes? Agglomeration is a situation in which all mobile factors have moved to one region in the search of higher factor remuneration. The size of the local market for a single firm is pivotal; it depends on, first, the overall market size in the region, and second, on the share for a single firms and, thus, on number of other firms in the region.

Following Krugman's [23] well-known thought experiment, let us assume that one entrepreneur/firm moves from region two to region one. This relocation has two effects: Since the entrepreneur moves with her expenditure, the overall size of the market in region one increases, which leads to higher nominal profits and more firms are attracted to the region under consideration—this is the so-called market size effect that fosters agglomeration. At the same time, the number of firms in region one increases as well, which reduces the market share for a single firm and which, thus, tends to lower nominal profits, and fewer firms are attracted to the region under consideration—this is the so-called competition effect that fosters an equal regional factor distribution.

However, mobility of entrepreneurs is not driven by nominal, but by real profits, and thus, the price index has to be taken into consideration as well. With more firms having moved into one region, more variants are available locally, i.e., without transport costs—the price index is lower and more entrepreneurs/firms are attracted to the region under consideration. This is the so-called price index effect that works in same direction as the market size effect; it fosters agglomeration.

Krugman [23] introduced the so-called "Tomahawk" diagram (see Fig. 4.2) that summarizes for a two-region model how the spatial distribution of economic activity looks like in the long-run and how it depends upon trade integration. The abscissa shows trade freeness ϕ, an index that is zero for prohibitively high trade costs and one for zero trade cost permitting free trade (see Sect. 4.2 for an algebraic definition); the ordinate shows the share of firms locating in region one with a share of 0.5 indicating equal distribution and a share of one (zero) agglomeration in region one (two); dashed (solid) lines indicate unstable (stable) long-run equilibria. In many NEG models, the

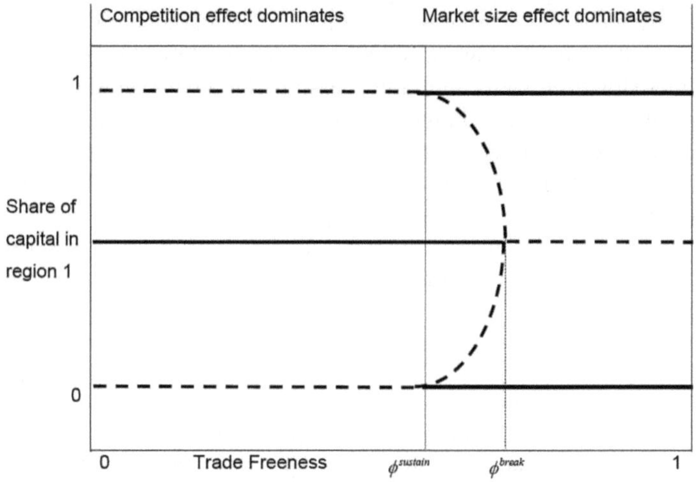

Fig. 4.2 Tomahawk diagram typically used in the NEG approach to identify the long-run equilibria and their local stability properties. It corresponds to a pitchfork bifurcation in the dynamic system theory of 1D nonlinear maps

symmetric equilibrium proofs to be stable (unstable) for low (high) values of trade freeness; with the opposite holding for the core-periphery equilibria that involve agglomeration in one region.

Three equilibria exist for the whole range of the trade freeness; however, the stability properties change in a remarkable way. Let us start in a situation with a low trade freeness and with an equal distribution of economic activity; intensifying interregional trade (which increases the trade freeness index) may trigger an agglomeration process that leads away from the symmetric equilibrium to one of the core-periphery equilibria—asymmetric patterns may emerge endogenously, once the agglomeration forces are stronger than the dispersion forces. This happens suddenly, in a "catastrophic" way, once the threshold of ϕ^{break} is trespassed. It is interesting to note that for a value ϕ above the break value, both core-periphery equilibria are (locally) stable; in which region the agglomeration will be actually found in the long-run depends upon initial conditions: if the initial industry share of region one is slightly above (below) 0.5, the endogenous agglomeration process will lead to an industrial core in region one (two). In this sense, history matters for the long run spatial distribution of economic activity. The last remarkable feature found for many NEG models is that for a value of ϕ between the so-called sustain and the so-called break value actually five equilibria coexist, three of which are (locally) stable. This introduces an element of irreversibility and hysteresis into agglomeration processes. Once ϕ^{break} is crossed, the symmetric equilibrium becomes (locally) unstable and the dynamic process leads to agglomeration in one of the regions. Reducing again the trade freeness, agglomeration persists until ϕ^{sustain} is crossed—thus, agglomeration processes are not exactly reversible and exhibit some degree of spatial hysteresis.

We finish this section by elaborating explicitly the relation between concepts from the NEG and the theory of dynamic systems.

First, where are dynamic processes involved? The market equilibria are instantaneously established; producing and shipping of commodities to other locations does not take time. However, factor mobility is assumed to evolve gradually over time—and the analytical core of NEG models is a dynamic equation for factor mobility.

Second, it is remarkable that the multiplicity of (core-periphery and interior) equilibria play a central role in the NEG paradigm (whereas elsewhere in economics unique equilibria are in the focus and multiplicity is something to be avoided).

Third, in NEG stories history matters, agglomeration processes happen in a catastrophic way and are at least to some extent irreversible. From a dynamic system's perspective this is the case, because multiple and (locally) stable equilibria coexist, each with a basin of attraction—in these circumstances, (historically given) initial conditions determine which of the coexisting equilibria will be attained in the long-run. It is interesting to note that in NEG models also unstable fixed points and boundary fixed points are important (in contrast to other economic approaches that often focus exclusively on stable interior fixed points). Therefore, a careful specification of boundary conditions is necessary (that may lead to piecewise defined models).

Fourth, note that the tomahawk diagram introduced above corresponds to a bifurcation diagram in dynamic system's theory.

Finally, we add a brief discussion on the temporal framework, i.e., whether the dynamic equation is specified in continuous or discrete time—a decision that matters because of differences in the implied dynamic phenomena. In contrast to the majority of NEG models, we set our models in discrete time on basis of the following observations: Economic processes, such as production and factor mobility typically involve lags that are not captured by a continuous time modeling, but easily captured in a discrete specification. Arguably, the "best" modeling might be a differential delay equation; however, that specification involves considerable analytic complexity. Simplifying to a pure differential equation loses dynamic aspects, whereas simplifying to a difference equation retains many of the dynamic phenomena. Therefore, we choose the latter approach.

The resulting analytic structure typically involves piecewise defined nonlinear difference equations; and we study cyclical and complex attractors, coexisting attractors and the (complex) structure of the basins of attraction.

4.2 A General Multi-regional Model

4.2.1 General Framework

We consider an economy composed of R regions (region r goes from 1 to R). There are two sectors, Agriculture A and Manufacturing M. While in the agricultural sector a homogeneous good is produced, manufacturing involves the production of N differentiated varieties (variety i goes from 1 to N). Finally, two types of agents exist: skilled workers (or entrepreneurs) E and unskilled workers (or simply workers) L, which are endowed with human capital and labor, respectively. Workers are immobile (but reallocation across sectors is possible), whereas entrepreneurs can move across regions. Regions could be grouped together into countries or trade blocs. We assume that there is no entrepreneurial migration across countries/trade blocs.

4.2.2 Consumers' Preferences

As in the standard footloose entrepreneur (FE) model, we assign Cobb-Douglas preferences over the choice between consumption of the agricultural good C_A and consumption of a composite of manufactured varieties C_M as

$$U = C_M^{\mu} C_A^{1-\mu}$$

and CES preferences across the manufactured varieties

$$C_M = \left(\sum_{i=1}^{N} c_i^{\frac{\sigma-1}{\sigma}} \right)^{\frac{\sigma}{\sigma-1}} ,$$

where c_i represents the consumption of the variety i, with $i = 1, \ldots, N$; σ the constant elasticity of substitution, with $\sigma > 1$ (the closer σ to 1, the lower is the degree of substitutability between varieties/the greater is the consumer's taste for variety); and μ and $1 - \mu$ the income shares devoted to the manufactured varieties and to the homogeneous agricultural good, respectively, with $0 < \mu < 1$.

The budget constraint of an individual (entrepreneur or worker) residing in region r is

$$\sum_{i=1}^{N} \tilde{p}_i c_i + p_A C_A = y ,$$

where p_A is the price of the agricultural good and $\tilde{p}_i$ is the price of the manufactured variety i at destination (inclusive of transport costs).

4.2.3 *Production*

The A sector is characterized by constant returns to scale and perfect competition. The production of 1 unit of output requires only 1 unit of L as input. The M sector, instead, is (Dixit-Stiglitz) monopolistically competitive involving increasing returns. It is modeled according to a few basic characteristics: firms with symmetric behavior produce differentiated varieties with the same production technology involving a fixed component, one entrepreneur, and a variable component, workers, with 1 unit of L required for each unit of the differentiated good (the assumption of a skilled labor input coefficient equal to 1 does not imply any loss of generality). Total cost for a firm i is

$$TC(q_i) = \pi_i + wq_i ,$$

where q_i is the output of firm i and w the wage rate and where the fixed cost component π_i represents the operating profit and the remuneration of the entrepreneur:

$$\pi_i = p_i q_i - wq_i ,$$

where p_i is the mill price or price at the origin (that disregards trade costs) fixed by firm i (see below). Given consumer's preference for variety and increasing returns, each firm will always produce a variety different from those produced by the other firms. Moreover, since one entrepreneur is required for each manufacturing firm (in the absence of economies of scope), the total number of firms/varieties, N, is always equal to the total number of entrepreneurs, $E = N$ (this is why we used the same subscript i for both). Denoting by $\lambda_{r,t}$ the share of entrepreneurs located in region r, the number of regional varieties produced in that period is

$$n_{r,t} = \lambda_{r,t} N = \lambda_{r,t} E \ .$$

4.2.4 Trade Costs

Trade between regions can be inhibited by transportation and/or tariffs (or non tariffs) barriers and/or other types of impediments/frictions. We use a broad definition of trade costs for manufactured commodities based on Samuelson's iceberg principle: denoting by T_{rs} the trade costs between region r and region s, with $T_{rs} \geq 0$, if one unit ships from r to s, then only $1/T_{rs}$ units arrive at destination. We also use the usual parameter transformation that introduces the concept of "trade freeness": $\phi_{rs} = T_{rs}^{1-\sigma}$, with $0 < \phi_{rs} \leq 1$, where $\phi_{rs} = 1$ implies no trade costs and $\phi_{rs} = 0$ impossibility or simply absence of trade. Moreover, we assume that no trade costs are incurred within each region, i.e., $T_{rr} = 1$. With R regions, we can construct the $R \times R$ trade cost matrix where for the entry T_{rs}, r represents the region of origin and s the region of destination of the commodity:

$$\mathbf{T} = \begin{bmatrix} 1 & T_{12} & \dots & T_{1R} \\ T_{21} & 1 & \cdots & T_{2R} \\ \vdots & & & \\ T_{R1} & T_{R2} & \cdots & 1 \end{bmatrix} .$$

By applying the transformation of iceberg trade costs into trade freeness, we can write

$$\mathbf{\Phi} = \begin{bmatrix} 1 & \phi_{12} & \dots & \phi_{1R} \\ \phi_{21} & 1 & \cdots & \phi_{2R} \\ \vdots & \vdots & \ddots & \vdots \\ \phi_{R1} & \phi_{R2} & \cdots & 1 \end{bmatrix} .$$

In the examples we give below, we assume symmetric trade costs, that is, they are the same in both directions $T_{rs} = T_{sr}$ and $\phi_{rs} = \phi_{sr}$. This implies also symmetric trade costs and trade freeness matrices.

Finally, we assume that no trade costs are incurred in the A-sector, implying perfect trade freeness for the agricultural product.

4.2.5 Short-Run General Equilibrium

A short-run general equilibrium (SRGE) is instantaneously realized within time t and it is defined for a given spatial distribution of entrepreneurs across the regions.

Considering the SRGE, we fix for time t the regional shares of entrepreneurs $(\lambda_{1,t}, \lambda_{2,t}, \ldots, \lambda_{R,t})$ and add the time subscript also in the definitions given above where necessary. In a SRGE all markets are cleared: supply equals demand for the agricultural good and each manufacturer meets the demand for its variety; and by Walras's law simultaneous equilibrium in the product markets implies equilibrium in the regional labor markets. With no trade costs, the price of the agricultural good is the same across regions; we set it equal to 1, representing the numéraire. From the agricultural market clearing condition, it follows $w = p_A = 1$.

Following profit maximization, firm i sets at the origin a price p_i on the basis of a perceived demand elasticity of $-\sigma$. Dropping the subscript i, under the assumption of symmetric firm behavior, we have

$$p = \frac{\sigma}{\sigma - 1} . \tag{4.1}$$

The overall demand for each variety (where we consider that the price of a variety produced in region r and sold in region s is $\widetilde{p} = pT_{rs}$) corresponds to

$$d_{r,t} = \left(\sum_{s=1}^{R} \mu Y_{s,t} P_{s,t}^{\sigma-1} T_{rs}^{1-\sigma} \right) p^{-\sigma} = \frac{\mu Y}{pE} \left(\sum_{s=1}^{R} \frac{Y_{s,t}}{\Delta_{s,t}} \phi_{rs} \right) , \tag{4.2}$$

where

$$P_{r,t} = \left[\sum_{s=1}^{R} \left(n_{s,t} p_{s,t} T_{rs} \right)^{1-\sigma} \right]^{\frac{1}{1-\sigma}} = \Delta_{r,t}^{\frac{1}{1-\sigma}} E^{\frac{1}{1-\sigma}} p$$

is the price index that consumers face in region r;

$$Y_{r,t} = L_r + n_{r,t} \pi_{r,t} \tag{4.3}$$

represents income and expenditure in region r; with L_r representing the (given) labor endowment of region r. Moreover we have defined

$$\Delta_{r,t} = \lambda_{1,t}\phi_{r1} + \lambda_{2,t}\phi_{r2} + \cdots + \lambda_{R,t}\phi_{rR} = \sum_{s=1}^{R} \lambda_{s,t}\phi_{rs} .$$

The entrepreneur remuneration in region r corresponds to

$$\pi_{r,t} = pq_{r,t} - q_{r,t} = \frac{pq_{r,t}}{\sigma} ,$$

where we obtain the latter term by using (4.1).

Since in equilibrium $d_{r,t} = q_{r,t}$, taking into account (4.2), the entrepreneur remuneration in region r can be expressed as follows:

$$\pi_{r,t} = \frac{\mu}{\sigma E}\left(\sum_{s=1}^{R} \frac{Y_{s,t}}{\Delta_{s,t}}\phi_{rs}\right).$$

After plugging (4.3) into the above expression and considering that $n_{r,t} = \lambda_{r,t}E$, we can write

$$\pi_{r,t} = \frac{\mu}{\sigma E}\left(\sum_{s=1}^{R} \frac{L_s + \lambda_{r,t}E\pi_{r,t}}{\Delta_{s,t}}\phi_{rs}\right). \tag{4.4}$$

Letting $r = 1, \ldots, R$, a system of R linear equations is generated from expression (4.4):

$$\begin{aligned}
\pi_{1,t} &= \frac{\mu}{\sigma E}\left(\sum_{s=1}^{R} \frac{L_s + \lambda_{1,t}E\pi_{1,t}}{\Delta_{s,t}}\phi_{1s}\right), \\
\pi_{2,t} &= \frac{\mu}{\sigma E}\left(\sum_{s=1}^{R} \frac{L_s + \lambda_{2,t}E\pi_{2,t}}{\Delta_{s,t}}\phi_{2s}\right), \\
&\vdots \\
\pi_{R,t} &= \frac{\mu}{\sigma E}\left(\sum_{s=1}^{R} \frac{L_s + \lambda_{R,t}E\pi_{R,t}}{\Delta_{s,t}}\phi_{Rs}\right).
\end{aligned} \tag{4.5}$$

This system can be expressed in matrix form as

$$\pi_t = \frac{\mu}{\sigma E}\mathbf{\Phi}\mathbf{C_t}\left(\mathbf{L} + E\mathbf{G}_t\pi_t\right), \tag{4.6}$$

where

$$\mathbf{C}_t = \begin{pmatrix} \frac{1}{\Delta_{1,t}} & 0 & \cdots & 0 \\ 0 & \frac{1}{\Delta_{2,t}} & \cdots & 0 \\ \vdots & \vdots & \ddots & \vdots \\ 0 & 0 & \cdots & \frac{1}{\Delta_{R,t}} \end{pmatrix}, \quad \mathbf{G}_t = \begin{pmatrix} \lambda_{1,t} & 0 & \cdots & 0 \\ 0 & \lambda_{2,t} & \cdots & 0 \\ \vdots & \vdots & \ddots & \vdots \\ 0 & 0 & \cdots & \lambda_{R,t} \end{pmatrix}, \quad \mathbf{L} = \begin{pmatrix} L_1 \\ L_2 \\ \vdots \\ L_R \end{pmatrix}.$$

The matrix **C** shows how the interaction of enterprise location and trade costs affects the regional price index; the matrix **G** gives the distribution of entrepreneurs across the R regions; and the vector **L** represents the regional labor endowments.

The solutions are

$$\pi_t^* = \begin{pmatrix} \pi_{1,t}^* \\ \pi_{2,t}^* \\ \vdots \\ \pi_{R,t}^* \end{pmatrix} = \left(\mathbf{I} - \frac{\mu}{\sigma}\mathbf{\Phi}\mathbf{C}_t\mathbf{G}_t\right)^{-1}\frac{\mu}{\sigma E}(\mathbf{\Phi}\mathbf{C}_t\mathbf{L}). \tag{4.7}$$

Finally $V_{r,t} = \pi_{r,t}/P^{\mu}_{r,t}$ is the real profit income perceived by an entrepreneur located in region r. As we shall see below, it represents a crucial variable in the entrepreneurial migration dynamics. Taking into account the above expressions, it can be formulated as a function of the entrepreneurial shares given in a SRGE: $V_{r,t} = V_{r,t}(\lambda_{1,t}, \lambda_{2,t}, \ldots, \lambda_{R,t})$.

4.2.6 From the Short-Run to the Long-Run

A shift from a short-run equilibrium t to the following $t+1$ occurs allowing for entrepreneurial migration based on an economic incentive. Specifically, our migration hypothesis involves a discrete time process centered on a comparison between the indirect utility obtained in region r and a weighted average of indirect utilities in all regions—a mechanism resembling the replicator dynamics:

$$\frac{M_{r,t+1} - \lambda_{r,t}}{\lambda_{r,t}} = \gamma_{rk}\left(\frac{V_{r,t} - \sum_{s=1}^{R}\lambda_{s,t}V_{s,t}}{\sum_{s=1}^{R}\lambda_{s,t}V_{s,t}}\right), \tag{4.8}$$

where $M_{r,t+1}$ denotes the share of entrepreneurs in region r disregarding boundary conditions; γ_{rs} is the speed of migration from region r to region k; $\gamma_{rk} \geq 0$, $r, k = 1, 2, \ldots, R$ and $r \neq k$; when $\gamma_{rk} = 0$ there is no migration from r to k. Moreover, we assume the same migration speed from region r to region k and from region k to region r, i.e., $\gamma_{rk} = \gamma_{kr}$. Finally, the boundary conditions on shares must hold: $0 \leq \lambda_{r,t} \leq 1$ and $\lambda_{1,t} + \lambda_{2,t} + \cdots + \lambda_{R,t} = 1$.

Letting $r = 1, \ldots R$, (4.8) generates an R-dimensional discrete time dynamic system. A long-run interior equilibrium fixed point of this system is defined by a vector $(\lambda_1^*, \lambda_2^*, \ldots, \lambda_R^*)$ such that

$$V_r(\lambda_1^*, \lambda_2^*, \ldots, \lambda_R^*) = \sum_{s=1}^{R} \lambda_s^* V_s(\lambda_1^*, \lambda_2^*, \ldots, \lambda_R^*)$$

for each r. Assuming regions with same labor endowments, i.e., letting $L_1 = L_2 = \cdots = L_R = L$, a fully symmetric equilibrium exists, corresponding to a uniform distribution of the economic activity across the regions:

$$\lambda_1^* = \lambda_2^* = \cdots = \lambda_R^* = \frac{1}{R}\,.$$

When $L_1 = L_2 = \cdots = L_R = L$, asymmetric equilibria—where the economic activity takes place in all the regions but is unevenly distributed—could also emerge that still enjoy some degree of symmetry; in the examples that we give below these equilibria exists in pairs and are positioned at the same distance from an invariant line.

Other stationary equilibria could emerge, taking into account the boundary conditions: when for at least one region r it holds $\lambda_r = 0$ (no manufacturing in that region), and for no region s it holds $\lambda_s = 1$, with $r \neq s$ and $r, s = 1, 2, \ldots, R$; we are in the presence of a border fixed point. The number of borders depending on the number of 0s in the vector of the stationary values: for example, $(\lambda_1^*, \ldots, 0, \ldots, \lambda_R^*)$ represents an equilibrium with one border, $(\lambda_1^*, \ldots, 0, \ldots, 0, \ldots, \lambda_R^*)$ an equilibrium with two borders, and so on. Again, letting $L_1 = L_2 = \cdots = L_R$, border symmetric equilibria may exist in correspondence of which the non-zero λ^*s are all equal to $1/(R-b)$, where b represents the number of borders. Asymmetric border equilibria also exist keeping some degree of symmetry: they exist in pairs and are positioned symmetrically with respect to an invariant line.

Core-periphery (CP) equilibria are stationary fixed points where for a region r it holds $\lambda_r = 1$, and for every other region s it holds $\lambda_s = 0$. In a CP equilibrium all the economic activity is agglomerated in one region. With no impediments to entrepreneurial migration, the number of existing CP equilibria corresponds to the number of regions, R.

Finally, when entrepreneurial migration between countries or trade blocs is not allowed, agglomeration-within-a-country could occur: that is, within a country (or within more than one) all the manufacture agglomerates only in one region. In this case border fixed points emerge in which, after a suitable renormalization of variables, also 1s appear in the vector of the stationary equilibrium solutions (see below).

We will not study here the local and global stability properties of the general system which includes (4.8) and the boundary conditions. We limit ourselves to the exploration of a few examples of models with a small number regions (3 or 4). In what follows, we introduce those models and in Sect. 4.3 we study their dynamic properties.

4.2.7 Example 1: A Model of a Custom Union in the Presence of an Outside Region

In [17, 18] we aimed to study the joint effect of regional integration and international liberalization. In those papers we put forward a model representing a three-region economy composed of a trade bloc (named the Union) and an outside region. In this model, the symmetric regions 1 and 2 are part of the Union, while region 3 is the outside trade partner. The mobile factor, entrepreneurs, moves only between regions 1 and 2, whereas the number of entrepreneurs located in region 3 is fixed. Denoting by $\widetilde{n}$ the share of entrepreneurs located in the trade block 1–2, $1-\widetilde{n}$ representing the share of entrepreneurs located in the outside region 3. This model can be derived from the more general one presented above by setting $R = 3$ and by changing suitably the main state variables. That is, we set

$$\lambda_{1,t} = x_t \widetilde{n}\ , \quad \lambda_{2,t} = (1-x_t)\widetilde{n}\ , \quad \lambda_{3,t} = 1-\widetilde{n},$$

where x_t and $1 - x_t$ represent the shares of entrepreneurs of the Union located in region 1 and in region 2, respectively. x_t is also the state variable of a one-dimensional dynamic model governing entrepreneurial mobility between regions 1 and 2 (see below). The boundary conditions become $0 \leq \lambda_{1,t} \leq \widetilde{n}$ and $\lambda_{1,t} + \lambda_{2,t} = \widetilde{n}$ or, equivalently, $0 \leq x_t \leq 1$. In [17] , in order to derive some analytic results, we considered the special case $\widetilde{n} = 1$, that is, in region 3 only the agricultural good is produced.

Trade costs take into account the stronger integration between 1 and 2. Moreover, in [17, 18] it is assumed that the trade distance between regions 1 and 3 and regions 2 and 3 is the same. Therefore, we set $T_{12} = T_S$, $T_{13} = T_{23} = T_L$ and $T_S < T_L$. The trade cost and trade freeness matrices become

$$\mathbf{T} = \begin{bmatrix} 1 & T_S & T_L \\ T_S & 1 & T_L \\ T_L & T_L & 1 \end{bmatrix} , \quad \mathbf{\Phi} = \begin{bmatrix} 1 & \phi_S & \phi_L \\ \phi_S & 1 & \phi_L \\ \phi_L & \phi_L & 1 \end{bmatrix} . \tag{4.9}$$

Another crucial difference between regions 1 and 2 and region 3 is the size difference: denoting by L the overall endowment of the three-region economy and with 2θ the share of workers living in the Union, we have: $L_1 = L_2 = \theta L$ and $L_3 = 1 - 2\theta L$.

For the case $\widetilde{n} < 1$, in order to find the solutions of the system (4.5), we have to substitute into (4.6) the trade freeness matrix reported in (4.9) and the following expressions:

$$\mathbf{C}_t = \begin{pmatrix} \frac{1}{\Delta_{1,t}} & 0 & 0 \\ 0 & \frac{1}{\Delta_{2,t}} & 0 \\ 0 & 0 & \frac{1}{\Delta_{3,t}} \end{pmatrix} , \quad \mathbf{G}_t = \begin{pmatrix} x_t\widetilde{n} & 0 & 0 \\ 0 & (1 - x_t)\widetilde{n} & 0 \\ 0 & 0 & 1 - \widetilde{n} \end{pmatrix} , \quad \mathbf{L} = \begin{pmatrix} \theta L \\ \theta L \\ 1 - 2\theta L \end{pmatrix} ,$$

where

$$\begin{aligned} \Delta_{1,t} &= x_t\widetilde{n} + (1 - x_t)\widetilde{n}\phi_S + (1 - \widetilde{n})\,\phi_L \ , \\ \Delta_{2,t} &= x_t\widetilde{n}\phi_S + (1 - x_t)\widetilde{n} + (1 - \widetilde{n})\,\phi_L \ , \\ \Delta_3 &= \widetilde{n}\phi_L + 1 - \widetilde{n} \ . \end{aligned}$$

For the case $\widetilde{n} = 1$, no profits are generated in region 3 and the vector of the solutions π_t^* only involves two entries.

Entrepreneurs only migrate between region 1 and region 2. After imposing $R = 3$, $\gamma_{12} = \gamma$ and $\gamma_{13} = \gamma_{23} = 0$, the migration process (4.8)—driven by entrepreneurial regional profitability—is reduced to a one-dimensional discrete time equation governed by the state variable x_t corresponding to

$$\frac{x_{t+1}^u - x_t}{x_t} = \gamma \left(\frac{V_{1,t} - \left[x_t V_{1,t} + (1 - x_t)\, V_{2,t} \right]}{x_t V_{1,t} + (1 - x_t)\, V_{2,t}} \right) ,$$

where γ is the migration speed between regions 1 and 2 (after dropping subscripts). Expressing x^u_{t+1} as a function of x_t, say $x^u_{t+1} = Z(x_t)$, one gets the unconstrained share of entrepreneurs of the Union located in region 1. Taking into account the constraints, $0 \leq x_t \leq 1$, the full dynamic model corresponds to the map

$$x_{t+1} = f(x_t) = \begin{cases} 0 & \text{if} \quad Z(x_t) < 0 \,, \\ Z(x_t) & \text{if} \quad 0 \leq Z(x_t) \leq 1 \,, \\ 1 & \text{if} \quad Z(x_t) > 1 \,. \end{cases} \tag{4.10}$$

Notice that, as long as $\widetilde{n} < 1$, agglomeration of the industrial sector can only occur within the Union. The size of the industry located in region 3 does not vary. We study the fixed points of the map (4.10) and its local and global dynamics in Sect. 4.3.3.

4.2.8 Example 2: A Two-Country Four-Region Model

4.2.8.1 The Case of Symmetric Regions

The focus of the model put forward in [16] are the dynamic processes governing a two-country four-region economy. In that paper, we introduced a specific geographical set-up: there are two countries/trade blocs, each of them composed of two regions; regions 1 and 2 are part of the first bloc (home, h) and regions 3 and 4 are part of the second bloc (foreign, f). The four regions are aligned from 1 to 4: thus, regions 2 and 3, which share a common border, enjoy a central position, whereas the position of regions 1 and 4 is marginal. One of the aims of the paper is indeed to study the issue of centrality vis-à-vis marginality, taking into account that central regions access larger markets whereas marginal regions are sheltered from competition. The border inhibits entrepreneurial migration; thus, mobility is allowed only within trade blocs between regions 1 and 2 and regions 3 and 4. The two blocs have the same endowment of entrepreneurs. Letting E the overall endowment of this factor of production, $E/2$ is located in each bloc. Moreover, each region has the same number of workers implying the same local market size. Letting L, the overall endowment of labor, $L/4$ is located in each region. As before, this model can be derived from the more general one presented above by setting $R = 4$ and by changing suitably the main state variables. That is, we set

$$\lambda_{1,t} = \frac{1 - x_t}{2} \,, \quad \lambda_{2,t} = \frac{x_t}{2} \,, \quad \lambda_{3,t} = \frac{y_t}{2} \,, \quad \lambda_{4,t} = \frac{1 - y_t}{2} \,,$$

where x_t and $1 - x_t$ are the shares of home entrepreneurs located in region 1 and region 2, respectively; and y_t and $1 - y_t$ are the shares of foreign entrepreneurs located in region 3 and region 4, respectively. x_t and y_t are also the state variables of a two-dimensional discrete time dynamic system governing entrepreneurial migration within home and foreign (see below). The boundary conditions become: $0 \leq \lambda_{1,t} \leq 1/2$, $\lambda_{1,t} + \lambda_{2,t} = 1/2$, $0 \leq \lambda_{3,t} \leq 1/2$ and $\lambda_{3,t} + \lambda_{4,t} = 1/2$ or, equivalently, $0 \leq x_t \leq 1$ and $0 \leq y_t \leq 1$. Notice that the one we have chosen is not the only possible choice of variables, x_t and y_t are related to the shares of entrepreneurs located in regions 2 and 3 to better stress the symmetric properties of the map (4.12). Other choices are possible serving a different purpose (see Sect. 4.2.8.2).

In [16] only direct trade is possible, that is, only contiguous regions can trade with each other. Therefore, taking also into account the specific geographical features of the economy, we set $T_{12} = T_{34} = T$ and $T_{23} = T_E$, where T represents domestic trade costs and T_E external trade costs. Given the absence of indirect trade—which implies infinite trade costs between noncontiguos regions,—we are able to fill the trade freeness matrix by setting $\phi = T^{1-\sigma}$, $\phi_E = T_E^{1-\sigma}$ and $\phi_{13} = \phi_{14} = \phi_{24} = 0$:

$$\mathbf{\Phi} = \begin{bmatrix} 1 & \phi & 0 & 0 \\ \phi & 1 & \phi_E & 0 \\ 0 & \phi_E & 1 & \phi \\ 0 & 0 & \phi & 1 \end{bmatrix} . \tag{4.11}$$

Given the above variables change and the assumptions on factor endowments and trade costs, in order to find the solutions of the system (4.5), we have to substitute into (4.6) the trade freeness matrix reported in (4.11) and the following expressions[2]:

$$\mathbf{C}_t = \begin{pmatrix} \frac{1}{\Delta_{1,t}} & 0 & 0 & 0 \\ 0 & \frac{1}{\Delta_{2,t}} & 0 & 0 \\ 0 & 0 & \frac{1}{\Delta_{3,t}} & 0 \\ 0 & 0 & 0 & \frac{1}{\Delta_{4,t}} \end{pmatrix}, \quad \mathbf{G}_t = \frac{1}{2}\begin{pmatrix} \frac{1-x_t}{2} & 0 & 0 & 0 \\ 0 & \frac{x_t}{2} & 0 & 0 \\ 0 & 0 & \frac{y_t}{2} & 0 \\ 0 & 0 & 0 & \frac{1-y_t}{2} \end{pmatrix}, \quad \mathbf{L} = \frac{L}{4}\begin{pmatrix} 1 \\ 1 \\ 1 \\ 1 \end{pmatrix},$$

where

$$\Delta_{1,t} = \frac{1-x_t}{2} + \frac{x_t}{2}\phi , \qquad \Delta_{2,t} = \frac{1-x_t}{2}\phi + \frac{x_t}{2} + \frac{y_t}{2}\phi_E ,$$

$$\Delta_{3,t} = \frac{x_t}{2}\phi_E + \frac{y_t}{2} + \frac{1-y_t}{2}\phi , \qquad \Delta_{4,t} = \frac{y_t}{2}\phi + \frac{1-y_t}{2} .$$

In this model entrepreneurs are allowed to move within both trade blocs, between regions 1 and 2 and regions 3 and 4. However, they cannot move between the two trade blocs (home and foreign); for example, migration cannot occur from 1 to 3 or in the opposite direction. After setting $R = 4$, $\gamma_{12} = \gamma_{34} = \gamma$ and $\gamma_{23} = \gamma_{13} = \gamma_{13} = 0$, the migration process (4.8) is reduced to a two-dimensional discrete time dynamic system, governed by the state variables x_t and y_t:

[2] We do not explicit the solutions of the models presented here. For these solutions, the interested reader can refer to the corresponding papers.

$$\frac{x_{t+1}^u - x_t}{x_t} = \gamma\left(\frac{V_{1,t} - \left[x_t V_{1,t} + (1-x_t)\, V_{2,t}\right]}{x_t V_{1,t} + (1-x_t)\, V_{2,t}}\right),$$

$$\frac{y_{t+1}^u - y_t}{y_t} = \gamma\left(\frac{V_{1,t} - \left[x_t V_{1,t} + (1-x_t)\, V_{2,t}\right]}{x_t V_{1,t} + (1-x_t)\, V_{2,t}}\right),$$

where $x_{t+1}^u = Z_h(x_t, y_t)$ is the unconstrained share of home entrepreneurs located in region 2, $y_{t+1}^u = Z_f(x_t, y_t)$ is the unconstrained share of foreign entrepreneurs located in region 3 and γ is the migration speed. Taking into account the constraints, $0 \le x_t \le 1$ and $0 \le y_t \le 1$, the full dynamic system corresponds to the map Z defined as follows:

$$x_{t+1} = \begin{cases} 0 & \text{if} \quad Z_h(x_t, y_t) < 0\,, \\ Z_h(x_t, y_t) & \text{if} \ \ 0 \le Z_h(x_t, y_t) \le 1\,, \\ 1 & \text{if} \quad Z_h(x_t, y_t) > 1\,, \end{cases}$$

$$y_{t+1} = \begin{cases} 0 & \text{if} \quad Z_f(x_t, y_t) < 0\,, \\ Z_f(x_t, y_t) & \text{if} \ \ 0 \le Z_f(x_t, y_t) \le 1\,, \\ 1 & \text{if} \quad Z_f(x_t, y_t) > 1\,. \end{cases} \tag{4.12}$$

Here it is worthwhile to notice that since the overall size of the industry in country h (or in country f) does not vary by assumption, agglomeration can only occur in one of the two regions within each country/trade bloc. No CP equilibria can emerge characterized by full agglomeration in one of the four regions and absence of industry in the other three. Similarly, as we shall see below, border fixed points may exist involving agglomeration in one country and spreading of manufacturing in the other one.

Finally, notice that due to the assumption of symmetric regions, the map (4.12) enjoys some degree of symmetry. Specifically, as we shall see below when we deal with its dynamic properties, it is symmetric with respect to some invariant line. Due to the assumed geography, according to which regions are positioned at different distances from each other, the symmetry is not full. Therefore, for example, the interior equilibrium that involves the condition $x^* = y^*$ is only partially symmetric: $x^* \neq 1/4$.

4.2.8.2 The Case of Asymmetric Regions

In [15] we put forward another four-region model with two important differences compared to the previous one: (i) we allow for indirect trade; (ii) the four regions have different labor endowments. As before, the paper has some focus on the dynamic processes governing a two-country (trade bloc) four-region economy. The geographical set-up is identical to that assumed in [16]: the four regions are aligned from 1 to 4; regions 1 and 2 are part of a trade bloc and regions 3 and 4 are part of another trade bloc. Regions 2 and 3 share a common border and enjoy a more central position.

Regions 1 and 4 are more difficult to reach, however, allowing for indirect trade, the manufactured good can be shipped from and to any region. Trade costs are not prohibitive but depend on distance. The other crucial difference compared to [16], i.e., unequal labor endowments, implies different local market sizes.[3]

We assume that regions 1 and 3 are smaller, in terms of market size, than regions 2 and 4. Supposing, for simplicity that region 1 is equal to region 3 and that region 2 is equal to region 4, it follows: $L_1 = L_3 = L_S$, $L_2 = L_4 = L_B$ and $L_S < L_B$. The aim of [15] is broader compared to [16], wanting to compare not only the importance of centrality vis-à-vis marginality—regions 2 and 3 are central and regions 1 and 4 are at the fringes—but also centrality vis-à-vis regional size—regions 2 and 4 are big and regions 1 and 3 are small. Moreover, by allowing for indirect trade we touched upon the issue of transit traffic through the central regions induced by commodity transportation. In [15] we changed the main state variables as follows:

$$\lambda_{1,t} = \frac{x_t}{2}\,, \quad \lambda_{2,t} = \frac{1-x_t}{2}\,, \quad \lambda_{3,t} = \frac{y_t}{2}\,, \quad \lambda_{4,t} = \frac{1-y_t}{2}\,,$$

where x_t and $1 - x_t$ are the shares of entrepreneurs of the first trade bloc located in the small region 1 and in the big region 2, respectively; and y_t and $1 - y_t$ are the share of entrepreneurs of the second trade bloc located in the small region 3 and in the large region 4, respectively. This specific choice allowed us to focus more on the two small regions. There are no substantial changes in the boundary conditions that, expressed in terms of the changed variables, are: $0 \leq x_t \leq 1$ and $0 \leq y_t \leq 1$.

Given the geographical set-up and taking into account that indirect trade is possible and trade costs are proportional to distance, we set the direct trade costs as $T_{12} = T_{34} = T_C$, $T_{23} = T_D$, and derived the indirect trade costs: $T_{13} = T_{24} = T_C T_D$, $T_{14} = T_C^2 T_D$. Thus, we are able to fill both the trade cost and the trade freeness matrices:

$$\mathbf{T} = \begin{bmatrix} 1 & T_C & T_C T_D & T_C^2 T_D \\ T_C & 1 & T_D & T_C T_D \\ T_C T_D & T_D & 1 & T_C \\ T_C^2 T_D & T_C T_D & T_C & 1 \end{bmatrix}, \quad \mathbf{\Phi} = \begin{bmatrix} 1 & \phi_C & \phi_C \phi_D & \phi_C^2 \phi_D \\ \phi_C & 1 & \phi_D & \phi_C \phi_D \\ \phi_C \phi_D & \phi_D & 1 & \phi_C \\ \phi_C^2 \phi_D & \phi_C \phi_D & \phi_C & 1 \end{bmatrix}. \tag{4.13}$$

Given the above variables change and the assumptions on factor endowments and trade costs, the solutions of the system (4.6) are found by substituting into (4.7) the trade freeness matrix in (4.13) and the following expressions:

$$\mathbf{C}_t = \begin{pmatrix} \frac{1}{\Delta_{1,t}} & 0 & 0 & 0 \\ 0 & \frac{1}{\Delta_{2,t}} & 0 & 0 \\ 0 & 0 & \frac{1}{\Delta_{3,t}} & 0 \\ 0 & 0 & 0 & \frac{1}{\Delta_{4,t}} \end{pmatrix}, \quad \mathbf{G}_t = \frac{1}{2} \begin{pmatrix} \frac{x_t}{2} & 0 & 0 & 0 \\ 0 & \frac{1-x_t}{2} & 0 & 0 \\ 0 & 0 & \frac{y_t}{2} & 0 \\ 0 & 0 & 0 & \frac{1-y_t}{2} \end{pmatrix}, \quad \mathbf{L} = \begin{pmatrix} L_S \\ L_B \\ L_S \\ L_B \end{pmatrix},$$

[3]Given the unitary wage rate, the number of immobile regional workers coincides with that part of local expenditure that does not change through time.

where

$$\Delta_{1,t} = \frac{x_t}{2} + \frac{1-x_t}{2}\phi_C + \frac{y_t}{2}\phi_C\phi_D + \frac{1-y_t}{2}\phi_C^2\phi_D ,$$
$$\Delta_{2,t} = \frac{x_t}{2}\phi_C + \frac{1-x_t}{2} + \frac{y_t}{2}\phi_D + \frac{1-y_t}{2}\phi_C\phi_D ,$$
$$\Delta_{3,t} = \frac{x_t}{2}\phi_C\phi_D + \frac{1-x_t}{2}\phi_D + \frac{y_t}{2} + \frac{1-y_t}{2}\phi_C ,$$
$$\Delta_{4,t} = \frac{x_t}{2}\phi_C^2\phi_D + \frac{1-x_t}{2}\phi_C\phi_D + \frac{y_t}{2}\phi_C + \frac{1-y_t}{2} .$$

The migration process driving the dynamics, after the due substitutions, coincides to that presented in (4.12), taking also into account that x_t and $1-x_t$ are now attached to the shares of entrepreneurs of region 1 and region 2. As before, agglomeration-within-a-country can occur. As we shall see in the discussion below, a crucial difference of this four-region model compared to the previous one is that for $L_S \neq L_B$, the map (4.12) is not symmetric. One of the consequences is that no (partial) symmetric equilibrium exists: $x^* \neq y^*$.

4.3 Dynamic Analysis of NEG Models

This section is based on the results presented in [15–18], related to analysis of local and global dynamics of several maps associated with the NEG models described in Sect. 4.2. Below we discuss peculiarities of such maps, itemize possible kinds of fixed points and their bifurcations, present examples of various attractors and their basins of attraction. Typical bifurcation scenarios observed in the NEG maps under variation of trade freeness parameters are also discussed.

As we have mentioned, in the considered NEG models the migration process of entrepreneurs resembles the *evolutionary replicator dynamics*. For the NEG models defined by *1D maps* (as, e.g., in [17, 18]) this leads to the following function defining the map:

$$Z(x) = x\left[1 + \gamma(1-x)\frac{\Omega(x)-1}{1+x(\Omega(x)-1)}\right] ,$$

where

$$\Omega(x) = \frac{V_1(x)}{V_2(x)}$$

is the ratio between real profits in region 1 and in region 2. The full dynamic model corresponds to the 1D piecewise smooth map f defined in (4.10). We present some results associated with dynamics of this map in Sect. 4.3.3.

For the NEG models defined by *2D maps* (as, e.g., in [15, 16]), the evolutionary replicator dynamics is defined by the following functions:

$$Z_h(x,y) = x\left[1 + \gamma(1-x)\frac{\Omega_h(x,y)-1}{1+x(\Omega_h(x,y)-1)}\right],$$
$$Z_f(x,y) = y\left[1 + \gamma(1-y)\frac{\Omega_f(x,y)-1}{1+x(\Omega_f(x,y)-1)}\right],$$

where

$$\Omega_h(x,y) = \frac{V_1(x,y)}{V_2(x,y)}, \quad \Omega_f(x,y) = \frac{V_3(x,y)}{V_4(x,y)}$$

are the ratio between real profits in region 1 and in region 2, and the ratio between real profits in region 3 and in region 4, respectively. The full dynamic model corresponds to the 2D piecewise smooth map Z defined in (4.12). We recall some results related to dynamics of map (4.12) in Sect. 4.3.4 (symmetric case) and Sect. 4.3.5 (nonsymmetric case).

4.3.1 Peculiarities of NEG Maps

As a first peculiarity of NEG maps we mention quite complicated expressions of the functions defining these maps, that allow to obtain only a few analytic results. Additionally to nonsmoothness, these maps are *noninvertible* for quite an essential part of the parameter space (see, e.g., [28], where properties of 2D noninvertible maps are studied). Therefore, various numerical tools and methods become of great help in the study of the dynamic properties of such maps. For example, a NEG map, besides attracting fixed points, can have other attractors, such as cycles of any period or chaotic attractors. Some of these attractors can coexist, so that varying one or several parameters different bifurcation scenarios can be observed depending on initial conditions, which we study with the help of 1D and 2D bifurcation diagrams (see, e.g., [24, 29] for a general bifurcation theory of nonlinear dynamic systems).

An important property of the NEG maps (4.10) and (4.12) is that besides standard attractors such maps can have attractors in Milnor sense (see [27]), which are caused by the 'flat' branches of the functions defining these maps. Let us first recall two different definitions of an attractor.

Definition 4.1 An *attracting set* A of a map F is a closed invariant set for which a neighborhood $U(A)$ exists such that $F(U(A)) \subset U(A)$ and $F^n(x) \to A$ as $n \to \infty$ for *any* $x \in U(A)$. An *attractor* A is an attracting set with a dense orbit.

Definition 4.2 A *Milnor attractor* is a closed invariant set $A \subset J$ such that the set $\rho(A)$, consisting of all the points $x \in J$ for which ω-limit set $\omega(x) \subset A$, has *strictly positive measure*, and there is no strictly smaller closed subset A' of A such that $\rho(A')$ coincides with $\rho(A)$ up to a set of measure zero.

Comparing these two definitions, one can note that according to the Definition 4.1, an attractor is a closed invariant set with a dense orbit, which has a *neighborhood each point of which is attracted to the attractor*, while an attractor in Milnor sense does not require the existence of such a neighborhood, but only a *positive measure set of points attracted to the attractor*. In fact, Definition 4.2 of an attractor obviously includes also attractors with an attracting neighborhood, as in Definition 4.1. For short, we say that a set is *M-attractor* if it is attracting in Milnor sense, *but not in a sense of the Definition* 4.1. In such a way, a M-attractor has no attracting neighborhood, but it has a basin of attraction of positive measure. For example, a fixed point which is locally repelling can be an M-attractor, with quite a large basin of attraction.

One more peculiarity of maps (4.10) and (4.12) is related to the fact that they have upper and lower borders at which the system function is not differentiable. Thus, rigorously speaking one can discuss only one-side (local) attractivity of invariant sets located at these borders. In the neighborhoods related to the flat branches of the system functions these invariant sets are obviously always locally superstable. For convenience, to make it shorter, we consider that the notion of stability of these sets refer to their local stability in the neighborhoods related to nonflat branches.

Next, as shown in [17, 18], an important property of the map (4.10) is related to its *symmetry with respect to* $x = 0.5$. Thus, any invariant set A of the map f (such as fixed points, cycles, chaotic attractors, basins of attraction, etc.) is either symmetric to itself with respect to $x = 0.5$, or there exists one more invariant set A' which is symmetric to A. The map (4.12) studied in [16] is *symmetric with respect to the main diagonal* of the phase plane, so that any invariant set A of Z is either symmetric with respect to this diagonal, or there exists one more invariant set A' symmetric to A. The map (4.12) studied in [15] is symmetric only for particular parameter values.

4.3.2 Fixed Points and Their Stability

As we already mentioned in Sect. 4.2, due to the specific analytic representation of maps (4.10) and (4.12) they always have so-called *Core-periphery (CP) fixed points*, which are

$$CP_0 : \ x = 0 \ , \qquad CP_1 : \ x = 1 \ ,$$

for the map (4.10) and

$$\begin{aligned} CP_{00} &: \ (x, y) = (0, 0) \ , \qquad & CP_{11} &: \ (x, y) = (1, 1) \ , \\ CP_{01} &: \ (x, y) = (0, 1) \ , \qquad & CP_{10} &: \ (x, y) = (1, 0) \ , \end{aligned}$$

for the map (4.12).

Besides CP fixed points the NEG maps can have *interior fixed points*. In particular, in the symmetric case, the map (4.10) always has an *interior symmetric fixed point*

$$S : \ x = 0.5$$

and, depending on parameters, it can also have *interior asymmetric fixed points*, that necessarily exist in pairs:

$$AS_a : \; x = a \, , \qquad AS'_a : \; x = 1 - a \, ,$$

(with $a \neq 0, \; a \neq 0.5$). When symmetric, the map (4.12) can have an *interior symmetric fixed point*

$$S_{aa} : \; (x, y) = (a, a) \, ,$$

(with $a \neq 0, \; a \neq 1$) and *interior asymmetric fixed points* (necessarily existing in pairs):

$$AS_{ab} : \; (x, y) = (a, b) \, , \quad AS_{ba} : \; (x, y) = (b, a) \, ,$$

(with $a, b \neq 0, a, b \neq 1$). In the asymmetric case, an *interior fixed point* of the map (4.12) is denoted as

$$IP_{ab} \; : (x, y) = (a, b) \, ,$$

with $a, b \neq 0, \, a, b \neq 1$.

The map (4.12) can also have *border fixed points*, which in the symmetric case exist in *pairs*

$$BP_{0a} : \; (x, y) = (0, a) \, , \; BP_{a0} : \; (x, y) = (a, 0) \, ,$$

and

$$BP_{1a} : \; (x, y) = (1, a) \, ; \, , \; BP_{a1} : \; (x, y) = (a, 1) \, ,$$

(with $a \neq 0, \; a \neq 1$), while in the asymmetric case the existence of a border fixed point does not depend on the existence of other border fixed points.

Note that the unit square $I^2 = [0, 1] \times [0, 1]$ of the phase plane of the map (4.12) is *invariant* under this map, as well as each of the borders of I^2, which we denote as follows:

$$\begin{aligned} I_{x0} &= \{(x, y) : y = 0\} \, , \quad I_{x1} = \{(x, y) : y = 1\} \, , \\ I_{0y} &= \{(x, y) : x = 0\} \, , \quad I_{1y} = \{(x, y) : x = 1\} \, . \end{aligned}$$

On each of these borders, the 2D map Z is reduced to the corresponding 1D map. These maps are helpful to understand the overall dynamics of Z.

It appears that any interior fixed point of the map (4.12)—when it exists—is an intersection point of the curves

$$\Omega_h = \left\{(x, y) \in I^2 : \Omega_h(x, y) = 1\right\} \, , \tag{4.14}$$

and

$$\Omega_f = \left\{(x, y) \in I^2 : \Omega_f(x, y) = 1\right\} \, . \tag{4.15}$$

On the other hand, for border fixed points, the following conditions hold:

$$BP_{0a} \in \left\{\Omega_f \cap I_{0y}\right\} \ , \ \ BP_{1a} \in \left\{\Omega_f \cap I_{1y}\right\} ,$$
$$BP_{a0} \in \left\{\Omega_h \cap I_{x0}\right\} \ , \ \ BP_{a1} \in \left\{\Omega_h \cap I_{x1}\right\} \ .$$

This information about fixed points helps to visualize their locations in the phase plane. Note that we allow for the existence of several fixed points of the same kind.

Stability conditions for the CP fixed points are easy to get:

$$CP_0 : \quad \Omega(0) < 1 \ , \quad CP_1 : \quad \Omega(1) > 1 \ ,$$

$$CP_{00} : \quad \left\{\Omega_h(0,0) < 1 \ , \ \Omega_f(0,0) < 1\right\} \ , \tag{4.16}$$

$$CP_{11} : \quad \left\{\Omega_h(1,1) > 1 \ , \ \Omega_f(1,1) > 1\right\} \ , \tag{4.17}$$

$$CP_{01} : \quad \left\{\Omega_h(0,1) < 1 \ , \ \Omega_f(0,1) > 1\right\} \ , \tag{4.18}$$

$$CP_{10} : \quad \left\{\Omega_h(1,0) > 1 \ , \ \Omega_f(1,0) < 1\right\} \ . \tag{4.19}$$

As we recall in the following sections, in the NEG models any CP fixed point changes its stability due to a *border-transcritical (BT) bifurcation*.[4] For the map (4.10) in a generic case this bifurcation results in the appearance/disappearance of an interior asymmetric fixed point, while interior symmetric fixed point can undergo either *pitchfork bifurcation (sub- or supercritical), or flip bifurcation*. For the map (4.12) a BT bifurcation of a CP fixed point in a generic case leads to the appearance/ disappearance of a border fixed point, which in turn may undergo a *flip bifurcation* leading to a 2-cycle belonging to the related border, a *fold bifurcation*—in which case the border fixed point disappears in a pair with another border fixed point,—or a *BT bifurcation* leading to an interior fixed point. Below we present examples of such bifurcations.

4.3.3 A 1D Symmetric NEG Map

We discuss here the dynamic properties of the 1D symmetric NEG map (4.10) derived by the model put forward in [17, 18] and presented in Sect. 4.2.7. Let us recall first some results related to the dynamics of this 1D symmetric NEG map following [18] (see also [17]). The map (4.10), the explicit expression of which can be found in [18], depends on 7 parameters, namely, $\widetilde{n}$, σ, μ, γ, θ, ϕ_L and ϕ_S. In Fig. 4.3 this map

[4]In short, a *border-transcritical bifurcation* of a fixed point of a piecewise smooth continuous map occurs when at the moment of the bifurcation this fixed point belongs to a border at which the system function is not differentiable, its one-side multiplier is equal to 1 and it merges with another fixed point. After the bifurcation, one fixed point disappears, while another one changes its stability.

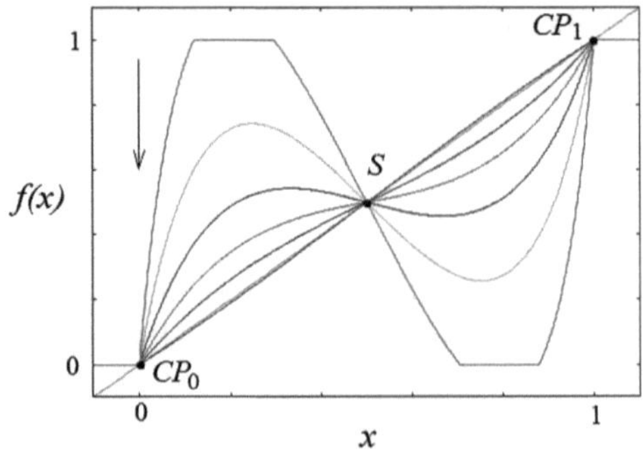

Fig. 4.3 The map f for $\sigma = 6$, $\mu = 0.45$, $\gamma = 10$, $\theta = 0.25$, $\phi_L = 0.1$, $\widetilde{n} = 0.8$ and $\phi_S = 0.11, 0.2, 0.3, 0.4, 0.5$ and 0.7 (the direction indicated by an *arrow* corresponds to increasing ϕ_S)

is shown for different values of the trade freeness parameter ϕ_S fixing the values of the other parameters.

The most important result is that the symmetric fixed point may lose stability not only through a subcritical pitchfork bifurcation, but also through a supercritical one. In [18] this result is proved analytically for the limiting case $\widetilde{n} = 1$; however, by continuity, a similar result must hold also for $0 < \widetilde{n} < 1$, and this conjecture is confirmed by numerical investigations.

Let first $\widetilde{n} = 1$. In Fig. 4.4 we show 1D bifurcation diagrams that illustrate for specific parameter values the possible scenarios; red circles mark pitchfork bifurcations, and green circles indicate BT bifurcations. Panel *(b)* represents the subcritical

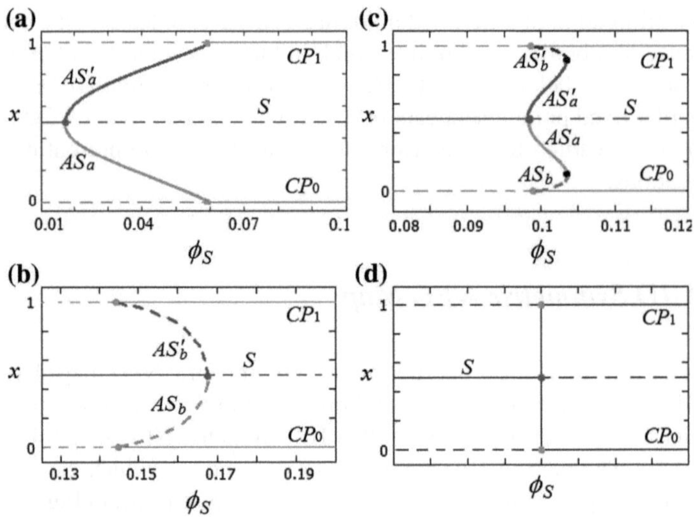

Fig. 4.4 1D bifurcation diagram for $\sigma = 2$, $\mu = 0.45$, $\gamma = 20$, $\widetilde{n} = 1$ and $\theta = 0.25$, $\phi_S \in [0.01, 0.1]$ in (**a**), $\theta = 0.4$, $\phi_S \in [0.08, 0.12]$ in (**b**), $\theta = 0.32$, $\phi_S \in [0.13, 0.19]$ in (**c**). These diagrams are related to parameter paths indicated in Fig. 4.5 by *horizontal lines with arrows*. Degenerate pitchfork bifurcation (which cannot occur in the map f) is shown schematically in (**d**)

pitchfork bifurcation that leads to catastrophic agglomeration and that is found in many standard NEG models: In that case, the pitchfork bifurcation gives rise to asymmetric fixed points that are unstable. They delimit the basins of attraction for the CP fixed points and the symmetric fixed point, all of which are (locally) stable between the break and sustain point values of the trade freeness parameter. Thus, stable fixed points coexist and the long-run pattern of regional industry location depends on parameters, as well as on the basin of attraction to which an initial condition belongs.

Instead, panel *(a)* depicts a supercritical pitchfork bifurcation that leads to smooth agglomeration: asymmetric stable fixed points are born after the bifurcation—the model, which is entirely based on standard NEG assumptions, is thus able to generate endogenously interior asymmetric outcomes (in which economic activity is neither symmetrically distributed between the two regions nor fully agglomerated in one of the regions).

What happens at the moment when the pitchfork bifurcation of the fixed point S and the transcritical bifurcation of the fixed point CP_0 and CP_1 occur simultaneously? One could expect a 1D bifurcation diagram like the one sketched in Fig. 4.4d. However, such a diagram is impossible because it would mean that a degenerate pitchfork bifurcation occurs, with $f(x) \equiv x$ in the complete interval [0, 1]. In fact, the true transition is as shown in Fig. 4.4c where fold bifurcations (marked with black circles) give rise to two stable asymmetric fixed points AS_a, AS'_a and two unstable asymmetric fixed points AS_b, AS'_b. Panel *(c)*, thus, shows that also for asymmetric stable equilibria coexistence with the CP equilibria is possible (and the related basins of attraction are delimited by additional asymmetric equilibria that are unstable).

In Fig. 4.5 a 2D bifurcation diagram is presented in the (ϕ_S, θ)-parameter plane for $\sigma = 2$, $\mu = 0.45$, $\gamma = 20$, $\widetilde{n} = 1$. Here different colors are related to different attracting cycles, namely, the red region S to the symmetric fixed point; the pink region AS to coexisting asymmetric fixed points AS_a and AS'_a; the blue region CP to

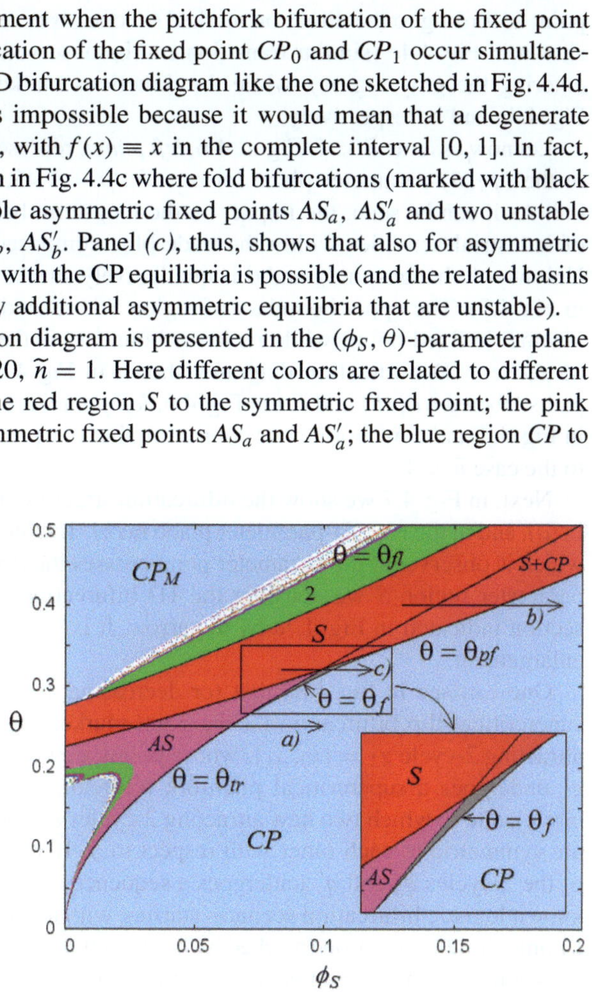

Fig. 4.5 2D bifurcation diagram in the (ϕ_S, θ)-parameter plane for $\sigma = 2$, $\mu = 0.45$, $\gamma = 20$, $\widetilde{n} = 1$. 1D bifurcation diagrams related to the *horizontal lines* marked *(a)*, *(b)* and *(c)* are shown in Fig. 4.4a–c, respectively. *Inset* presents an enlargement of the indicated *window*

the fixed points CP_0 and CP_1; the region marked as $S + CP$ to coexisting attracting symmetric and CP fixed points; the gray region shown also in an inset (bounded by the fold bifurcation curve $\theta = \theta_f$ that can be obtained only numerically) is related to coexisting attracting fixed points CP_0, CP_1 and AS_a, AS'_a; the blue region CP_M to the fixed points CP_0 and CP_1 which are M-attractors; the green region to 2-cycles; the other colors correspond to cycles of periods $k \leq 30$ and white region is related to higher periodicity or to chaotic attractors.

In the limiting case $\widetilde{n} = 1$ the bifurcation curves of the fixed points S, CP_0 and CP_1 can be obtained analytically, in particular, the flip bifurcation boundary $\theta = \theta_{fl}$, the pitchfork bifurcation boundary $\theta = \theta_{pf}$ and the BT bifurcation boundary $\theta = \theta_{tr}$ (see [18] for the explicit expressions of these boundaries). These curves are shown in Fig. 4.5, in particular, one can see that the curves $\theta = \theta_{pf}$ and $\theta = \theta_{tr}$ intersect each other, and if (ϕ_S, θ)-parameter point moves through the intersection point according to the direction marked *(c)* one observes the 1D bifurcation diagram shown in Fig. 4.4c (similar transition can be observed also for any parameter path entering the gray region), while the directions marked *(a)* and *(b)* are related to Fig. 4.4a and b, respectively.

Let now assume $\widetilde{n} < 1$. Figure 4.6, *left panel*, represents a 2D bifurcation diagram in the (ϕ_S, θ)-parameter plane for $\sigma = 6$, $\mu = 0.45$, $\gamma = 10$, $\phi_L = 0.1$ and $\widetilde{n} = 0.8$. As it can be seen, the impact on the long-term behavior of x—the distribution of industrial activities within the trade bloc composed of regions 1 and 2 labeled Union—of changes in ϕ_S and θ is qualitatively similar to the case $\widetilde{n} = 1$ (as shown in Fig. 4.5). The upper boundary of region S is related to the flip bifurcation of the symmetric fixed point, while its lower boundary is the pitchfork bifurcation curve. Moreover, 1D bifurcation diagrams shown in Fig. 4.6, *right panel*, related to the paths labeled *(a)*, *(b)* and *(c)* in Fig. 4.6, *left panel*, are quite similar to those shown in Fig. 4.4 and corroborate our conjecture that our previous results can be extended to the case $\widetilde{n} < 1$.

Next, in Fig. 4.7 we show the bifurcation structure in the $(\phi_S, \widetilde{n})$-parameter plane in *(a)*, and in the $(\phi_L, \widetilde{n})$-parameter plane in *(b)*. To comment the bifurcation scenario which is observed if the parameter point crosses the flip bifurcation boundary of the parameter region S we consider the 1D bifurcation diagram related to the cross-section indicated in Fig. 4.7a by the arrow. It is shown in Fig. 4.8 together with an enlargement.

One can see in Fig. 4.8a that for decreasing ϕ_S the fixed point S undergoes a supercritical flip bifurcation (at the point marked by a black circle) leading to an attracting 2-cycle $g_2 = \{x_0, x_1\}$, whose points are symmetric with respect to S. Then g_2 undergoes a supercritical pitchfork bifurcation (at the point marked by a red circle), due to which two new attracting 2-cycles q_2 and q'_2 are born, points of which are symmetric to each other with respect to S. If we continue to decrease ϕ_S each of the 2-cycles q_2 and q'_2 undergoes a sequence of bifurcations following the well-known logistic bifurcation scenario starting with a cascade of flip bifurcations up to a homoclinic bifurcation (marked by blue points) of 2-cycle g_2 (see Fig. 4.8b). Thus, we see that the map f can have coexisting attracting cycles and chaotic attractors.

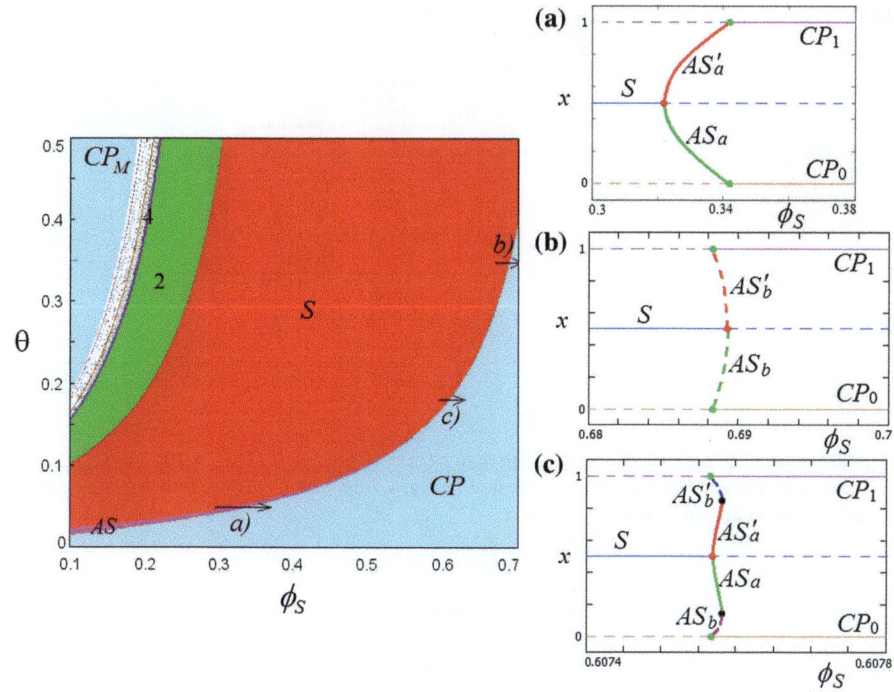

Fig. 4.6 *Left panel* 2D bifurcation diagram in the (ϕ_S, θ)-parameter plane for $\phi_L = 0.1$, $\widetilde{n} = 0.8$, $\sigma = 6$, $\mu = 0.45$, $\gamma = 10$. *Right panel* 1D bifurcation diagrams related to *parameter paths* marked **a**, **b** and **c**, respectively, in the 2D diagram on the *left*

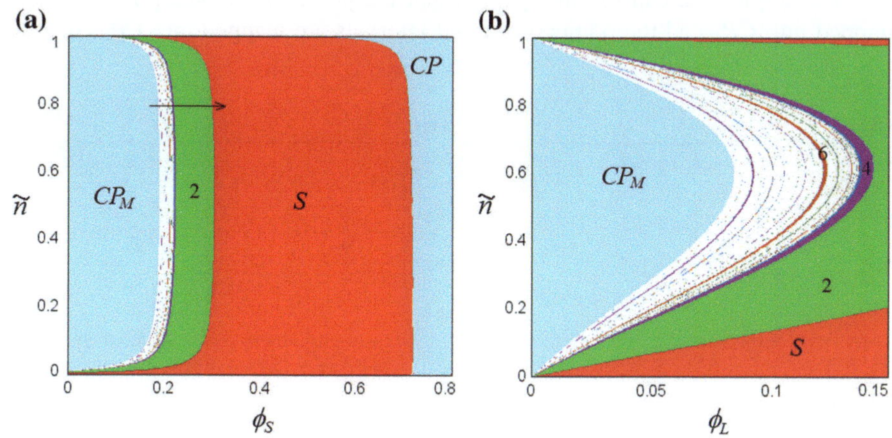

Fig. 4.7 2D bifurcation diagrams in $(\phi_S, \widetilde{n})$- and $(\phi_L, \widetilde{n})$-parameter plane for $\sigma = 6$, $\mu = 0.45$, $\gamma = 10$, $\theta = 0.25$ and $\phi_L = 0.01$ in (**a**), $\phi_S = 0.15$ in (**b**)

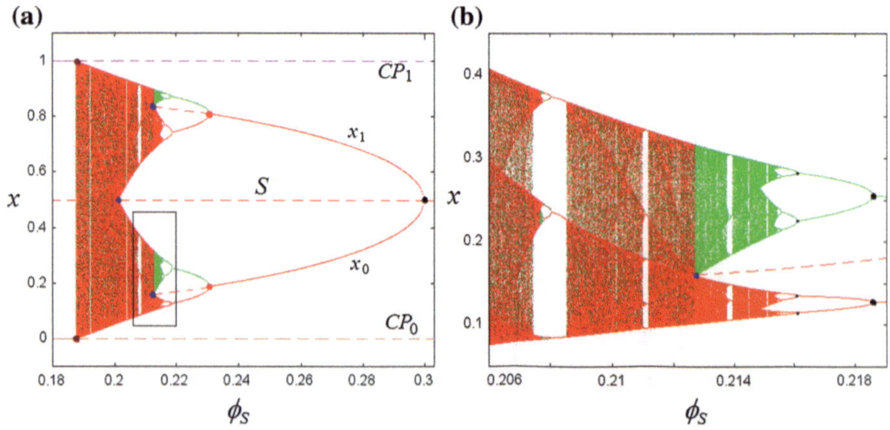

Fig. 4.8 In **a** 1D bifurcation diagram of the map f for $\sigma = 6$, $\mu = 0.45$, $\gamma = 10$, $\theta = 0.25$, $\phi_L = 0.01$, $\tilde{n} = 0.8$ related to cross-section indicated in Fig. 4.7 by a *black arrow*. In **b** An enlargement of the *window* indicated in (**a**)

Let us comment now a bifurcation marked in Fig. 4.8a by brown circles. It is a contact bifurcation of a one-piece chaotic attractor, bounded by the critical points of the map f denoted c and c', with its basin confined by the fixed point CP_0 and CP_1. Such a contact occurs if a parameter point crosses the boundary of the region CP_M (see Fig. 4.7). In Fig. 4.9a the map f is shown at the moment of such a contact defined by the condition $c = 0$ or $c' = 1$. After this bifurcation the locally repelling fixed points CP_0 and CP_1 become M-attractors.

For example, one can see in Fig. 4.9b that the points of the green intervals are mapped into CP_1 and the points of the red intervals are mapped into CP_0. In fact,

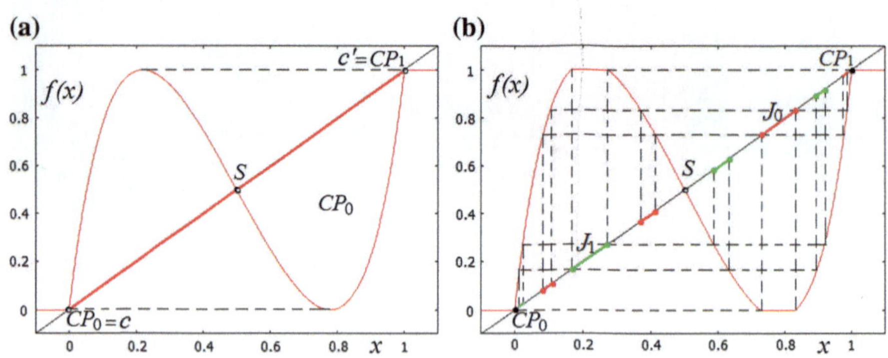

Fig. 4.9 The map f at the moment (**a**) in and (**b**) after the contact bifurcation of the chaotic attractor $A = [c, c']$ with its basin confined by the fixed points CP_0 and CP_1. Here $\sigma = 6$, $\mu = 0.45$, $\gamma = 10$, $\theta = 0.25$, $\phi_L = 0.01$, $\tilde{n} = 0.8$ and $\phi_S = 0.187626$ (the related point is indicated in Fig. 4.8a by a *brown circle*) in (**a**), $\phi_S = 0.18$ in (**b**)

the interval J_1 related to the left flat branch of f and all its preimages, a few of which are shown in Fig. 4.9b, constitute the stable set of the fixed point CP_1. There is a sequence of preimages of J_1 accumulating at CP_0, and there is also a sequence of preimages of J_1 accumulating at CP_1. The same can be said about sequences of preimages of J_2 related to the second flat branch of f. Thus, in any (one-side) neighborhood of CP_0 or CP_1 there is a positive measure set of points which first escape from this neighborhood and then eventually are mapped to CP_0, as well a symmetric positive measure set of points mapped to CP_1. Clearly, not all the point of I are mapped to CP_0 or CP_1: a chaotic repellor, separating the basins of the CP fixed points, remains in I, which is a Cantor set formed by all the repelling cycles and their preimages, as well as uncountably many aperiodic orbits.

4.3.4 2D Symmetric NEG Map

Now let us consider the 2D symmetric NEG map (4.12) presented in Sect. 4.2.8.1 and studied in [16]. This map depends on 7 parameters. In all the numerical simulations we fix $E = 100$, $L = 400$, $\mu = 0.5$ and study how the dynamics depends on the parameters γ, σ, ϕ, and ϕ_E.

The map is symmetric with respect to the main diagonal $L_d = \{(x, y) : x = y\}$, thus, this diagonal is invariant under the map Z. So, one can consider a 1D map which is a restriction of map Z to L_d :

$$z_d : x \mapsto \begin{cases} 0 & \text{if } z_d(x) < 0 , \\ z_d(x) & \text{if } 0 \le z_d(x) \le 1 , \\ 1 & \text{if } z_d(x) > 1 , \end{cases}$$

where an explicit expression of $z_d(x)$ can be found in [16]. In a similar way 1D maps $z_{(0)}$ and $z_{(1)}$ associated with the invariant borders I_{x0} and I_{1y} can be obtained (due to the symmetry of Z an analogous reduction holds on the border I_{0y} and I_{x1}, respectively).

To give an example of dynamics generated by the 1D map z_d we show in Fig. 4.10a a 1D bifurcation diagram of z_d for $\gamma = 5$, $\sigma = 2$, $\phi_E = 0.1$, $0 < \phi < 1$, while Fig. 4.10b, c present examples of the map z_d for various values of ϕ. In Fig. 4.10a a fold bifurcation can be recognized leading (for decreasing ϕ) to attracting and repelling fixed points denoted S_a and S_b, shown by solid and dashed lines, respectively, which in terms of the map Z are associated with the symmetric interior fixed points S_{aa} and S_{bb}. If we continue to decrease ϕ, the repelling fixed point S_b merges quite soon with the fixed point CP_1 due to the BT bifurcation (this bifurcation occurs for $\phi \approx 0.135$), then the fixed point S_a undergoes a flip bifurcation. The other two BT bifurcations indicated in Fig. 4.10a occur for the fixed point CP_0 for increasing ϕ (these bifurcations occur at $\phi \approx 0.1891$ and $\phi \approx 0.9472$).

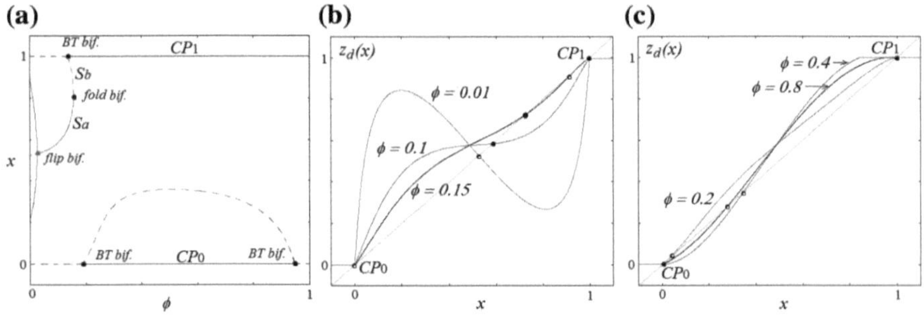

Fig. 4.10 **a** 1D bifurcation diagram of the map z_d for $\gamma = 5$, $\sigma = 2$, $\phi_E = 0.1$, $0 < \phi < 1$; the map z_d for $\phi = 0.01, 0.1, 0.15$ is shown in (**b**) and for $\phi = 0.2, 0.4, 0.8$ in (**c**)

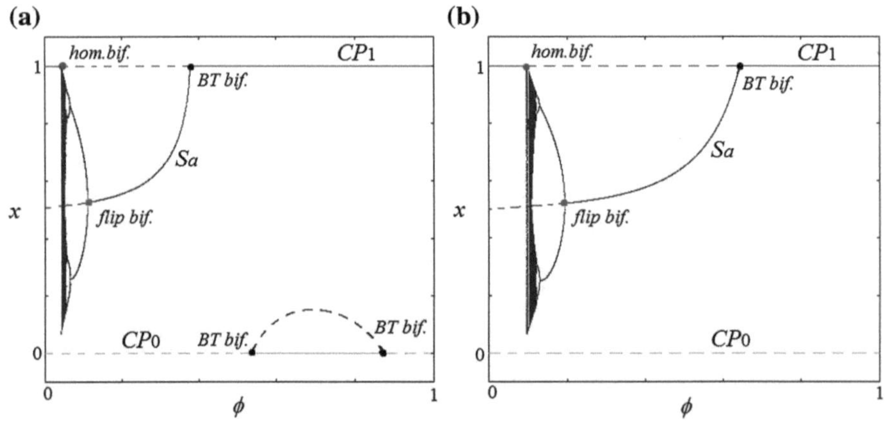

Fig. 4.11 1D bifurcation diagrams of the map z_d for $\gamma = 5$, $\phi_E = 0.1$, $0 < \phi < 1$ and $\sigma = 3$ in (**a**), $\sigma = 8$ in (**b**)

Figure 4.11 shows two more 1D bifurcation diagrams of the map z_d, for $\sigma = 3$ and $\sigma = 8$, where one observes for decreasing ϕ the logistic bifurcation scenario up to the contact of a chaotic attractor with its basin boundary defined by the fixed point CP_1 and its preimage. This bifurcation (called also *final bifurcation*) is caused by the *homoclinic bifurcation* of the fixed point S_a, after which the fixed point CP_1 becomes an M-attractor, while the chaotic attractor is transformed into a chaotic repellor.

Figures 4.12 and 4.13 illustrate how we can use the 1D maps z_d, $z_{(0)}$ and $z_{(1)}$ to describe the dynamics of the map Z. For the considered parameter values Z has four attracting fixed points, namely, CP_{11}, S_{aa}, BP_{c1} and BP_{1c}. Their basins of attraction shown in Fig. 4.12 are separated by the stable sets of the saddle fixed points denoted AS_{ef}, AS_{fe}, BP_{1d} and BP_{d1}. In this figure as well as in the others, attracting, repelling and saddle fixed points are marked by black, white and gray circles, respectively. The curves Ω_h and Ω_f given in (4.14) and (4.15) are also drawn. In Fig. 4.13 the

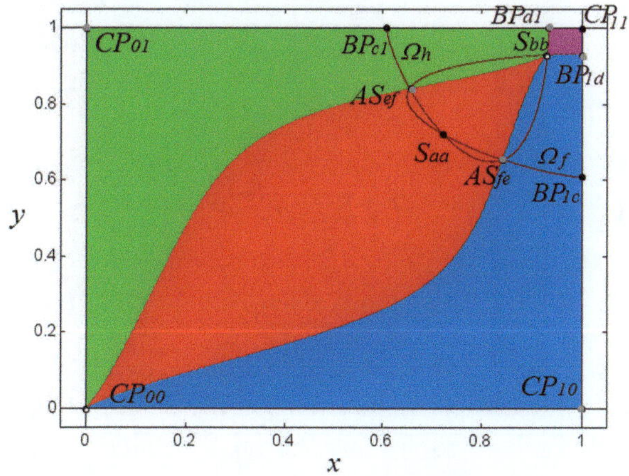

Fig. 4.12 Basins of attraction of fixed points S_{aa}, BP_{1c}, BP_{c1} and CP_{11} of the map Z for $\gamma = 5$, $\sigma = 2$, $\phi_E = 0.1$, $\phi = 0.15$

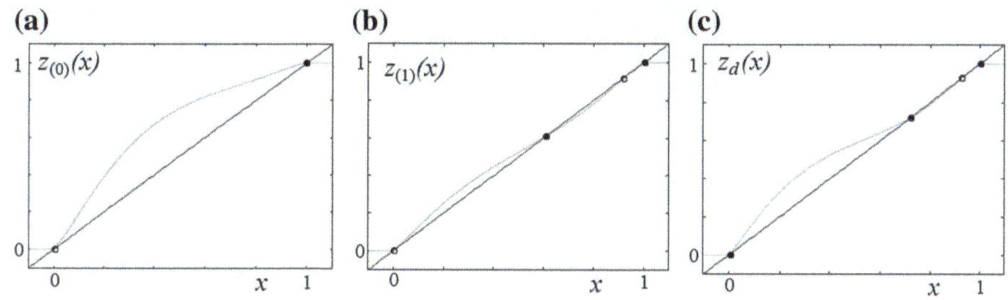

Fig. 4.13 The map $z_{(0)}$ in (**a**), $z_{(1)}$ in (**b**) and z_d in (**c**) for $\gamma = 5$, $\sigma = 2$, $\phi_E = 0.1$, $\phi = 0.15$

corresponding maps z_d, $z_{(0)}$ and $z_{(1)}$ are shown. One can deduce that an initial point $(x_0, y_0) \in I_{0y}$ (see the map $z_{(0)}$) is attracted to the saddle fixed point CP_{01} (similarly, an initial point $(x_0, y_0) \in I_{x0}$ is attracted to CP_{10}), while an initial point $(x_0, y_0) \in I_{x1}$ (see the map $z_{(1)}$) with $0 < x_0 < d$ is attracted to BP_{c1}, and an initial point $(x_0, y_0) \in I_{x1}$ with $d < x_0 < 1$ is attracted to CP_{11} (analogous conclusions hold for an initial point $(x_0, y_0) \in I_{1y}$). Finally, an initial point $(x_0, y_0) \in L_d$ (see the map z_d) with $0 < x_0 < b$ is attracted to S_{aa}, and with $b < x_0 < 1$ is attracted to CP_{11}.

Figures 4.14 and 4.15 illustrate one more example. As can be seen in Fig. 4.14, the map Z has an attracting 2-cycle located on the main diagonal (compare with the map z_d in Fig. 4.15c) with a basin of attraction shown in red, and two M-attracting fixed points, $BP_{c1} \in I_{x1}$ and $BP_{1c} \in I_{1y}$ (compare with the map $z_{(1)}$ in Fig. 4.15b) with green and blue basins, respectively. Boundaries of these basins are proper segments of I_{0y}, I_{x1}, I_{x0}, I_{1y} and their preimages. An initial point $(x_0, y_0) \in I_{x1}$ is attracted to BP_{c1}, an initial point $(x_0, y_0) \in I_{1y}$ is attracted to BP_{1c}, while the dynamics on I_{x0} and I_{0y} are more complicated: as it follows from the dynamics of the map $z_{(0)}$ (see

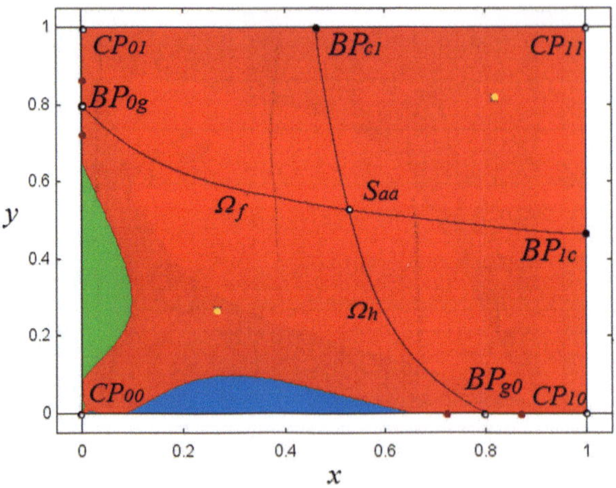

Fig. 4.14 An attracting 2-cycle (*yellow circles*) and its basin shown in *red*, and M-attracting fixed points BP_{c1} and BP_{1c} with *green* and *blue* basins, respectively, for $\gamma = 5$, $\sigma = 2$, $\phi_E = 0.1$, $\phi = 0.01$

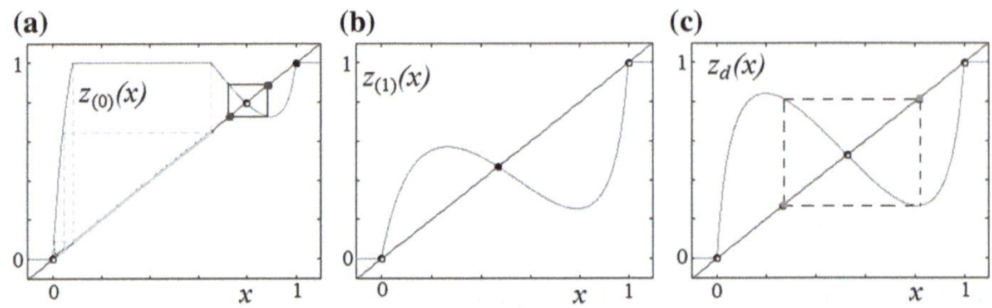

Fig. 4.15 The 1D maps $z_{(0)}$ in (**a**), $z_{(1)}$ in (**b**) and z_d in (**c**) for $\gamma = 5$, $\sigma = 2$, $\phi_E = 0.1$, $\phi = 0.01$

in Fig. 4.15a), an initial point $(x_0, y_0) \in I_{0y}$ belonging to the segment shown green (related to the flat branch of $z_{(0)}$) or to its preimages is mapped into CP_{01}, while other initial points of I_{0y} (except for the preimages of the fixed point BP_{0g}) are attracted to a 2-cycle. Similarly, an initial point $(x_0, y_0) \in I_{x0}$ belonging to the blue segment or its preimages is mapped to CP_{10} while other initial points of I_{x0} (except for the preimages of the fixed point BP_{g0}) are attracted to a 2-cycle belonging to I_{x0}.

Figure 4.16 presents a 2D bifurcation diagram and its enlargement in the (ϕ, γ)-parameter plane, together with the bifurcation curves BT_{d0}, BT_{d1}, $BT_{(0)1}$ and $BT_{(1)0}$ (their explicit expressions can be found in [16]). The horizontal lines with double arrows indicate stability regions of the related CP fixed points. In particular, it can be seen that the blue region bounded by BT_{d0} and $BT_{(1)0}$ is associated with four coexisting attracting CP fixed points. The dark gray region marked CP_M corresponds to CP fixed points which are M-attractors. The region shown in yellow is related to attracting border fixed points BP_{x1}, BP_{1x}. The red region marked S is associated

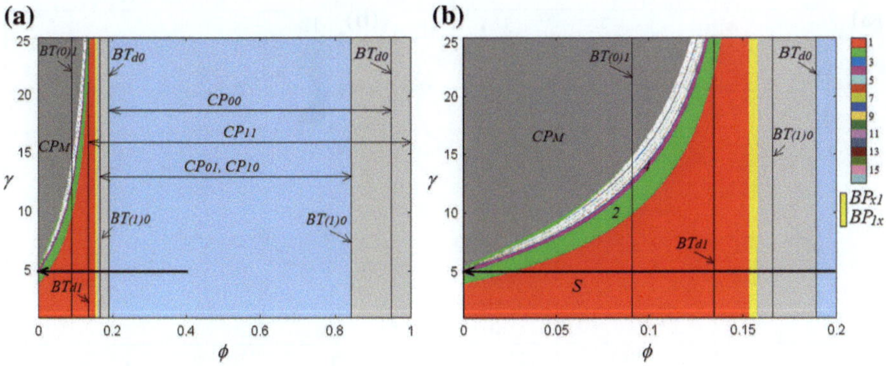

Fig. 4.16 2D bifurcation diagram in (**a**) and its enlargement in (**b**) in the (ϕ, γ)-parameter plane for $\sigma = 2$, $\phi_E = 0.1$

with attracting symmetric interior fixed point S_{xx}. It can be seen that this fixed point undergoes a flip bifurcation leading to an attracting 2-cycle (its stability region is shown in green). Then, when decreasing ϕ, a region related to an attracting 4-cycle is recognizable (it is shown in magenta), as well as other periodicity regions. Here, a white region is related to either higher periodicity or chaotic attractors. Given that the 2D bifurcation diagram in Fig. 4.16 is obtained for only one initial point, $(x_0, y_0) = (0.5, 0.51)$, coexistence of attractors other than the *CP* fixed points cannot be seen in this figure. In order to study such a coexistence, we consider below a 1D bifurcation diagram related to the cross-section of the 2D diagram for $\gamma = 5$ indicated by the thick arrow.

First in Fig. 4.17a we present a 1D bifurcation diagram (x, y) versus ϕ for $\gamma = 5$, $\sigma = 2$, $\phi_E = 0.1$ and $0 < \phi < 1$, where only fixed points are shown, namely, the branches related to the border fixed points BP_{x0}, BP_{x1}, BP_{0y}, BP_{1y} are shown in

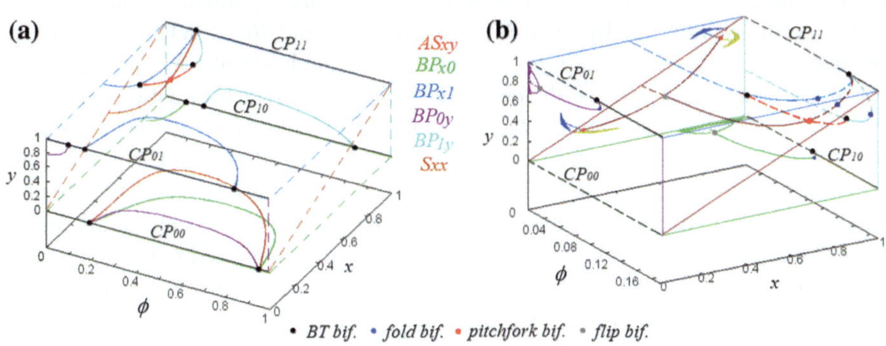

Fig. 4.17 In **a** Fixed points of the map Z in the (x, y, ϕ)-space for $0 < \phi < 1$. In **b** 1D bifurcation diagram (x, y) versus ϕ for $0 < \phi < 0.17$. The other parameters are fixed as $\gamma = 5$, $\sigma = 2$, $\phi_E = 0.1$

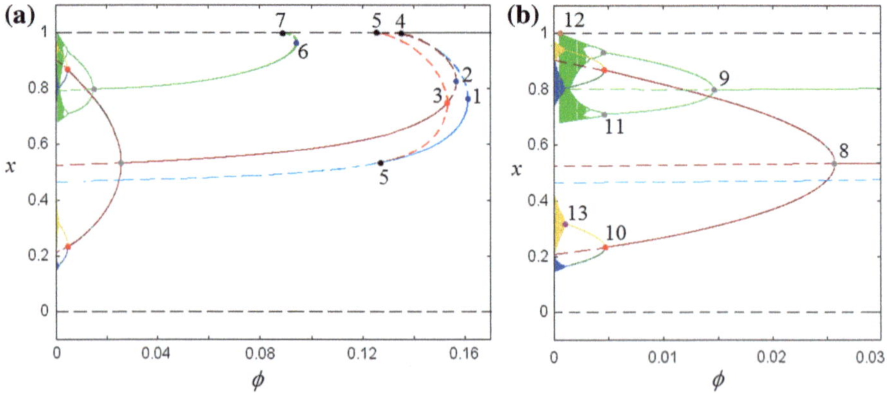

Fig. 4.18 Projection of the 1D bifurcation diagram shown in Fig. 4.17b to the (x, ϕ)-plane. Here $\gamma = 5$, $\sigma = 2$, $\phi_E = 0.1$ and $0 < \phi < 0.17$ in (**a**), $0 < \phi < 0.03$ in (**b**)

green, blue, magenta and light blue, respectively, while the branches of interior asymmetric and symmetric fixed points AS_{xy} and S_{xx} are shown in red and brown, respectively. Border-transcritical and pitchfork bifurcations that these fixed points undergo, are indicated by black and red circles. Next, in Fig. 4.17b a complete 1D bifurcation diagram (x, y) versus ϕ is presented for $0 < \phi < 0.17$. Its projection on the (x, ϕ)-plane is shown in Fig. 4.18. In these figures dashed lines are related to repelling or saddle fixed points, while attracting fixed points are shown by solid lines. Besides border-transcritical and pitchfork bifurcations, fold and flip bifurcations are also indicated.

Let us comment now on the bifurcation sequence observed for decreasing ϕ at fixed $\sigma = 2$, $\phi_E = 0.1$, $\gamma = 5$ (see the thick arrow in Fig. 4.16), and how basins of attraction of coexisting attractors change due to these bifurcations. We begin with the value $\phi = 0.4$ at which the map Z has four coexisting attracting CP fixed points. Their basins of attraction shown in Fig. 4.19a are separated by the stable sets of the border saddle fixed points BP_{0b}, BP_{b0}, BP_{c1} and BP_{1c}. When ϕ decreases, the basins of CP_{00}, CP_{10} and CP_{01} decrease while the basin of CP_{11} increases (see Fig. 4.19b where $\phi = 0.25$). If the parameter point crosses the bifurcation curve BT_{d0} (at $\phi \approx 0.1891$), the fixed point CP_{00} loses stability, so that between the curves BT_{d0} and $BT_{(1)0}$, the map Z has three coexisting attracting fixed points, CP_{11}, CP_{10} and CP_{01}. Then crossing $BT_{(1)0}$ (at $\phi \approx 0.1655$), the fixed points CP_{10} and CP_{01} become saddles so that between the curves $BT_{(1)0}$ and BT_{d1} only CP_{11} is attracting among the CP fixed points. However, in this parameter range other fixed points appear.

In fact, at $\phi \approx 0.1613$ (the parameter point enters the yellow region in Fig. 4.16) a fold bifurcation occurs in the map $z_{(1)}$ (see the bifurcation marked 1 in Fig. 4.18a) leading to one attracting and one repelling fixed points which in terms of the map Z are associated with two pairs of border asymmetric fixed points, namely, two attracting fixed points, BP_{c1}, BP_{1c}, and two saddle fixed points, BP_{d1}, BP_{1d}. Thus,

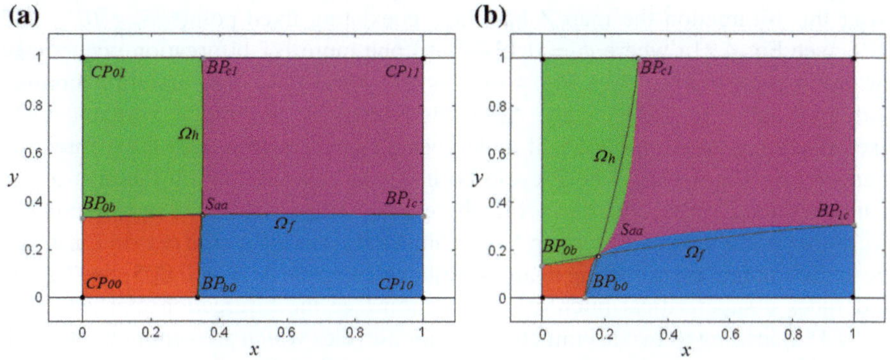

Fig. 4.19 Basins of attraction of CP fixed points CP_{00}, CP_{11}, CP_{01} and CP_{10} for $\sigma = 2$, $\phi_E = 0.1$, $\gamma = 5$ and $\phi = 0.4$ in (**a**), $\phi = 0.25$ in (**b**)

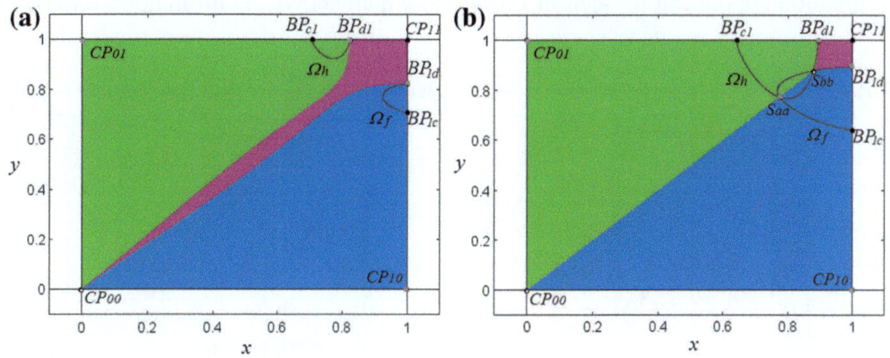

Fig. 4.20 Basins of attraction of fixed points CP_{11}, BP_{c1} and BP_{1c} for $\sigma = 2$, $\phi_E = 0.1$, $\gamma = 5$ and $\phi = 0.16$ in (**a**), $\phi = 0.155$ in (**b**)

after this bifurcation, the map Z has three coexisting attracting fixed points, CP_{11}, BP_{c1} and BP_{1c}. An example of their basins is shown in Fig. 4.20a where $\phi = 0.16$.

If we continue to decrease ϕ, a fold bifurcation in the map z_d occurs at $\phi \approx 0.1565$ (see the bifurcation marked 2 in Fig. 4.18a) leading to a pair of symmetric interior fixed points, a saddle fixed point S_{aa} and a repelling fixed point S_{bb}. An example of the phase portrait soon after this bifurcation is shown in Fig. 4.20b where $\phi = 0.155$. Next bifurcation is the subcritical pitchfork bifurcation of S_{aa} occurring at $\phi \approx 0.1529$ (the parameter point enters the red region in Fig. 4.16; see also the bifurcation marked 3 in Fig. 4.18a) after which this fixed point becomes attracting and two asymmetric interior saddle fixed points are born, denoted AS_{ef} and AS_{fe}. An example of the basins of attraction of coexisting fixed points CP_{11}, BP_{c1}, BP_{1c} and S_{aa} is shown in Fig. 4.12 where $\phi = 0.15$.

Next bifurcation occurs when the parameter point crosses the curve BT_{d1} (at $\phi \approx 0.135$) so that CP_{11} loses stability merging simultaneously with the fixed points S_{bb}, BP_{d1} and BP_{1d} due to a BT bifurcation (see the bifurcation marked 4 in Fig. 4.18a).

After this bifurcation the map Z has three coexisting fixed points, S_{aa}, BP_{c1} and BP_{1c} (see Fig. 4.21a where $\phi = 0.134$) until one more BT bifurcation occurs (see the bifurcation marked 5 in Fig. 4.18a) as BP_{c1} and BP_{1c} lose stability merging with AS_{ef} and AS_{fe}, respectively. After this bifurcation the unique attractor is the fixed point S_{aa}. At $\phi \approx 0.0256$, it undergoes a flip bifurcation (see the bifurcation marked 8 in Fig. 4.18b). The 2-cycle born due to this bifurcation belongs to the main diagonal L_d. See, e.g., Fig. 4.14, where $\phi = 0.01$, which shows an attracting 2-cycle coexisting with M-attracting fixed points BP_{c1} and BP_{1c} (the reason for which these fixed points are attracting again is explained below). At $\phi \approx 0.005$, the 2-cycle undergoes a supercritical pitchfork bifurcation (see the bifurcation marked 10 in Fig. 4.18a) leading to two asymmetric interior 2-cycles which are symmetric to each other with respect to L_d (see Fig. 4.21b where $\phi = 0.004$). Then each of these 2-cycles undergoes a Neimark-Sacker bifurcation leading to closed invariant attracting curves which after destruction lead to chaotic attractors. The pitchfork and Neimark-Sacker bifurcation curves can be seen in Fig. 4.22b which shows 2D bifurcation diagrams

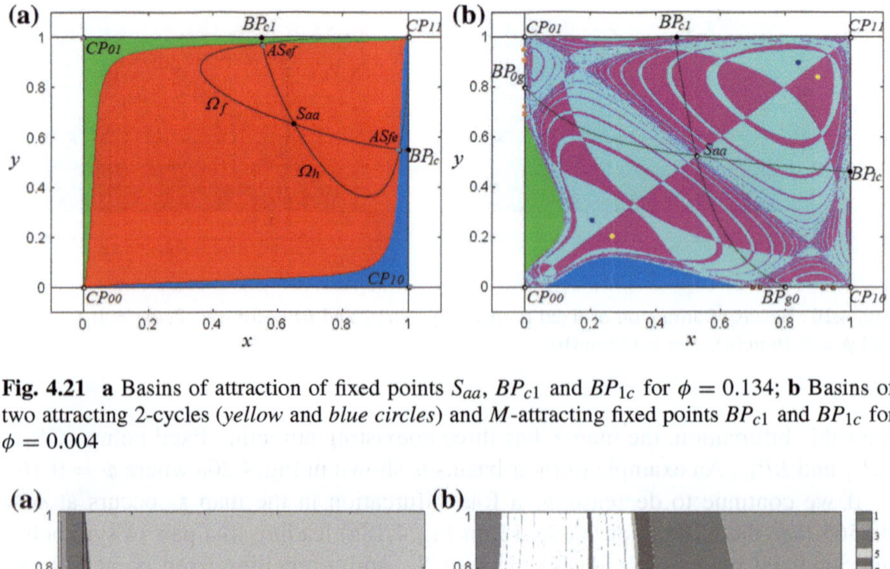

Fig. 4.21 **a** Basins of attraction of fixed points S_{aa}, BP_{c1} and BP_{1c} for $\phi = 0.134$; **b** Basins of two attracting 2-cycles (*yellow* and *blue circles*) and M-attracting fixed points BP_{c1} and BP_{1c} for $\phi = 0.004$

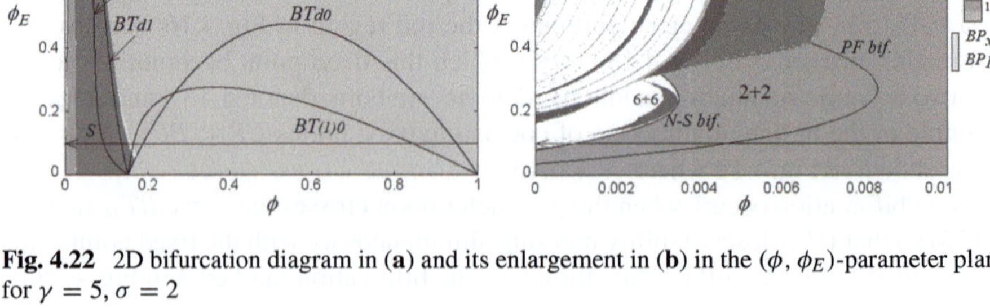

Fig. 4.22 2D bifurcation diagram in (**a**) and its enlargement in (**b**) in the (ϕ, ϕ_E)-parameter plane for $\gamma = 5, \sigma = 2$

in the (ϕ, ϕ_E)-parameter plane for $\gamma = 5$, $\sigma = 2$. In the meantime, at $\phi \approx 0.0936$, the map $z_{(0)}$ undergoes a fold bifurcation (see the bifurcation marked 6 in Fig. 4.18a) leading to attracting and repelling fixed points, AS_g and AS_h, respectively, which in terms of the map Z are associated with two pairs of border asymmetric fixed points, saddles BP_{0g}, BP_{g0}, and repellors BP_{0h}, BP_{h0}. Soon after, at $\phi \approx 0.0911$ (the parameter point crosses $BT_{(0)1}$) the fixed points BP_{0h} and BP_{h0} merge with CP_{01} and CP_{10}, respectively, due to a BT bifurcation (see the bifurcation marked 7 in Fig. 4.18a), so that saddles CP_{01} and CP_{10} become repelling fixed points. Then the fixed point AS_g of the map $z_{(0)}$ undergoes a cascade of flip bifurcations (two of which are marked 9 and 11 in Fig. 4.18b) following the logistic bifurcation scenario up to the homoclinic bifurcation of AS_g (see the bifurcation marked 12 in Fig. 4.18b). In the meantime at $\phi \approx 0.0515$ (when the fixed point AS_g is still stable) the flat branch with $x = 1$ appears in the definition of the map $z_{(0)}$ inside the interval [0, 1], so that its fixed point CP_1 becomes an M-attractor (see, e.g., Fig. 4.15a). In terms of the map Z this leads to the stabilization of fixed points BP_{c1} and BP_{1c} which also become M-attractors (see, e.g., Fig. 4.14 or Fig. 4.21b). So, for $\phi \lessapprox 0.0515$ all the mentioned above attractors coexist with the M-attracting fixed points BP_{c1} and BP_{1c}.

To see how the described above scenario changes if other parameters are varied, we show in Fig. 4.22 a 2D bifurcation diagram in the (ϕ, ϕ_E)-parameter plane for $\gamma = 5$, $\sigma = 2$. As before, the arrow indicates a cross-section for $\phi_E = 0.1$ along which the bifurcation sequence described above is observed. In this parameter plane the pitchfork bifurcation curve of the 2-cycle leading to two coexisting 2-cycles is shown. One can also see the Neimark-Sacker bifurcation curve and a 1 : 3 resonance region which is associated with two coexisting 6-cycles. Basins of two attracting 6-cycles and M-attracting fixed points BP_{c1} and BP_{1c} are shown in Fig. 4.23, where $\phi = 0.003$, $\phi_E = 0.2$.

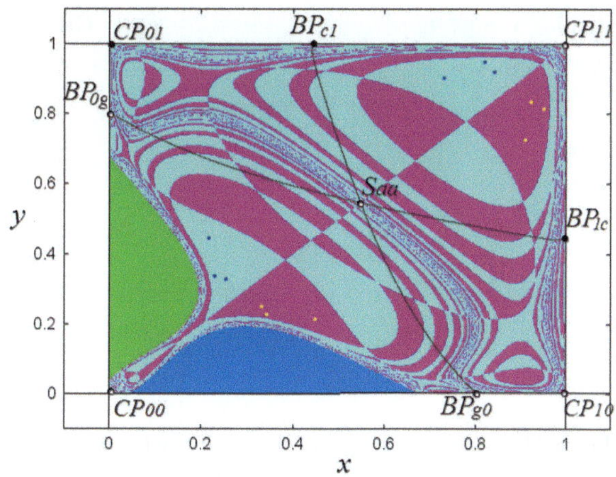

Fig. 4.23 Basins of two attracting 6-cycles (shown in *light blue* and *magenta*) and M-attracting fixed points BP_{c1} and BP_{1c} (shown in *green* and *dark blue*, respectively) for $\gamma = 5$, $\sigma = 2$, $\phi = 0.003$, $\phi_E = 0.2$

4.3.5 2D Nonsymmetric NEG Map

Now we consider the asymmetric map (4.12) presented in Sect. 4.2.8.2 and studied in [15] representing an economy with four regions characterized by different labor endowments. This map depends on 8 parameters: the values of the parameters μ, σ, E, γ and L_B are fixed as $\mu = 0.7$, $\sigma = 3$, $E = 100$, $\gamma = 20$, $L_B = 400$, the parameter L_S can take the values 200, 300, and $0 < \phi_C, \phi_D < 1$ are parameters to be varied.

First, in Fig. 4.24 we show the stability regions of the CP fixed points bounded by the BT bifurcation curves for $L_S = 200$. In the curves of Fig. 4.24a '*vs*' refers to 'vertical stability' associated with multiplier $\lambda_v(x, y) = 1$, '*hs*' means 'horizontal stability' associated with multiplier $\lambda_h(x, y) = 1$, and the numbers indicate the related CP fixed point. For example, the curve marked '*vs00*' corresponds to the condition $\lambda_v(0, 0) = 1$ that holds for $\Omega_f(0, 0) = 1$. A CP fixed point is stable if the parameter point is located on the right-hand side of both bifurcation curves, *vs* and *hs*, where the stability conditions (4.16)–(4.19) are satisfied. For the considered parameter values there are four regions corresponding to different combinations of the stable CP fixed points. Namely, in the light-blue area all four CP fixed points are stable; in the dark-gray area, the fixed points CP_{00}, CP_{01} and CP_{10} are stable; in the light-grey area CP_{00} and CP_{01} are stable; and, finally, in the green and the yellow areas, only one CP fixed point, CP_{01} or CP_{00} respectively, is stable. The 1D bifurcation diagram ϕ_D versus (x, y) shown in Fig. 4.24b corresponds to the parameter path for $\phi_C = 0.8$, $0.1 < \phi_D < 0.8$, indicated by a red arrow in Fig. 4.24a. In Fig. 4.24b, numbered red points indicate BT bifurcations, tick solid lines stable fixed points, while dashed and dotted lines correspond to saddle and repelling fixed points, respectively.

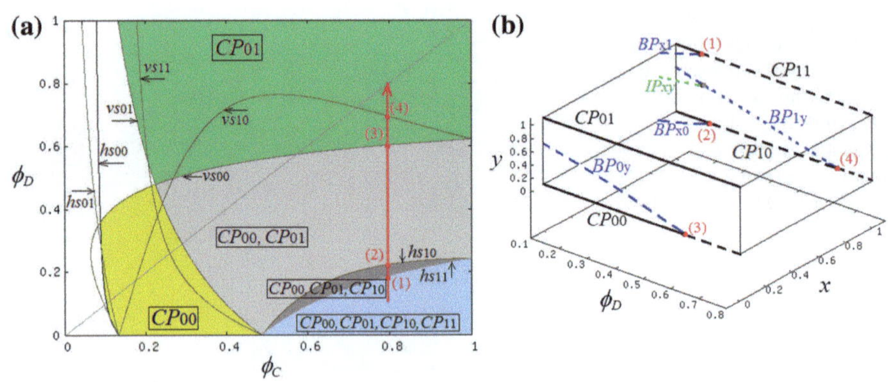

Fig. 4.24 In **a** stability regions of the CP fixed points in the (ϕ_C, ϕ_D)-parameter plane; In **b** 1D bifurcation diagram ϕ_D versus (x, y) for $\phi_C = 0.8$ related to the parameter path indicated in (**a**) by a *red arrow*. Here $L_s = 200$

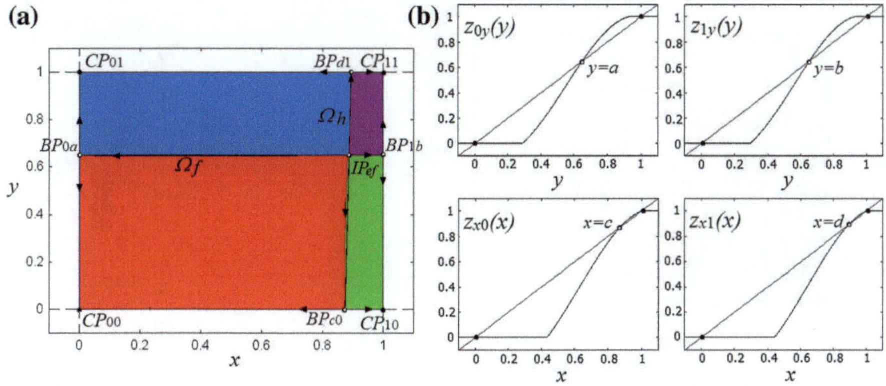

Fig. 4.25 In **a** basins of four coexisting attracting CP fixed points for $\phi_C = 0.8$, $\phi_D = 0.1$; in **b** 1D maps defined on the invariant borders of I^2. Here $L_s = 200$

An example of the basins of four coexisting attracting CP fixed points is shown in Fig. 4.25a, where $L_S = 200$, $\phi_C = 0.8$, $\phi_D = 0.1$. These basins are separated by the stable invariant sets of the saddle border fixed points BP_{0a}, BP_{1b}, BP_{c0} and BP_{d1}, whose origin is the repelling interior fixed point IP_{ef}. In Fig. 4.25b, we show the 1D maps z_{x0}, z_{x1}, z_{0y} and z_{1y} to which map Z is reduced to on the related invariant borders of I^2.

When increasing ϕ_D, the fixed point CP_{11} loses stability and becomes a saddle via a BT bifurcation (the point BP_{x1} merges with CP_{11} and at the bifurcation $\lambda_h(1, 1) = 1$, see the point (1) in Fig. 4.24). After this bifurcation three attracting CP fixed points coexist, CP_{00}, CP_{01} and CP_{10}. If ϕ_D keeps on increasing, the fixed point IP_{xy} undergoes a BT bifurcation merging with the border point BP_{1y} (at this bifurcation $\Omega_f \cap \Omega_h = IP_{xy} = BP_{1y}$ and $\lambda_h(1, y) = 1$), and then a BT bifurcation occurs for point BP_{x0}; at the same time the fixed point CP_{10} becomes a saddle (see the point (2) in Fig. 4.24). After this bifurcation only two attracting CP fixed points coexist, CP_{00} and CP_{01}. Then, the point CP_{00} loses stability and becomes a saddle via a BT bifurcation of the fixed point BP_{0y} (see point (3) in Fig. 4.24), after which only one attracting fixed point is left, namely, CP_{01}.

Let us investigate now the complete bifurcation structure of the (ϕ_C, ϕ_D)-parameter plane. The parameter region which is left uncolored in Fig. 4.24a is associated with more complex dynamics. In Fig. 4.26a we present 2D bifurcation diagram in the (ϕ_C, ϕ_D)-parameter plane where regions related to attracting cycles of different periods $n \leq 23$ are shown by different colors (periods of the largest regions are indicated by numbers), white region is associated either with cycles of higher periods, or with chaotic attractors, pink and red regions correspond to attracting border and interior fixed points, respectively, and dark gray region is related to M-attracting CP fixed points denoted CP_M. To clarify which bifurcations occur at the transitions from one area to another we show in Fig. 4.26b a 1D bifurcation diagram ϕ_C versus (x, y) (note that values of ϕ_C are shown decreasing) for fixed $\phi_D = 0.1$, which corresponds

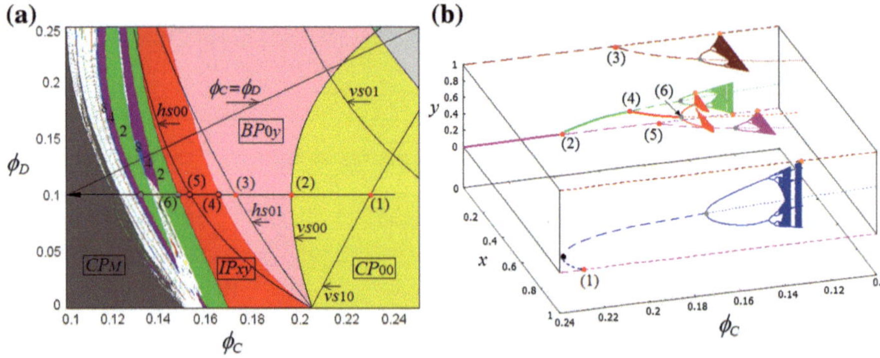

Fig. 4.26 In **a** 2D bifurcation diagram in the (ϕ_C, ϕ_D)-parameter plane; In **b** bifurcation diagram ϕ_C versus (x, y) for $\phi_D = 0.1$ related to the parameter path indicated in (**a**) by a *long black arrow*. Note that values of ϕ_C are shown decreasing. Here $L_s = 300$

to the parameter path indicated by the horizontal arrow in Fig. 4.26a. Given that only one initial point is used to produce Fig. 4.26a, coexisting attractors cannot be seen in this figures. To clarify this issue we show below several examples of basins of coexisting attractors.

Let us first comment on the bifurcations occurring for parameter values indicated by numbers (1)–(6) in Fig. 4.26. For decreasing ϕ_C first, at $\phi_C \approx 0.2295$, a BT bifurcation of the saddle fixed point CP_{10} occurs in the vertical direction (the related parameter point is indicated by (1)) at which CP_{10} becomes repelling merging with a repelling border fixed point BP_{1y} which is born a bit before, at $\phi_C \approx 0.2389$, due to a fold bifurcation (see the black circle in Fig. 4.26b) in a pair with a saddle border fixed point BP'_{1y}. Then at $\phi_C \approx 0.1962$ the attracting fixed point CP_{00} becomes locally a saddle via a BT bifurcation in the vertical direction (in Fig. 4.26 this bifurcation is marked by (2)) that leads to the appearance of an attracting border fixed point BP_{0y}. If we continue to decrease ϕ_C, at $\phi_C \approx 0.173$ the saddle fixed point CP_{01} undergoes a BT bifurcation in the horizontal direction (see the point (3) in Fig. 4.26) and becomes repelling leading to the appearance of a saddle border fixed point BP_{x1}. Then at $\phi_C \approx 0.1659$ the attracting border fixed point BT_{0y} undergoes a BT bifurcation in the horizontal direction (see the point (4)) leading to an attracting interior fixed point IP_{xy}. A BT bifurcation of CP_{00} in horizontal direction occurring at $\phi_C \approx 0.1534$ (see the point (5)) transforms the saddle fixed point CP_{00} into a repelling one and gives birth to a saddle border fixed point BP_{x0}. First period-doubling bifurcations of the border and interior fixed points are indicated in Fig. 4.26 by gray circles. Period-doubling bifurcation of the interior fixed point occurs at $\phi_C \approx 0.1481$. Each border 2-cycle undergoes a cascade of period-doubling bifurcations leading to chaos, while an interior 2-cycle undergoes a Neimark-Sacker bifurcation resulting in appearance of 2-cyclic attracting closed invariant curves (we show below an example of such an attractor). Orange circles in Fig. 4.26 indicate homoclinic

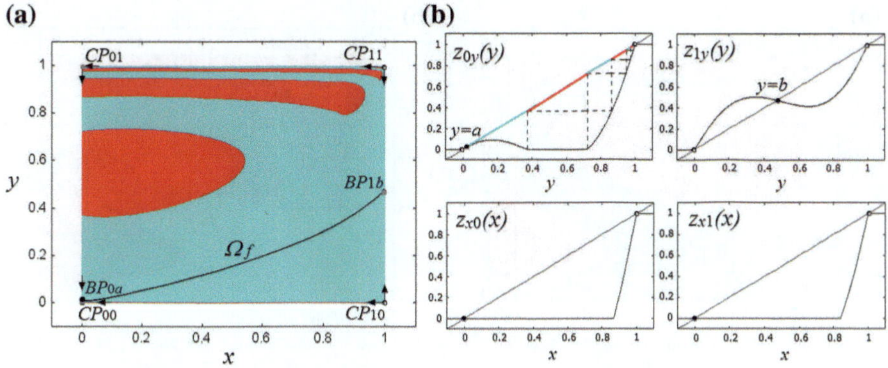

Fig. 4.27 In **a** coexisting attracting fixed point BP_{0a} (with *blue* basin) and M-attracting fixed point CP_{00} (with *red* basin) for $\phi_C = 0.196$, $\phi_D = 0.1$; in **b** 1D maps defined on the invariant borders of I^2. Here $L_s = 300$

bifurcations leading to disappearance of chaotic attractors on the related borders of I^2, as well as disappearance of an interior chaotic attractor.

Let us clarify the described above bifurcation sequence presenting several examples of basins of coexisting attractors.

For $0.1962 \lessapprox \phi_C \lessapprox 0.2295$ (see the range of ϕ_C between the points (2) and (1) in Fig. 4.26) map Z has a unique attractor which is fixed point CP_{00}, while for $0.1659 \lessapprox \phi_C \lessapprox 0.1962$ (see the range between the point (4) and (2)) map Z has coexisting attracting fixed point BP_{0a} and M-attracting fixed point CP_{00}: Fig. 4.27a shows basins of attraction of these fixed points at $\phi_C = 0.196$, and in Fig. 4.27b the 1D maps defined on the related borders of I^2 are presented. It becomes clear why fixed point CP_{00} is an M-attractor if one looks at the graph of map z_{0y} defined on border I_{0y}: even if locally the fixed point $y = 0$ is already repelling and initial points from its right neighborhood are attracted to fixed point $y = a$, it has quite a large basin of attraction shown in Fig. 4.27b in red. Indeed, any point of the interval associated with the flat branch of z_{0y}, as well as any its preimage, is mapped into $y = 0$ in a finite number of steps. Respectively, initial points belonging to the red islands Fig. 4.27a are mapped into CP_{00} in a finite number of steps (in particular, points from the largest red island are mapped into CP_{00} in one step).

For $0.1534 \lessapprox \phi_C \lessapprox 0.1659$ (see the range between the point (5) and (4) in Fig. 4.26) map Z has coexisting attracting fixed point IP_{ef} and two M-attracting fixed points, CP_{00} and BP_{0a}. An example of their basins is shown in Fig. 4.28a where $\phi_C = 0.159$.

For $0.1481 \lessapprox \phi_C \lessapprox 0.1534$ (see the range between the point (6) and (5) in Fig. 4.26) the map Z has one coexisting attracting fixed point IP_{ef} and three M-attracting fixed points, CP_{00}, BP_{0a} and BP_{c0}. An example of their basins is shown in Fig. 4.28b where $\phi_C = 0.15$.

As already mentioned, for decreasing ϕ_C the fixed point IP_{ef} undergoes a period-doubling bifurcation leading to an interior 2-cycle which then undergoes a Neimark-

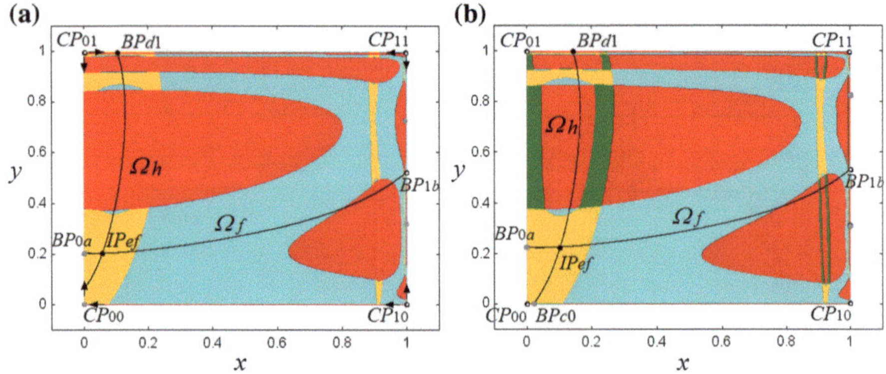

Fig. 4.28 In **a** coexisting attracting fixed point IP_{ef} (with *orange* basin), M-attracting fixed points CP_{00} and BP_{0a} (with *red* and *blue* basins, respectively) for $\phi_C = 0.159$, $\phi_D = 0.1$; In **b** coexisting M-attracting fixed points CP_{00}, BP_{0a} and BP_{c0} (with *red*, *blue* and *dark-green* basins, respectively) and attracting fixed point IP_{ef} (with *orange* basin) for $\phi_C = 0.15$

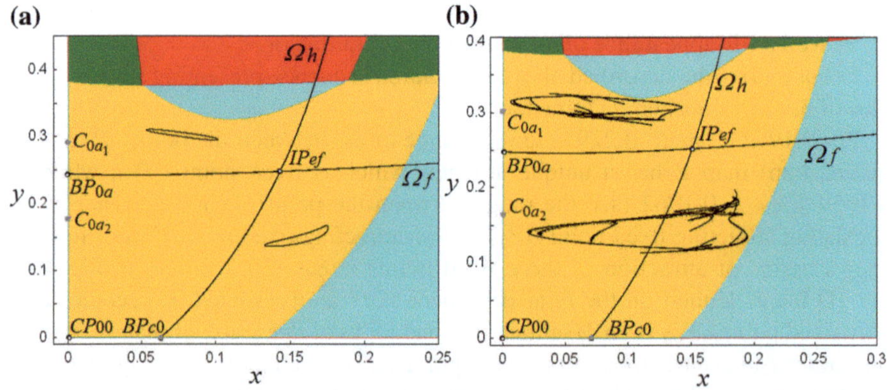

Fig. 4.29 In **a** an enlarged part of the phase plane with coexisting M-attracting fixed points CP_{00} and BP_{c0} (with *red* and *dark-green* basins, respectively), M-attracting 2-cycle $\{C_{0a_1}, C_{0a_2}\}$ (with *blue* basin) and 2-cyclic closed invariant curves (with *orange* basin) for $\phi_C = 0.1398$. In **b** An enlarged part of the phase plane with coexisting M-attracting fixed points CP_{00} and BP_{c0}, M-attracting 2-cycle $\{C_{0a_1}, C_{0a_2}\}$ and 2-cyclic chaotic attractor for $\phi_C = 0.1375$. Here $L_s = 300$

Sacker bifurcation. As a result of this bifurcation 2-cyclic attracting closed invariant curves are born. Figure 4.29a presents an enlarged part of the basin of such an attractor for $\phi_C = 0.1398$, as well as basins of coexisting M-attracting fixed points CP_{00}, BP_{c0} and M-attracting 2-cycle $\{C_{0a_1}, C_{0a_2}\}$ belonging to the border I_{0y}. If we continue to decrease ϕ_C 2-cyclic closed invariant curves are transformed in a chaotic attractor. An example of 2-cyclic chaotic attractor is shown in Fig. 4.29b for $\phi_C = 0.1375$, which is near its final bifurcation occurring due to a contact of the attractor with its basin boundary.

4.4 Conclusion and Further Research Topics

In this contribution, we presented the basic ideas of the New Economic Geography approach and a general multi-regional model. We showed how methods of dynamic system's theory can be applied to analyze these models. Thus, this contribution is also an introduction into a broader field of research that we pursued with several coauthors: In a two-region framework, we reformulated various NEG models in discrete time and studied how the dynamic properties change (see [7, 8, 20]); we analyzed the implication for regional dynamics of various policy options, such as tax policy (see [13, 14]) and productive public expenditures (see [9–11]); and we introduced first nature asymmetries between the regions and more specific expectation formation hypotheses for the firm relocation decision (see [2–4]). Most recently, we studied multi-regional NEG models and their dynamic properties (for an overview see: [12]; for specific analyses see [15–19]) and addressed questions of regional integration and international globalization.

References

1. Abreu, M.: Neoclassical growth models. In: Fischer, M., Nijkamp, P. (eds.) Handbook of Regional Science. Springer, Amsterdam (2013)
2. Agliari, A., Commedatore, P., Foroni, I., Kubin, I.: Border collision bifurcations in a footloose capital model with first nature firms. Comput. Econ. **38**, 349–366 (2011)
3. Agliari, A., Commedatore, P., Foroni, I., Kubin, I.: Expectations and industry location: a discrete time dynamical analysis. Decis. Econ. Finan. **37**, 3–26 (2014)
4. Agliari, A., Commedatore, P., Foroni, I., Kubin, I.: Agglomeration dynamics and first nature asymmetries. Math. Comput. Simul. **108**, 81–98 (2015)
5. Baldwin, R.E., Forslid, R., Martin, P., Ottaviano, G., Robert-Nicoud, F.: Economic Geography and Public Policy. Princeton University Press, Princeton (2003)
6. Caselli, F.: Accounting for cross-country income differences. In: Aghion, P., Durlauf, S. (eds.) Handbook of Economic Growth. Elsevier, Amsterdam (2005)
7. Commedatore, P., Currie, M., Kubin, I.: Footloose entrepreneurs, taxes and subsidies. Spat. Econ. Anal. **3**, 115–141 (2008)
8. Commedatore, P., Currie, M., Kubin, I.: Chaotic footloose capital, nonlinear dynamics. Psychol. Life Sci. **11**, 267–289 (2007)
9. Commedatore, P., Kubin, I., Petraglia, C.: Productive public expenditure in a new economic geography model. Economie Internationale **114**, 133–160 (2008)
10. Commedatore, P., Kubin, I., Petraglia, C.: Footloose capital and productive public services. In: Salvadori, N., Commendatore, P., Tamberi, M. (eds.) Geography, Structural Change and Economic Development: Theory and Empirics. Edward Elgar, Cheltenham (2009)
11. Commedatore, P., Kubin, I., Petraglia, C.: R&d public expenditure, knowledge spillovers and agglomeration: comparative statics and dynamics. In: Bischi, G.I., Chiarella, C., Gardini, L. (eds.) Nonlinear Dynamics in Economics, Finance and the Social Sciences. Springer, Berlin (2010)
12. Commendatore, P., Filosofo, V., Grafenender-Weissteiner, T., Kubin, I.: Towards a multiregional neg framework: comparing alternative modelling strategies. In: Commendatore, P., Kayam, S., Kubin, I. (eds.) Complexity and Geographical Economics: Topics and Tools. Springer, Cham (2015)

13. Commendatore, P., Kubin, I.: Source versus residence: a comparison from a new economic geography perspective. Forthcoming in Papers in Regional Science
14. Commendatore, P., Kubin, I.: Taxation, public expenditures and agglomeration. Economica Politica **3**, 357–386 (2013)
15. Commendatore, P., Kubin, I., Mossay, P., Sushko, I.: The role of centrality and market size in a 4-region asymmetric new economic geography model. Submitted to J. Evol. Econ
16. Commendatore, P., Kubin, I., Mossay, P., Sushko, I.: Dynamic agglomeration patterns in a two-country new economic geography model with four regions. Chaos Solitons Fractals **79**, 2–17 (2015)
17. Commendatore, P., Kubin, I., Petraglia, C., Sushko, I.: Economic integration and agglomeration in a customs union in the presence of an outside region. Working Paper 146, Vienna University of Economics and Business Administration, Department of Economics (2012)
18. Commendatore, P., Kubin, I., Petraglia, C., Sushko, I.: Regional integration, international liberalisation and the dynamics of industrial agglomeration. J. Econ. Dyn. Control **48**, 265–287 (2014)
19. Commendatore, P., Kubin, I., Sushko, I.: Typical bifurcation scenario in a three region identical new economic geography model. Math. Comput. Simul. **108**, 63–80 (2015)
20. Currie, M., Kubin, I.: Chaos in the core periphery model. J. Econ. Behav. Organ. **60**, 252–275 (2006)
21. Feenstra, R.F.: Advanced International Trade. Princeton University Press, Princeton (2004)
22. Forslid, R., Ottaviano, G.I.P.: An analytically solvable core periphery model. J. Econ. Geogr. **3**, 229–240 (2003)
23. Krugman, P.: Increasing returns and economic geography. J. Polit. Econ. **99**, 483–499 (1991)
24. Kuznetsov, Y.: Elements of Applied Bifurcation Theory. Springer, New York (1995)
25. Le Gallo, J., Fingleton, B.: Regional growth and convergence empirics. In: Fischer, M., Nijkamp, P. (eds.) Handbook of Regional Science. Springer, Amsterdam (2014)
26. Martin, P., Rogers, C.A.: Industrial location and public infrastructure. J. Int. Econ. **39**, 335–351 (1995)
27. Milnor, J.: On the concept of attractor. Commun. Math. Phys. **99**, 177–195 (1985)
28. Mira, C., Gardini, L., Barugola, A., Cathala, J.C.: Chaotic Dynamics in Two-Dimensional Noninvertible Maps. World Scientific, Singapore (1996)
29. Wiggins, S.: Introduction to Applied Nonlinear Dynamical Systems and Chaos. Springer, New York (2003)

Chapter 5
Dynamic Modeling in Renewable Resource Exploitation

Fabio Lamantia, Davide Radi and Lucia Sbragia

Abstract This chapter reviews some fundamental models related to the exploitation of a renewable resource, an important topic when dealing with regional economics. The chapter starts by considering the growth models of an unexploited population and then introduces commercial harvesting. Still maintaining a dynamic perspective, an analysis of equilibrium situations is proposed for a natural resource under various market structures (monopoly, oligopoly and open access). The essential dynamic properties of these models are explained, as well as their main economic insights. Moreover, some key assumptions and tools of intertemporal optimal harvesting are recalled, thus providing an interesting application of the theory of optimal growth.

5.1 Introduction

In this chapter we review some fundamental models about the exploitation of a single-species renewable resource.

We start by considering the growth models of an unexploited population, and then we introduce a commercial harvesting. Still maintaining a dynamic perspective, we first comment on the steady state levels of the natural resource under various market

F. Lamantia (✉)
Department of Economics, Statistics and Finance, University of Calabria,
3C Via P. Bucci, 87036 Rende, CS, Italy
e-mail: fabio.lamantia@unical.it; lamantia@unical.it

F. Lamantia
Economics—School of Social Sciences,
The University of Manchester Arthur Lewis Building, Manchester, UK

D. Radi
School of Economics and Management, LIUC - Università Cattaneo,
22 C.so Matteotti, 21053 Castellanza, VA, Italy
e-mail: dradi@liuc.it

L. Sbragia
Department of Economics, Durham University Business School, Durham, UK
e-mail: lucia.sbragia@durham.ac.uk

G.I. Bischi et al. (eds.), *Qualitative Theory of Dynamical Systems, Tools and Applications for Economic Modelling*, Springer Proceedings in Complexity,
DOI 10.1007/978-3-319-33276-5_5

structures (monopoly, oligopoly and open access). We explain the essential dynamic properties of these models as well as the main economic insights. We then review the key models of intertemporal optimal harvesting, which constitute an interesting application of the theory of optimal growth. In this case, a single agent has the exclusive right to harvest the resource and has to conceive an optimal extraction plan over an infinite planning period. We also consider the situation where many agents have at the same time the right to harvest from the same common stock of the resource, model led by dynamic optimization problems with strategic interactions, i.e., dynamic games. Starting from the 70s the game theoretic approach has been applied to study the exploitation of the fisheries and, more generally, of renewable resources. This is possible because the use of this common pool natural resource by many harvesters gives rise to numerous externalities, namely, resource stock externalities, crowding externalities and mesh externalities, just to cite some. We review in this chapter two well-known examples of dynamic games in fisheries. Finally, we end the chapter by showing how to build up an evolutionary fishery model with different time scales. The example in this chapter deals with labeling to promote the adoption of "environmentally-friendly" technologies versus standard technologies in the fishing industry.

For expositional purposes, we only consider one resource and disregard multi-species models. We simplified the treatment to provide a quick overview on the modeling of renewable resources, to show different mathematical tools at work and to suggest an accessible set of examples to a general readership. Since there are many specialized books on mathematical bioeconomics with an in-depth review of the subject, we refer the reader to these books for a more extensive analysis of the topic. Here we suggest [11] as an excellent example.

5.2 Basic Models

A general dynamic model to describe the time evolution of the stock of a renewable natural resource subject to harvesting is given, in continuous time, by the differential equation

$$\dot{x}(t) := \frac{\mathrm{d}x(t)}{\mathrm{d}t} = f(x(t)) - h(t), \tag{5.1}$$

where $x(t)$ represents the amount of fish available in a given region at a certain time t, $f(x(t))$ is the instantaneous growth rate of the resource and $h(t)$ is the instantaneous harvesting rate. Discretizing equation (5.1), i.e., replacing the time derivative $\mathrm{d}x/\mathrm{d}t$ with the difference quotient $(x(t+\Delta t) - x(t))/\Delta t$ and further measuring time such that $\Delta t = 1$, one obtains the discrete version of equation (5.1)[1]:

$$x' := x(t+1) = x(t) + f(x(t)) - h(t). \tag{5.2}$$

[1] In this chapter $'$ denotes the unit-time advancement operator, that is $x' = x(t+1)$.

In this case, time is punctuated at discrete intervals (or periods) whose duration is determined by the characteristics of the system considered (days, months, seasons, years) and $f(x(t))$ is the growth of the resource in a unit of time. Notice that $f(x(t))/x(t)$ represents the specific growth rate of the resource in unit time; $h(t)$ denotes the amount of resource removed in unit time (in the case of fish resources, it represents the landings made through fishing).

The function $f(.)$, which represents the law of evolution of the unexploited resource (the "natural growth"), allows to calculate, on the basis of the amount of this resource at time t, the amount which will be present in the successive instant of time (continuous time) or in the following period $t+1$ (discrete time). The function $f(.)$ takes into account the biological characteristics of the species and of the natural environment. The laws of motion (5.1) or (5.2) incorporate harvesting through a function $h(t)$. Harvesting decisions are usually made on the basis of social and/or economic considerations (profit maximization; conservation of the resource) within the constraints imposed by the current legislation.

In continuous, as well as in discrete time, according to equation (5.1) or (5.2), respectively, the resource is in a steady state (or equilibrium) when the values of $x(t)$ are such that $f(x(t)) = h(t)$. This equation, known as equilibrium condition, states that the population level remains constant over time if and only if the net growth in the unit time is equal to the amount of the resource removed per unit time. Obviously, if at time t it is $f(x(t)) > h(t)$ [$<$] then the population will grow [decrease] at the current time or in the following period, depending on whether time is continuous or discrete.

5.2.1 Unexploited Population Dynamics

If $h(t) = 0$, i.e., in the absence of harvesting, equations (5.1) or (5.2) describe the dynamics of the natural resource, which essentially depends on the biological properties of the population considered and the characteristics of the environment in which it lives. The simplest model assumes that in a population $x(t)$, $\alpha x(t)$ individuals are born and $mx(t)$ die at each unit of time, so that in a time interval of length Δt, the natural growth is given by

$$x(t+\Delta t) = x(t) + rx(t)\Delta t, \tag{5.3}$$

where $r = \alpha - m$ is the intrinsic growth rate. From (5.3), one obtains the continuous dynamics

$$\frac{\mathrm{d}x}{\mathrm{d}t} = rx(t) \tag{5.4}$$

by letting $\Delta t \to 0$ and the discrete dynamics

$$x(t+1) = x(t)\,(1+r) \tag{5.5}$$

by replacing $\Delta t = 1$ in (5.3).

Equations (5.4) and (5.5) are clearly linear in x. If $r < 0$, then the resource converges to the extinction in the long run at exponential speed. If $r = 0$, then the population is maintained at a constant value, equal to the initial population level. Finally, if $r > 0$, then (5.4) or (5.5) provide for an exponential growth of the population. This constitutes one of the basic principles of ecology, according to which a population, in an environment characterized by virtually unlimited quantities of vital resources and space, reproduces at a rate directly proportional to the actual number of individuals.

If the environment is characterized by finite available living resources, then (5.4) and (5.5) can be regarded as an approximation of the law of population growth in the initial stage, when the population size is small relatively to space and food availability.

When the natural environment can only offer a finite amount of living resources the death rate m is not constant, but it increases with the population size, for example $m = \beta x(t)$. This can be seen as the result of overcrowding, which causes shortages of food and living space.

In this case, the specific growth rate becomes $r = \alpha - \beta x(t)$. The point at which $r = 0$ is given by $K = \alpha/\beta$, which represents a point of balance, since it is characterized by zero growth, and it is called *carrying capacity*.

As a consequence, models (5.4) or (5.5) are nonlinear and represent the well-known equations of a logistic growth, introduced for the first time in [29]. In continuous time and in discrete time the system assumes, respectively, the following forms

$$\frac{\mathrm{d}x}{\mathrm{d}t} = \alpha x(t) - \beta\,(x(t))^2 \tag{5.6}$$

and

$$x(t+1) = x(t) + \alpha x(t) - \beta\,(x(t))^2\,. \tag{5.7}$$

Equations (5.6) and (5.7) have steady states at $x_0 = 0$ and $x_K = K$ which correspond, respectively, to the extinction (or nonexistence) of the species and to natural carrying capacity (see Fig. 5.1a).

From the definition of equilibrium, it follows that if the system is located, at a certain instant, in one of these points, there it remains also in subsequent periods. However, if we perturb the two equilibria they show different behaviors: in the case of $x_0 = 0$, a small increase in the resource will be amplified, and then the subsequent values $x(t)$ move away from that equilibrium; in the case of the equilibrium $x_K = K$, endogenous forces of the system will tend to dampen any small displacements, and bring back the value of $x(t)$ to its original equilibrium K, being $f(x) > x$ on the left of $x_K = K$ and $f(x) < x$ on the right. This explains why the carrying-capacity is considered "natural" equilibrium value of a species in its habitat.[2]

[2]New dynamic phenomena can be observed when the species has a unimodal growth function, i.e., it has maximum growth at an intermediate value of the population. In fishery models, this typically occurs when the population has the tendency to decrease when its level drops below a

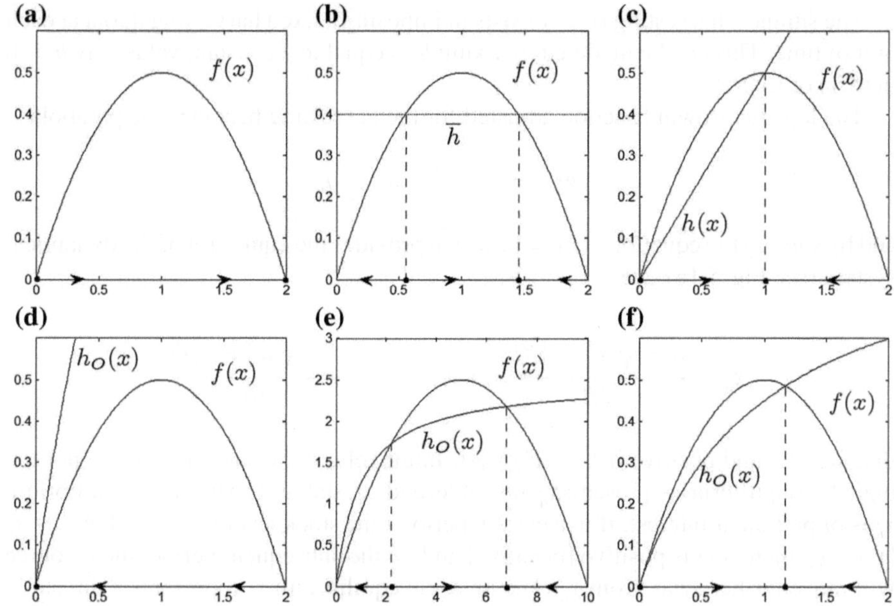

Fig. 5.1 **a** Dynamics of fish without harvesting; **b** Dynamics of fish with constant harvesting $h = \overline{h}$; **c** Dynamics of fish with harvesting $h(x) = qEx$; **d** Dynamics of fish with harvesting $h_O(x) = ax/(2[x+\gamma])$, $h'_O(0) > f'(0)$ and $h_O(\alpha/(2\beta)) > \alpha^2/(4\beta)$; **e** Dynamics of fish with harvesting $h_O(x) = ax/(2[x+\gamma])$, $h'_O(0) > f'(0)$ and $h_O(\alpha/(2\beta)) < \alpha^2/(4\beta)$; **f** Dynamics of fish with harvesting $h_O(x) = ax/(2[x+\gamma])$ and $h'_O(0) < f'(0)$

The instability of the extinction equilibrium entails that, if for any reason the population was reduced to a few units, its reproductive capacity would allow, after a certain number of periods, to return to the natural equilibrium, i.e., to the carrying-capacity level K.

In biology there are many growth functions that describe the behavior of different types of species. However, we can say that the extinction and the carrying capacity are steady states for most of them.[3]

5.2.2 Basic Harvesting Modeling

Consider again (5.1) or (5.2) with $h(t) > 0$. Function $h(t)$ can take many forms according to the socio-economic assumptions that one can make.

(Footnote 2 continued)
certain threshold, known as the *critical depensation level*. We do not deepen further this point here for the sake of time.

[3]For instance this happens under Gompertz growth, considered later on in an example of this chapter.

The simplest harvesting rule consists in imposing a fixed harvesting quota at each unit of time. This can be modeled by fixing $h(t)$ equal to a constant value, say $\overline{h} > 0$ in (5.1) or (5.2).[4]

The logistic growth function adjusted for the harvesting becomes the parabola

$$g(x) = -\beta x^2 + \alpha x - \overline{h}$$

and by solving the equation $g(x) = 0$, which provides the equilibria of the dynamical system (see Fig. 5.1b), we get

$$x_{\overline{h}} = \frac{\alpha - \sqrt{\alpha^2 - 4\beta\overline{h}}}{2\beta} \quad \text{and} \quad x_K = \frac{\alpha + \sqrt{\alpha^2 - 4\beta\overline{h}}}{2\beta},$$

that are real and positive if $\overline{h} < \alpha^2/(4\beta)$. Intuitively, if the share taken $\overline{h}$ is not too high, two equilibria are present: $x_{\overline{h}}$, unstable, and x_K, stable. The unstable equilibrium $x_{\overline{h}}$ is of particular interest: if in a certain period t the stock level x is such that $x > x_h$ [$x < x_h$], then $g(x)$ is positive [negative] and, in the subsequent period, the resource will move further away from $x_{\overline{h}}$. The unstable equilibrium $x_{\overline{h}}$ represents a threshold value, such that if the value of the stock x becomes lower than $x_{\overline{h}}$ at a certain time (for example due to an unauthorized harvesting or a higher mortality), the dynamics of the system will lead to negative values, i.e., towards the extinction in a finite time; analogously, if an external shock keeps the resource above the value $x_{\overline{h}}$, then the system will evolve spontaneously towards the stable equilibrium x_K. For this reason, the unstable equilibrium $x_{\overline{h}}$ is also said "survival threshold".

Notice that $x_K < K$, where $K = \alpha/\beta$ is the carrying capacity of the unexploited population. Both equilibrium values $x_{\overline{h}}$ and x_K depend on the size of $\overline{h}$, and as $\overline{h}$ increases $x_{\overline{h}}$ increases and x_K decreases. This means that, the equilibrium value of the species decreases and the survival threshold decreases as well, making the system more vulnerable. When the parameter $\overline{h}$ reaches the value $\overline{h} = \alpha^2/(4\beta)$ the two equilibrium points merge into each other and the parabola $g(x)$ becomes there tangent to the abscissa axis. A further increase in $\overline{h}$ leads to the disappearance of any equilibrium. So, if the harvesting exceeds $\alpha^2/(4\beta)$, which is a (fold) bifurcation value, the only possible evolution is the one that leads to extinction.

We turn now to the case where a constant effort is imposed, as suggested in [25]. By "fishing effort" we mean a parameter that represents the number of fishing vessels sent in the water, the number of hours or days spent fishing, the number of hooks employed in long-line fishing and so on. Under constant effort, harvesting is assumed to be proportional to the current stock of fish present in the water basin considered, i.e.,

$$h = qEx, \tag{5.8}$$

[4]To simplify notation, unless otherwise stated, in the following we write $x(t) = x$ and $h(t) = h$.

where E denotes the fishing effort. This functional form leads to the so called Gordon-Schaefer fishery model, see [17, 25]. The additional parameter q is referred to as the *catchability* coefficient, and it models the degree of sophistication of the harvesting technology. Obviously, the higher is q, the more productive the harvesting is. With this type of harvesting, the growth function becomes:

$$g_E(x) = -\beta x^2 + x(\alpha - qE).$$

The function $g_E(x)$ is, again, a parabola that passes through the origin of the axes and it reaches its maximum value at the resource level given by $x = (\alpha - qE)/(2\beta)$. By solving equation $g_E(x) = 0$, which provides the equilibria of the dynamical system, we get the following steady states

$$x_0 = 0 \quad \text{and} \quad x_E = \frac{\alpha - qE}{\beta}. \tag{5.9}$$

Let us start from an unexploited population, so that the stable equilibrium is the carrying capacity $K = \alpha/\beta$ (see Fig. 5.1c). Suppose that a gradual increment in fishing effort occurs. If $qE < \alpha$, the equilibrium x_E is positive and stable, while the extinction equilibrium $x_0 = 0$ is unstable. However, with an increase in the fishing effort, the equilibrium level x_E decreases. If the fishing effort E and/or the level of technology used q are increased to the value $qE = \alpha$, the two equilibria coincide; further increases of these parameters will cause a (stability-exchange) bifurcation in which the equilibrium x_E becomes unstable and negative, and the extinction equilibrium x_0 becomes stable. This means that if qE exceeds the parameter α, which is the birth rate of the species considered, the only possible evolution in the long run leads to the extinction of the resource.

Under constant effort harvesting, the long-term harvesting (the "sustainable yield"), given by $Y = qEx_E$, can be written as

$$Y = \begin{cases} \frac{\alpha qE - q^2E^2}{\beta} & \text{if } 0 < qE < \alpha, \\ 0 & \text{if } qE \geq \alpha. \end{cases}$$

Notice that the maximum production Y is obtained for an effort level $E_{\text{MSY}} = \alpha/(2q)$, at which it corresponds to the production

$$Y_{\text{MSY}} = \frac{\alpha^2}{4\beta} \tag{5.10}$$

known as the *Maximum Sustainable Yield* (MSY). Usually, an effort level such that $E > E_{\text{MSY}}$ corresponds to a situation of "overfishing", whereas the less common case with $E < E_{\text{MSY}}$ is referred to as "underfishing".

If we assume that the price of landed fish is constant[5] and equal to p and that the average and marginal cost of effort is constant at c, the profit for a representative fisherman in terms of effort E is given by

$$\pi = pqEx - cE. \tag{5.11}$$

If only one fisherman has the right to harvest the resource, then the sustainable profit (5.11) becomes

$$\pi_E = pqEx_E - cE = \frac{-pE^2q^2 + \alpha pqE}{\beta} - cE,$$

which is maximized if the fisherman exerts the *Maximum Economic Yield* (MEY) effort level

$$E_{\text{MEY}} = \frac{\alpha}{2q} - \frac{c\beta}{2pq^2}.$$

Since the MEY effort E_{MEY} is lower than the MSY effort E_{MSY}, the single owner is willing to harvest below the MSY, that is, the single owner will always underfish. However, [17] observes that if the harvesting rights do not belong exclusively to a single owner but the resource is open access, then new vessels will enter the fishery until it remains profitable, that is, until the aggregate effort is such that $\pi = 0$. This condition defines the so called *bionomic equilibrium*

$$x_{\text{BE}} = \frac{c}{qp}. \tag{5.12}$$

By equating x_{BE} to x_E in (5.9), the bionomic equilibrium effort level is obtained:

$$E_{\text{BE}} = \frac{\alpha}{q} - \frac{c\beta}{pq^2}.$$

Hence, when the resource is characterized by open access, overfishing occurs when $p > 2c\beta/(q\alpha)$, that is, provided that the selling price of the resource is sufficiently high. Note that under open access the total fishing effort is twice the level exerted by a single owner, $E_{\text{BE}} = 2E_{\text{MEY}}$.

5.3 Commercial Harvesting

Assume that total landings are sold in a market with a linear inverse demand function of the form

[5] As fish is considered a staple food for the large majority of consumers, in fisheries models price is often assumed to be constant, see, e.g. [11, 13].

$$p = a - h, \tag{5.13}$$

where p is the maximum selling price and a is the choke-off price. Harvesting cost takes the form (see [11, 28])

$$C(x, h) = \gamma \frac{h^2}{x}. \tag{5.14}$$

This cost function can be obtained from a harvesting function of a Cobb-Douglas type with effort and resource stock as inputs, and with total factor productivity ρ

$$h(x, E) = \rho x^{\lambda} E^{\mu}, \tag{5.15}$$

from which $E = \rho^{-1/\mu} x^{-\alpha/\mu} h^{1/\mu}$. Moreover, by assuming that the "production function" $h(x, E)$ is homogeneous of degree one with $\lambda = \mu = \frac{1}{2}$ and that total cost of fishing is proportional to exerted effort, i.e., $C = \delta E$, then we obtain[6]

$$C = \delta \rho^{-2} x^{-1} h^2 = \gamma \frac{h^2}{x}.$$

Without loss of generality, we assume that $\rho = 1$, so that γ can be interpreted as a measure of the inefficiency of the harvesting activity.

If a single exploiter maximizes his current profit

$$\pi = h \left[p - \gamma \frac{h}{x} \right]$$

we get the harvesting function

$$h = h_O(x) = \frac{ax}{2\left[x + \gamma\right]} \quad \text{with} \quad x \geq 0, \tag{5.16}$$

which depends on the current stock level x. Thus, a dynamical system in continuous or discrete time is obtained by substituting (5.16) in (5.1) or (5.2), respectively.

When $\gamma = 0$, i.e., when the cost for harvesting is negligible, harvesting reduces to $h = \overline{h} = a/2$. In this case, the model with constant harvesting is retrieved, for which we have already commented the main properties. When $\gamma > 0$, $h_O(x)$ is strictly increasing and strictly concave with $h_O(0) = 0$ and $\lim_{x \to \infty} h_O(x) = \frac{a}{2}$. A qualitative analysis of the model shows that the main (nondegenerate) cases are (see Fig. 5.1d,e,f):

[6]A production function (or harvesting function) of Cobb-Douglas type with fishing effort and fish biomass as production inputs and $\lambda = \mu = 1/2$, is used by several authors, see, e.g., [11, 13, 28], as it captures two fundamental aspects of fishing activity such as gear saturation and congestion. In particular, gear saturation is expressed by decreasing marginal return to fishing effort while congestion is expressed by decreasing marginal return to stock (or fish biomass).

1. $h'_O(0) > f'(0)$ and $h_O(\alpha/(2\beta)) > \alpha^2/(4\beta)$: no positive equilibrium exists. In this case, the selling price is too high and the myopic optimization of profits π leads to the convergence to the extinction equilibrium $x_0 = 0$;
2. $h'_O(0) > f'(0)$ and $h_O(\alpha/(2\beta)) < \alpha^2/(4\beta)$: two positive equilibria exist, denoted by x_h and x_K, with $x_h < x_K$. The smaller fixed point, x_h, is unstable and represents a survival threshold for the resource; on the other hand, x_K is stable and represents the carrying capacity modified for harvesting;
3. $h'_O(0) < f'(0)$: one positive equilibrium exists for all parameter values. In this case, the system is more robust, as extinction of the resource is unlikely to happen.

This model can be extended to include N identical agents. The profit of fisherman i becomes

$$\pi_i = h_i\left[p - \gamma\frac{h_i}{x}\right] = h_i\left[a - h_i - \sum_{j=1;j\neq i}^{N} h_j - \gamma\frac{h_i}{x}\right].$$

Since profits are strictly concave in harvesting, the first order condition ensures the existence of a profit maximizing harvesting level:

$$h_O(x) = \max_{h_i} \pi_i.$$

In addition, as all agents are assumed to be identical, we can write the harvesting of a single exploiter as follows

$$h = h_O(x) = \frac{ax}{(N+1)\,x + 2\gamma} \quad \text{with} \quad x \geq 0.$$

The time evolution of a fish stock harvested by a number of fishing vessels that changes over time is studied in [27]. We refer the reader to this paper for the modeling approach and the analysis of the entry of new vessels in the competition, which [27] assumes to be proportional to the profits gained by the fishing activity.

Another important contribution to the oligopolistic models of a fishery is the one in [24]. The novelty of this paper consists in the introduction of an imperfectly competitive international commercial fishing market which means that the model to solve is a duopoly with two demand functions. In the model the fish stock evolves in continuous time following the logistic equation

$$\dot{x} = \alpha x(1 - \beta x).$$

The natural resource is harvested by two countries in an open-access sea and sold in two markets: home and foreign. The inverse demand functions for fish in the two countries are linear and decreasing and given by

$$p_1 = a_1 - b_1(h_{11} + h_{21}), \quad p_2 = a_2 - b_2(h_{12} + h_{22}),$$

where h_{ij} represents the amount of fish harvested by country i and sold in country j $(i, j = 1, 2)$ and a_i and b_i are positive constants for $i = 1, 2$. The harvesting costs are similar to (5.14), that is

$$C_i = \gamma_i \frac{(h_{ii} + h_{ij})^2}{x} + c_i,$$

where γ_i and c_i are positive constants with $i = 1, 2$, which represent, respectively, a measure of country's i inefficiency to harvest and its fixed costs of harvesting. The profit of country i is then given by

$$\pi_i = p_i h_{ii} + p_j h_{ij} - C_i.$$

If the two countries behave as Cournot duopolists, by solving the first order conditions it is possible to compute the best response functions[7] as

$$\begin{aligned}
a_1 - 2b_1 h_{11} - b_1 h_{21} - 2\gamma_1 \frac{(h_{11} + h_{12})}{x} &= 0, \\
a_2 - 2b_2 h_{12} - b_2 h_{22} - 2\gamma_1 \frac{(h_{11} + h_{12})}{x} &= 0, \\
a_1 - b_1 h_{11} - 2b_1 h_{21} - 2\gamma_2 \frac{(h_{21} + h_{22})}{x} &= 0, \\
a_2 - b_2 h_{12} - 2b_2 h_{22} - 2\gamma_2 \frac{(h_{21} + h_{22})}{x} &= 0.
\end{aligned}$$

The solution of the system of the four best response functions gives the Nash equilibrium harvesting strategies from which it is possible to derive the equilibrium total harvest of country 1 as

$$H_{1\mathrm{NE}} = h_{11\mathrm{NE}} + h_{12\mathrm{NE}}$$

and the equilibrium total harvest of country 2 as

$$H_{2\mathrm{NE}} = h_{21\mathrm{NE}} + h_{22\mathrm{NE}}.$$

Finally, the equilibrium total harvest of the two countries corresponds to

$$H = H_{1\mathrm{NE}} + H_{2\mathrm{NE}} = \frac{(Dx + E)\,x}{Ax^2 + Bx + C}$$

with

[7] Here, the best response function (also known as reaction function) gives the optimal output for a country given the output of another country.

$$A = 3b_1^2 b_2^2, \quad B = 4b_1 b_2(b_1 + b_2)(\gamma_1 + \gamma_2), \quad C = 4\gamma_1\gamma_2(b_1 + b_2)^2,$$
$$D = 2b_1 b_2(a_1 b_1 + a_2 b_2), \quad E = 2\,(b_1 + b_2)\,(a_1 b_1 + a_2 b_2)(\gamma_1 + \gamma_2).$$

The differential equation governing the change of the fish stock in the presence of commercial fishing is thus given by

$$\dot{x} = \alpha x(1 - \beta x) - \frac{(Dx + E)\,x}{Ax^2 + Bx + C},$$

and the study of its long run evolution shows that the natural resource has always one steady state that corresponds to its extinction and then one or two positive steady states depending on the values of the parameters. If two positive steady states exist, then extinction can be reached only from a certain set of initial stock levels (and from the remaining ones the natural resource will reach a positive level), otherwise it is the only possible outcome. In order to avoid this the government can only affect the technical parameter for the fishing firms, i.e., their cost functions. The reader can consult [24] for details on the analysis.

5.4 Intertemporal Optimal Harvesting

5.4.1 Single Exploiter and Constant Prices

The simplest and most renown fishery model that employs the intertemporal optimization technique has been proposed in [12] and can be stated as follows

$$\max_{h \geq 0} \int_0^{+\infty} e^{-\delta t} R(h, x) \mathrm{d}t \qquad \text{s.t.} \qquad \begin{cases} \dot{x} = f(x) - h, \\ x(0) = x_0, \end{cases} \tag{5.17}$$

where $R(h, x)$ is the instantaneous profit from selling h units of fish in the market, δ is the discount factor and x is the stock available at time t.[8] If one considers the logistic growth function $f(x) = \alpha x - \beta x^2$, then the state ODE in (5.17) becomes

$$\dot{x} = f(x) - h = \alpha x - \beta x^2 + \gamma x - \gamma x - h = (\alpha + \gamma)\,x - \beta x^2 - \gamma x - h.$$

[8]Notice that (5.17) can be regarded as an optimal growth model. However, two important features present in the optimal fishery model differentiate it from a standard growth model. First, the type of resource suggests a "production" function that does not satisfy the Inada conditions. Second, the profit function depends, in general, not only on the harvesting h, but also on the level of the resource x.

If $R(h, x) = R(h)$ the profit does not depend on the level of the biomass and the model (5.17) has the same formulation as a capital accumulation model.[9]

In the fishery model the "utility" (profit) does not depend only on consumption (harvesting) but also on the level of the stock. A simple motivation for the profit R to depend both on h and x is related to the cost of harvesting: if the fish stock is abundant, then harvesting should be cheap and vice versa when the fish stock is low. For instance, assume that the instantaneous profit is given by

$$R(h, x) = \left[p - c(x)\right] h, \tag{5.18}$$

where p is the (constant) selling price of fish, $c(x)$ is the marginal cost of harvesting, which depends on the level of the resource, and h is the harvesting. We assume that the harvesting h is given by the Gordon-Schaefer equation (5.8) where q is normalized to one without loss of generality: $h = Ex$. If the average cost of effort is constant and equal to θ, then the total cost of fishing is

$$\theta E = \theta \frac{h}{x},$$

i.e., in (5.18) we have $c(x) = \theta / x$. Since it is realistic to assume that fishing effort is non-negative and below an upper bound (due to capacity constraints of the fleets), the control E must be chosen such that $0 \leq E \leq E_{\max}$. Summing up, the optimal fishery model with a constant price and stock-dependent marginal costs can be formulated as follows

$$\max_{0 \leq E \leq E_{\max}} \int_0^{+\infty} e^{-\delta t} Ex\,(p - c(x))\, \mathrm{d}t \qquad \text{s.t.} \qquad \begin{cases} \dot{x} = f(x) - Ex, \\ x(0) = x_0, \end{cases} \tag{5.19}$$

where $Ex\,(p - c(x))$ represents the instantaneous profit.

The current-value Hamiltonian for problem (5.19) is given by

$$H^c = Ex\,(p - c(x)) + \mu\,(f(x) - Ex) = Ex\,(p - c(x) - \mu) + \mu f(x). \tag{5.20}$$

Notice that (5.20) is linear in the control variable E: to maximize the current-value Hamiltonian (5.20) one requires to set $E = 0$ whenever $p - c(x) - \mu < 0$ and $E = E_{\max}$ whenever $p - c(x) - \mu > 0$. If we call $s(x) = p - c(x) - \mu$ the *switching function*, the optimal control then looks like this

$$E = \begin{cases} 0 & \text{if } s(x) < 0, \\ E_{\max} & \text{if } s(x) > 0. \end{cases} \tag{5.21}$$

[9] As remarked in the previous note, the fishery "production" function $f(x) = (\alpha + \gamma)x - \beta x^2$ does not satisfy the Inada conditions, in particular, $\lim_{x \to 0^+} f'(x) \neq +\infty$ and $\lim_{x \to +\infty} f'(x) \neq 0^+$.

When $s(x) = 0$ we obtain $\mu = p - c(x)$, and we can compute the time derivative of μ which gives

$$\dot{\mu} = -c'(x)\dot{x} = -c'(x)(f(x) - Ex).$$

Moreover, an optimal μ has to satisfy also the equation of motion of the costate variable, which is given by

$$\begin{aligned}\dot{\mu} &= -\frac{\partial H^c}{\partial x} + \mu\delta = -E\left[\underbrace{p - c(x) - \mu}_{=0}\right] + Exc'(x) - \mu f'(x) + \mu\delta \\ &= Exc'(x) - \mu\left(f'(x) - \delta\right) = Exc'(x) - (p - c(x))\left(f'(x) - \delta\right),\end{aligned}$$

where the last line follows from $p - c(x) - \mu = 0$ in the interval. By equating the two expressions for $\dot{\mu}$

$$-c'(x)(f(x) - Ex) = Exc'(x) - (p - c(x))\left(f'(x) - \delta\right),$$

we obtain

$$f'(x^*) - \frac{c'(x^*)f(x^*)}{p - c(x^*)} = \delta. \tag{5.22}$$

Equation (5.22) is called the *Modified Golden Rule* of the fishery (with constant price). The particular level of resource x^* which solves (5.22) is called *singular solution*. Being x^* constant, it follows that $\dot{x} = f(x^*) - Ex^* = 0$, from which we get that the optimal effort, when $s(x) = 0$, is $E^* = f(x^*)/x^*$ and called the *singular control*. The optimal controls constitute the so-called *Most Rapid Approach* harvesting: when the resource is abundant (i.e., $x > x^*$), then it is optimal to harvest as much as possible (i.e., $E = E_{\max}$) until the resource converges to x^* and then apply always the singular control. Analogously, when the resource is scarce (i.e., $x < x^*$), then it is optimal not to harvest (i.e., $E = 0$) until the resource grows to the level x^*, to which there remains to employ the singular control. Thus, if we want to express the optimal control E^* as a function of the state of the system (stock of resource), we can write it in feedback form as follows

$$E^* = \begin{cases} 0 & \text{if } x < x^*, \\ \frac{f(x^*)}{x^*} & \text{if } x = x^*, \\ E_{\max} & \text{if } x > x^*. \end{cases}$$

5.4.2 Single Exploiter and Nonconstant Prices

We now revisit the previous example and allow for nonconstant prices, along the lines of [11]. The problem can be stated as follows

$$\max_{h\geq 0}\int_0^{+\infty} e^{-\delta t}h\left[p(h)-c\left(x\right)\right]\mathrm{d}t \qquad \text{s.t.} \qquad \begin{cases}\dot{x}=f(x)-h,\\ x(0)=x_0,\end{cases} \tag{5.23}$$

where h is the control variable,[10] $p(h)$ is the inverse demand function that depends on the harvest and $\pi(h)=h\left(p(h)-c\left(x\right)\right)$ is the instantaneous profit. The current-value Hamiltonian for this problem is

$$H^c = h\left(p(h)-c\left(x\right)\right)+\mu\left(f(x)-h\right). \tag{5.24}$$

With positive harvesting $h>0$, the optimal control must satisfy the condition

$$\frac{\partial H^c}{\partial h}=p(h)+hp'(h)-c(x)-\mu=0,$$

from which it follows

$$\mu=p(h)+hp'(h)-c(x)=\pi'(h). \tag{5.25}$$

The costate equation is given by

$$\dot{\mu}=-\frac{\partial H^c}{\partial x}+\mu\delta=hc'(x)+\mu\left(\delta-f'(x)\right). \tag{5.26}$$

Differentiating (5.25) with respect to time, one gets $\dot{\mu}=\pi''(h)\dot{h}$ that equated to (5.26) gives

$$hc'(x)+\mu\left(\delta-f'(x)\right)=\pi''(h)\dot{h},$$

so that, by using again (5.25), the dynamics of the optimal harvesting must satisfy the condition

$$\dot{h}=\frac{hc'(x)+\pi'(h)\left(\delta-f'(x)\right)}{\pi''(h)}. \tag{5.27}$$

This last ODE, the stock dynamics $\dot{x}=f(x)-h$ and the proper transversality condition constitute the necessary conditions that the optimal solution couple (harvesting and resource dynamics) must satisfy. From the conditions $\dot{h}=\dot{x}=0$, it is possible to obtain a steady state of the system x^*, which satisfies the "modified golden rule" condition

$$\frac{f(x^*)c'(x^*)+\pi'(f(x^*))\left(\delta-g'(x^*)\right)}{\pi''(f(x^*))}=0,$$

that can be rewritten as

[10] In the previous example we have $h=Ex$. Here we reason directly in terms of harvesting h for the sake of simplicity.

$$f'(x^*) - \frac{c'(x^*)f(x^*)}{\pi'(f(x^*))} = \delta, \tag{5.28}$$

which is the analogous of (5.22) with nonconstant price.

For instance, if we assume that the price of fish follows the linear inverse demand function

$$p(h) = a - bh,$$

that the cost of harvesting is independent on the stock ($c(x) = c$), and the fish stock grows following a logistic function

$$f(x) = x(\alpha - \beta x), \tag{5.29}$$

then the optimal control problem (5.23) becomes

$$\max_{h\geq 0} \int_0^{+\infty} e^{-\delta t} \left[mh - bh^2\right] \mathrm{d}t \qquad \text{s.t.} \qquad \begin{cases} \dot{x} = x(\alpha - \beta x) - h, \\ x(0) = x_0, \end{cases}$$

where $m = a - c$ represents the unitary markup and $\pi(h) = mh - bh^2$ is the instantaneous profit. The correspondent current-value Hamiltonian is thus given by

$$H^c = mh - bh^2 + \mu(x(\alpha - \beta x) - h),$$

and by applying the above stated FOCs we obtain the following nonlinear system of ODEs that the solution candidate must satisfy:

$$\begin{cases} \dot{x} = x(\alpha - \beta x) - h, \\ \dot{h} = \frac{\pi'(h)(\delta - f'(x))}{\pi''(h)} = -\frac{(m-2bh)(\delta-\alpha+2\beta x)}{2b}. \end{cases} \tag{5.30}$$

Notice, in particular, that H^c is concave with respect to x and h. In the (x, h) plane, the nullcline $\dot{x} = 0$ is represented by the parabola $f(x)$ in (5.29); the nullcline $\dot{h} = 0$ is represented by two lines: one horizontal line of equation $h = m/(2b)$, which maximizes instantaneous profits being there $\pi'(h) = 0$, and one vertical line of equation $x = (\alpha - \delta)/(2\beta)$, which corresponds to the modified golden rule stock, since in this case (5.28) reduces to $f'(x^*) = \delta$ being $c'(x) = 0$. Equilibria of the ODE system (5.30) are obtained by solving $\dot{x} = \dot{h} = 0$ with respect to x and h. Hereafter, for the sake of simplicity we set $b = \frac{1}{2}$.

When $x = (\alpha - \delta)/(2\beta)$ (vertical nullcline $\dot{h} = 0$), a unique equilibrium exists

$$E_1 = (x_1, h_1) = \left(\frac{\alpha - \delta}{2\beta}, \frac{(\alpha - \delta)(\alpha + \delta)}{4\beta}\right),$$

which is meaningful provided that $\alpha \geq \delta$. As already observed, x_1 in E_1 is the (modified) golden rule level (5.28).

When $h = m/(2b) = m$ (horizontal nullcline $\dot{h} = 0$), from the first equation in (5.30) we get the following two equilibrium values (x, h):

$$E_2 = \left(\frac{\alpha - \sqrt{\alpha^2 - 4m\beta}}{2\beta}, m \right), \quad E_3 = \left(\frac{\alpha + \sqrt{\alpha^2 - 4m\beta}}{2\beta}, m \right).$$

The first coordinate in E_2 and E_3 are real numbers provided that $\alpha^2 - 4m\beta \geq 0$. When $\alpha^2 - 4m\beta < 0$, i.e., when

$$m > \frac{\alpha^2}{4\beta}, \tag{5.31}$$

E_2 and E_3 are not meaningful. From an economic point of view, the harvesting that would maximize the instantaneous profits ($h = m$) is "out-of-reach", as it is greater than the maximum value that the parabola (5.29) can take, $f(\alpha/(2\beta)) = \alpha^2/(4\beta)$, which is the maximum sustainable yield (MSY) already considered in (5.10).[11] Thus, the model has different number of equilibria and different dynamic properties according to the sign of $\alpha^2 - 4m\beta$.

If $\alpha^2 - 4m\beta < 0$, i.e., if condition (5.31) holds, then the unique equilibrium of the ODE (5.30) is E_1. The Jacobian matrix of (5.30) is

$$J = \begin{bmatrix} \frac{\partial \dot{x}}{\partial x} & \frac{\partial \dot{x}}{\partial h} \\ \frac{\partial \dot{h}}{\partial x} & \frac{\partial \dot{h}}{\partial h} \end{bmatrix} = \begin{bmatrix} \alpha - 2\beta x & -1 \\ 2\beta(h - m) & -\alpha + 2\beta x + \delta \end{bmatrix},$$

from which we calculate

$$J(E_1) = \begin{bmatrix} \delta & -1 \\ \frac{\alpha^2 - 4m\beta - \delta^2}{2} & 0 \end{bmatrix}.$$

Clearly $\mathrm{Tr}(J) = \delta > 0$ and $\mathrm{Det}(J) = (\alpha^2 - 4m\beta - \delta^2)/2$. Notice that $\mathrm{Det}(J) < 0$ is equivalent to $m > (\alpha^2 - \delta^2)/(4\beta)$. Because of (5.31), we can conclude that when system (5.30) admits the unique equilibrium E_1, then E_1 is a saddle point.[12] For any initial condition $x(0) \neq (\alpha - \delta)/(2\beta)$, the optimal trajectory belongs to the stable manifold of the saddle point E_1, to which the solution converges in the long run. Notice also that in this case, the optimal time path is analogous to the one with

---

[11] Notice that $\alpha/(2\beta)$ is indeed the golden rule level of the stock. To be more precise, from $\dot{x} = 0$, we get that at equilibrium there is $\overline{h} = x(\alpha - \beta x)$, so that instantaneous profit is $\overline{\pi} = m\overline{h} - b\overline{h}^2 = m[x(\alpha - \beta x)] - b[x(\alpha - \beta x)]^2$. Instantaneous profit $\overline{\pi}(x)$ is maximized by $x = \alpha/(2\beta)$ if $m > \alpha^2/(4\beta)$, since $\overline{\pi}'(\alpha/(2\beta)) = 0$ and $\overline{\pi}''(\alpha/(2\beta)) < 0$. However, for $m < \alpha^2/(4\beta)$, $\overline{\pi}(x)$ has minimum at $x = \alpha/(2\beta)$, whereas $\overline{\pi}(x)$ is maximized by the "golden rule" levels given by the state values in E_2 and E_3, namely, by $x = \alpha \pm \sqrt{\alpha^2 - 4m\beta}/(2\beta)$.

[12] This equilibrium is a solution of system (5.30) and, being constant, satisfies a transversality condition of the form $\lim_{t\to+\infty} \pi'(h_1)e^{-\delta t} = 0$, which implies $\lim_{t\to+\infty} H^c e^{-\delta t} = 0$.

constant prices. However, with constant prices the optimal control is a combination of bang-bang controls with the singular control, whereas with nonconstant prices the optimal control moves along the stable manifold of the saddle point.

Figure 5.2 shows this case, where the red curves are the nullclines. Different colors correspond to different signs of the vector field (5.30), namely, $\dot{x} > 0$ and $\dot{h} > 0$ (pink region), $\dot{x} > 0$ and $\dot{h} < 0$ (blue region), $\dot{x} < 0$ and $\dot{h} < 0$ (yellow region), $\dot{x} < 0$ and $\dot{h} > 0$ (white region). The equilibrium E_1, which is given by the intersection between the parabola and the vertical line, is a saddle point.

If $\alpha^2 - 4m\beta > 0$, then all three equilibria exist. After E_2 and E_3 are created, through a fold bifurcation for (5.30) at $m = \alpha^2/(4\beta)$ (see Fig. 5.3a), E_1 remains a saddle point as long as $m > (\alpha^2 - \delta^2)/(4\beta)$.

At E_2, the Jacobian matrix is given by

$$J(E_2) = \begin{bmatrix} \sqrt{\alpha^2 - 4m\beta} & -1 \\ 0 & -\sqrt{\alpha^2 - 4m\beta} + \delta \end{bmatrix}$$

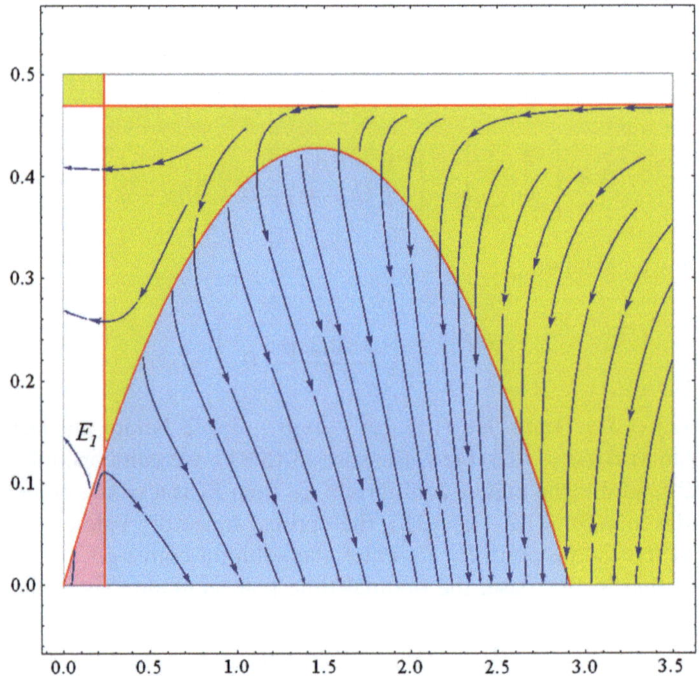

Fig. 5.2 Typical phase portrait of dynamical system (5.30) when $\alpha^2 - 4m\beta < 0$. *Red curves* are the nullclines, which separate regions where the vector field has different signs: $\dot{x} > 0$ and $\dot{h} > 0$ (*pink region*), $\dot{x} > 0$ and $\dot{h} < 0$ (*blue region*), $\dot{x} < 0$ and $\dot{h} < 0$ (*yellow region*), $\dot{x} < 0$ and $\dot{h} > 0$ (*white region*)

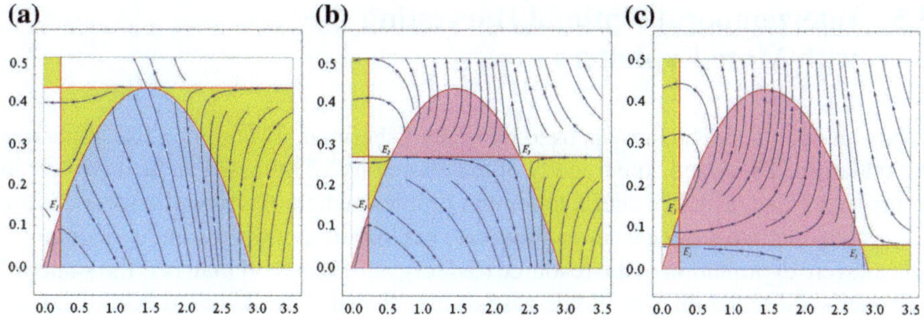

Fig. 5.3 Phase portraits of dynamical system (5.30) with $\alpha^2 - 4m\beta \geq 0$. **a** $\alpha^2 - 4m\beta = 0$, at the fold bifurcation with the creation of equilibria E_2 and E_3; **b** $m \in \big((\alpha^2 - \delta^2)/(4\beta), \alpha^2/(4\beta)\big)$; **c** $m \in \big(0, (\alpha^2 - \delta^2)/(4\beta)\big)$. Again, *red curves* are the nullclines, which separate regions where the vector field has different signs: $\dot{x} > 0$ and $\dot{h} > 0$ (*pink region*), $\dot{x} > 0$ and $\dot{h} < 0$ (*blue region*), $\dot{x} < 0$ and $\dot{h} < 0$ (*yellow region*), $\dot{x} < 0$ and $\dot{h} > 0$ (*white region*)

and the eigenvalues are the entries along the diagonal. We note that $-\sqrt{\alpha^2 - 4m\beta} + \delta > 0$ for $m \in \big((\alpha^2 - \delta^2)/(4\beta), \alpha^2/(4\beta)\big)$, hence, for $m \in \big((\alpha^2 - \delta^2)/(4\beta), \alpha^2/(4\beta)\big)$, E_2 is an unstable node and E_1 is a saddle point (see Fig. 5.3b). When $m \in \big(0, (\alpha^2 - \delta^2)/(4\beta)\big)$, E_2 is a saddle point and E_1 is an unstable node (Fig. 5.3c). At $m = (\alpha^2 - \delta^2)/(4\beta)$ a transcritical bifurcation occurs at which E_2 and E_1 exchange their stability properties.

At E_3, the Jacobian matrix is given by

$$J(E_3) = \begin{bmatrix} -\sqrt{\alpha^2 - 4m\beta} & -1 \\ 0 & \sqrt{\alpha^2 - 4m\beta} + \delta \end{bmatrix}$$

with eigenvalues, again, on the diagonal. Since $\sqrt{\alpha^2 - 4m\beta} + \delta > 0$ for all $m \in \big(0, \alpha^2/(4\beta)\big)$, we conclude that E_3, when exists, is a saddle point.

Summing up, for $m \in \big(\alpha^2/(4\beta), +\infty\big)$, with $\alpha > \delta$, the only equilibrium is the saddle point E_1. This equilibrium constitutes the optimal equilibrium of the system and the stable manifold of the equilibrium is the optimal combination of state and control to maximize profits, i.e., it provides the feedback control (see Fig. 5.2).

When $m \in \big((\alpha^2 - \delta^2)/(4\beta), \alpha^2/(4\beta)\big)$ the saddle point E_1 coexists with the unstable node E_2 and the saddle point E_3. Depending on the initial conditions of the system, the optimal trajectory is either along the stable manifold of the saddle point E_1 or along the stable manifold of the saddle point E_3 (see Fig. 5.3b).

When $m \in \big(0, (\alpha^2 - \delta^2)/(4\beta)\big)$, the saddle point E_2 coexists with the unstable node E_1 and the saddle point E_3: when $x(0) > (\alpha - \delta)/(2\beta)$ it is optimal to harvest $h = m$ in perpetuity and the system converge to E_3; when $x(0) < (\alpha - \delta)/(2\beta)$, then it is convenient to stay along the stable manifold of the saddle point E_2 and to let the resource grow until it is sustainable to harvest $h = m$ in perpetuity (see Fig. 5.3c). Notice that in this scenario, the modified golden rule does not identify the optimal biomass level since it is optimal to harvest according to the golden rule.

5.5 Intertemporal Optimal Harvesting with More Exploiters

When an open access fishery is used by many fishermen, its common use generates what Hardin [18] called "the tragedy of the commons". This case can be modeled, in terms of a game, as a Prisoner's Dilemma problem where the two prisoners are the fishermen, the two strategies are a moderate exploitation of the resource (cooperate) and a severe exploitation of the resource (defect), and the payoff matrix is the standard Prisoner's Dilemma matrix. In this case of the "Fisherman's Dilemma Game", both fishermen have as a dominant strategy a severe exploitation of the resource, and these are played in the Nash equilibrium with negative consequences on the fish population.

Here we present two main contributions to the application of a game-theoretic approach with dynamic optimization to the fisheries. The first one is the difference game proposed in [23]. This model focuses on what the authors call the dynamic externality, which is the bioeconomic loss that arises when a single dynamic population is exploited by a finite number of fishers. The second model is the differential game considered in [14]. This model extends the Gordon-Schaefer fishery model by introducing a market externality, that is, the price of landed fish is no more constant, but it depends on the quantity harvested by all fishermen, so that interactions in the marketplace do matter.

In [23], the resource stock develops over (discrete) time according to the biological growth rule given by

$$x' = x^{\alpha}, \quad 0 < \alpha < 1,$$

with carrying capacity normalized to 1 if no exploitation of the fishery occurs.

The resource is harvested by two countries denoted by $i = 1, 2$. Each country receives a utility from the present consumption of fish h_i in the form

$$U_i(h_i) = \log (h_i) .$$

The objective of each player is to maximize the sum of his own discounted utilities over an infinite horizon.

Three different equilibrium concepts, namely, the Cournot-Nash equilibrium, the cooperative equilibrium and the Stackelberg equilibrium are applied to compute the equilibrium harvest of the two countries and then study their impact on the evolution of the resource.

When each country acts as a Cournot duopolist in a dynamic framework, it takes the policy of the other participant as given while trying to maximize the sum of his own discounted utilities. To solve the problem, discrete-time dynamic programming is used as a maximization technique. The Cournot-Nash polices are first computed for a finite horizon, then, by letting the horizon tend to infinity, the infinite horizon Cournot-Nash polices are obtained and, consequently, the corresponding size of the fish population.

If there is only one future period, country 1's maximization problem can be written as

$$\max_{0\le h_1\le x-h_2}\left(\log h_1 + r\log\left(\frac{1}{2}(x-h_1-h_2)^\alpha\right)\right),$$

where r is country 1's discount factor and there is an underlying assumption that if there were no future periods, each country would get an equal share of all the remaining fish.

By solving the first order condition, country 1's reaction function is computed as

$$h_1 = \frac{x-h_2}{r\alpha+1}, \tag{5.32}$$

and by repeating a similar procedure for country 2 its reaction function is given by

$$h_2 = \frac{x-h_1}{d\alpha+1}, \tag{5.33}$$

where d is country 2's discount factor.

By solving the system of (5.32) and (5.33), the Cournot-Nash equilibrium is obtained as

$$\begin{aligned} h^1_{1\mathrm{NE}} &= \frac{\alpha d}{\alpha d+\alpha r+\alpha^2 dr}x, \\ h^1_{2\mathrm{NE}} &= \frac{\alpha r}{\alpha d+\alpha r+\alpha^2 dr}x \end{aligned}$$

with a remaining stock given by

$$x - h^1_{1\mathrm{NE}} - h^1_{2\mathrm{NE}} = \frac{\alpha^2 dr}{\alpha d+\alpha r+\alpha^2 dr}x.$$

The problem can be extended to consider a two-period horizon under the assumption that country 1 reacts to the second country in the present supposing that in the future the one-period horizon Cournot-Nash solution, given above, will prevail. By computing the best response functions for both countries and by solving their system, the Cournot-Nash equilibrium for the two-period horizon case can be found. The process can be repeated for an n-period horizon, and then by taking the limit to infinity, the equilibria for the infinite horizon problem are derived as

$$\begin{aligned} h_{1\mathrm{NE}} &= \frac{d\alpha\,(1-d\alpha)\,x}{1-(1-d\alpha)\,(1-r\alpha)}, \\ h_{2\mathrm{NE}} &= \frac{r\alpha\,(1-r\alpha)\,x}{1-(1-d\alpha)\,(1-r\alpha)}, \\ x - h_{1\mathrm{NE}} - h_{2\mathrm{NE}} &= \frac{\alpha^2 drx}{\alpha d+\alpha r-dr\alpha^2}. \end{aligned}$$

Now the long run evolution of the fish stock can be studied by computing its steady state, that is,

$$x_{\mathrm{NE}} = \left[\frac{\alpha^2 dr}{\alpha d + \alpha r - dr\alpha^2}\right]^{\frac{\alpha}{1-\alpha}} .$$

If countries are symmetric, that is, they have the same rate of time preferences ($d = r$), the Nash equilibrium over an infinite horizon for identical players is given by

$$h_{1\mathrm{NE}} = h_{2\mathrm{NE}} = \frac{1-\alpha r}{2-\alpha r} x$$

and the correspondent steady state of the resource is

$$x_{\mathrm{NE}_{\mathrm{Sym}}} = \left(\frac{\alpha r}{2 - r\alpha}\right)^{\frac{\alpha}{1-\alpha}} ,$$

which is smaller than the natural steady state.

In the cooperative case, countries combine their harvest so as to maximize the discounted sum of both countries' utilities. The identical rate of time preference serves as the common discounting factor and the problem to solve becomes

$$\max_h \sum_{t=0}^{\infty} r^t (2 \log h) \qquad \text{s.t.} \qquad x' = (x - 2h)^\alpha .$$

In the case of a cooperative resource management, the equilibrium consumption of the resource over an infinite horizon is

$$h_{\mathrm{C}} = \frac{1-\alpha r}{2} x$$

and the corresponding steady state of the resource is given by

$$x_{\mathrm{C}} = (\alpha r)^{\frac{\alpha}{(1-\alpha)}} .$$

Finally, if we assume that country 1 is more sophisticated than country 2, he acts as a leader and country 2 as a follower. Similarly to the Cournot case, a leader-follower game is played with one, two and n future periods, and then by taking the limit to infinity, the equilibrium consumptions for the infinite horizon problem are obtained as

$$h_{1\mathrm{S}} = (1 - \alpha d)x,$$
$$h_{2\mathrm{S}} = \alpha d(1 - \alpha r)x$$

with the remaining stock given by

$$x - h_{1S} - h_{2S} = (\alpha r)\,(\alpha d)\,x.$$

The long run steady state of the resource corresponds to

$$x_S = [(\alpha d)\,(\alpha r)]^{\frac{\alpha}{(1-\alpha)}}.$$

Two are the main results that need to be highlighted. Firstly,

$$h_{\mathrm{NE_{sym}}} < h_C,$$

which means that in the cooperative case

> the two countries will consume, for each level of population of fish, smaller quantities of fish, but will be able to achieve a higher "permanent" catch. Hence, the conflict implicit in the duopoly problem leads both countries to overconsume, with less left for future generations ([23, p. 253]).

Secondly, the resource steady states can be ranked in the following order

$$x_C > x_{\mathrm{NE_{Sym}}} > x_{\mathrm{S_{Sym}}}.$$

Turning to the game proposed in [14], the natural evolution of the fish stock is described by the Gompertz growth function

$$\dot{x} = x(\alpha - \rho \ln x),$$

and the economic sub-model assumes that there are two players denoted by $i = 1, 2$, who harvest by using a constant fishing effort function, that is, the harvesting rate of player i is given by the Gordon-Schaefer equation (5.8). The marginal (and average) cost of effort for player i is constant and given by c_i, and each country has a different time preference described by the discount rate r_i. The inverse demand function of the landed fish is assumed to be hyperbolic (isoelastic) of the form

$$p(h_1 + h_2) = \frac{1}{h_1 + h_2} \quad \Leftrightarrow \quad p(E_1 x + E_2 x) = \frac{1}{(E_1 + E_2)x} \text{ with } q_i = 1.$$

After transforming the state equation by using the state transformation: $z = \ln x$, the problem for country i can be written as

$$\max_{E_i > 0} \left\{ J_i = \int_0^\infty \left[\frac{1}{h_1 + h_2} - c_i \right] E_i e^{-r_i t} \mathrm{d}t \right\} \tag{5.34}$$
$$\text{s.t.} \begin{cases} \dot{z} = \alpha - \rho z - E_1 - E_2, \\ z(0) = \ln x_0. \end{cases}$$

This dynamic game can be easily solved as a static game because in state-separable games the determination of Nash optimal controls can be done separately from the determination of the state variables, see [16] for details on this point and, in general, on differential games.

We solve the model for a Nash and a Stackelberg solutions. For the Nash solution, player 1 solves the static problem

$$\max_{E_1>0}\left\{J_1=\left[\frac{E_1}{E_1+E_2}-c_1E_1\right]\right\}.$$

By computing the first order condition, we derive the best response function

$$E_2-c_1\left(E_1+E_2\right)^2=0. \tag{5.35}$$

Player 2 solves a similar problem and reacts to player 1 by using

$$E_1-c_2\left(E_1+E_2\right)^2=0. \tag{5.36}$$

The equilibrium of the game is obtained by solving the system of the two best response functions (5.35) and (5.36), which gives

$$E_{1\mathrm{NE}}=\frac{c_2}{(c_1+c_2)^2},\qquad E_{2\mathrm{NE}}=\frac{c_1}{(c_1+c_2)^2}. \tag{5.37}$$

For the Stackelberg solution, player 1 is assumed to be the leader and player 2 acts as a follower. Player 2 reaction function $E_2\left(E_1\right)$ is used by player 1 in his profit maximization problem:

$$\max_{E_1>0}\left\{J_1=\left[\frac{E_1}{E_1+E_2\left(E_1\right)}-c_1E_1\right]\right\}\quad\Leftrightarrow\quad\max_{E_1>0}\left\{J_1=\left[E_1^{0.5}c_2^{0.5}-c_1E_1\right]\right\},$$

which solves for

$$E_{1\mathrm{S}}=\frac{c_2}{4c_1^2},\qquad E_{2\mathrm{S}}=\frac{2c_1-c_2}{4c_1^2}. \tag{5.38}$$

By replacing the equilibrium efforts of both players (5.37) and (5.38) in (5.34) and by solving the integral, it is possible to compute the equilibrium payoffs for all players under the two cases.[13]

A comparison of the above results shows that

1. In both the Nash and the Stackelberg cases, the player with the smaller average cost is able to choose higher catch rates than his opponent.
2. In both the Nash and the Stackelberg cases, with equal discount rates, the player with the smaller unit cost is able to gain a higher payoff than his opponent. This

---

[13]They can also be computed when countries have equal discount factors: $r_1=r_2=r$.

means that a more efficient technology eliminates any information disadvantage in the sense of Stackelberg followership.
3. If the follower has a cost advantage over the leader, in the Stackelberg solution the total catch is smaller, the price is higher, as well as the profits, than in the correspondent Nash solution.

Given the equilibrium solutions (5.37) and (5.38), it is also interesting to study what happens to the long run evolution of the resource under the two different information scenarios. If countries act simultaneously, the evolution of the fish stock is described by

$$\dot{x} = x\,(\alpha - \rho \ln x) - \frac{1}{c_1 + c_2} x$$

with steady states given by

$$x^* = 0 \quad \text{and} \quad x^* = \exp\left(\frac{\alpha - \frac{1}{c_1+c_2}}{\rho}\right).$$

If country 1 becomes a leader and country 2 is a follower, the dynamics of the fisheries is given by

$$\dot{x} = x\,(\alpha - \rho \ln x) - \frac{1}{2c_1} x$$

with steady states corresponding to

$$x^* = 0 \quad \text{and} \quad x^* = \exp\left(\frac{\alpha - \frac{1}{2c_1}}{\rho}\right).$$

Note that in both cases the positive steady state is stable, since $\frac{\mathrm{d}\dot{x}}{\mathrm{d}x}(x^*) = -\rho < 0$. Moreover, if the leader has a cost disadvantage ($c_1 > c_2$), then the Stackelberg solution has to be preferred to the Nash solution in terms of the long run evolution of the fish stock.

5.6 Evolutionary Adoption of Harvesting Technology

In this section, we briefly consider some evolutionary fishery models recently proposed in the specialized literature. These models can be considered as an extension of the evolutionary oligopoly models, see [1, 8, 15, 21] for some examples. In general, in environmental economics, evolutionary models have been employed to study cooperative versus noncooperative behaviors in the exploitation of a fishery (see [3]), the establishment of protected areas (see [4, 9]) and related issues.

Here we recall a model recently proposed in [22] to describe a fishery where two different harvesting technologies can be employed: a standard one and one that is more environmentally friendly, that is, less intensive. In this model the resource is assumed to be common-pool, that is, N harvesters have the right to exploit the resource, but new exploiters can not enter the fishery. These exploiters are assumed to select one of the two harvesting technologies that are available according to a profit-driven adaptive mechanism based on the evolutionary selection rule, known as replicator dynamics.[14] In particular, fishermen can decide to employ a less efficient but more "environmentally-friendly" fishing technology if the loss in efficiency is counterbalanced by a higher price that consumers might be willing to pay for the green product. In practice, the use of the green technology is indicated by an eco-label of the product, such as the dolphin-safe labeling. The choice between the two types of technology depends only on agents' assessment of the expected profits and not on ethical or environmental concerns, as agents are assumed to be selfish profit maximizers. Exploiters have to make two choices over time: the technology to adopt and, given that, the quantity to harvest.

As before, let us assume that the resource follows a logistic growth such as (5.6) or (5.7) and is subjected to harvesting. Mathematically, the dynamics of the resource can be modeled as

$$\dot{x} = x(\alpha - \beta x) - h(x) \tag{5.39}$$

in continuous time or as

$$x(t+1) = x(t) + \alpha x(t) - \beta (x(t))^2 - h(x(t)) \tag{5.40}$$

in discrete time. Function $h(x)$ in (5.39) and in (5.40) represents the total harvest, which is instantaneous in (5.39) and at regular interval of time in (5.40). Total harvesting represents the aggregate landings by the N agents who have access to the common pool. In the following we obtain $h(x)$, which has the same functional form in continuous or in discrete time, by game-theoretic considerations.

The two available technologies are denoted, respectively, by s for the standard one and by c for the more environmental friendly. Technology s is more efficient than technology c and these different in efficiency is reflected in the different catchability coefficients q_s and q_c with $q_s > q_c$. At any time period, the N harvesters can be split in two groups according to the technology they use, with n_c denoting the "clean" agents and $n_s = N - n_c$ representing the "standard" agents.

Finally, we assume that fishermen adopt a Cobb-Douglas harvesting function with CRS (constant returns to scale) of the form

$$h_i(x) = \sqrt{q_i E_i x}, \tag{5.41}$$

[14] Differently from the modeling where harvesting efforts are controlled, here the regulator does not enforce any restraint.

where E_i denotes his/her current harvesting effort and $i = s, c$. The total cost of fishing for a type i agent, denoted by C_i, is proportional to the effort plus a fixed cost, i.e., $C_i = c_i + \gamma E_i$. In terms of harvested quantity, the total cost can be written as

$$C_i(h(x)) = c_i + \gamma \frac{(h(x))^2}{q_i x}, \quad i \in \{s, c\}.$$

Note that, given the previous assumptions, the marginal cost for "clean" harvesting is higher than for standard harvesting.

We assume that the landed harvest is sold at a constant price a_i with $i = s, c$ and that $a_c > a_s$, as consumers might be willing to pay more for the product obtained through a more environmentally-friendly technology.[15]

We assume that fishermen maximize just their profit at any instant of time if harvesting can take place continuously, or they maximize their expected one-period profit if they can harvest only at discrete instants of time. The profit of a representative fisherman employing technology $i \in \{s, c\}$ is given by

$$\pi_i(x) = a_i h_i(x) - c_i - \gamma \frac{(h_i(x))^2}{q_i x}. \tag{5.42}$$

By solving the optimality conditions[16] and employing symmetry among all players adopting the same technology, the Nash equilibrium harvesting strategy for a type i agent can be written as follows:

$$h_i^{\mathrm{NE}}(x) = \frac{a_i q_i}{2\gamma} x, \quad i \in \{s, c\}, \tag{5.43}$$

with $h_i^{\mathrm{NE}}(x) \geq 0$ whenever $x \geq 0$. Profits (5.42) computed at the Nash equilibrium (5.43) are non-negative and given by $\pi_i^*(x) = a_i^2 q_i/(4\gamma)x > 0$.

In what follows we introduce a dynamic mechanism that allows agents to move from the adoption of one technology to the other one. In particular, agents tend to switch from one strategy to the other if they expect this change to be profitable for them, according to the paradigms of the evolutionary game theory. By following [26], we assume that, at any time, agents harvest the Nash equilibrium quantity (5.43) and the corresponding profits (5.42), evaluated at the Nash equilibrium and denoted by $\pi_i^*(x)$, $i \in \{s, c\}$, are taken as fitness measures for the adoption of one of the two technologies. We first develop the model with a continuous-time switching, then we consider a discrete-time switching. Lastly, we relax the assumption that agents are able to change their technology continuously, and we introduce a time lag after

---

[15] This assumption of perfectly elastic demand for the resource is particularly well justified whenever the resource is a staple food for the consumers or several substitutes to the resource are traded in the market, see also [11] on this point.

[16] More precisely, by solving the first order condition $\partial\pi_i/\partial h_i = 0$, which is also sufficient for being $\partial^2\pi_i/\partial h_i^2 = -2\gamma/(q_i x) < 0$.

which agents can switch to the other technology, thus, obtaining a hybrid system (with continuous-time growth and discrete-time harvesting).

5.6.1 The Case of Instantaneous Switching

First, consider the case in which agents can revise their strategy (i.e., the employed technology) continuously and instantaneously. The most common evolutionary model, on which [22] focuses the analysis, is the replicator dynamics in continuous-time. Let us denote by $r(t) = r = n_s/N$ the fraction of agents using the "standard" technology (and $1 - r$ the complementary fraction of agents using the "clean" technology). The replicator equation is one possible way to model that from $\pi_s^*(x) > \pi_c^*(x)$ it follows that $\dot{r} > 0$ and vice versa (see [20, 30] for further details on the foundation of the evolutionary model). Mathematically, the dynamical system with resource and replicator dynamics is given by the following system of ODEs

$$\begin{cases} \dot{x} = x(\alpha - \beta x) - N(rh_s^{\mathrm{NE}}(x) + (1-r)h_c^{\mathrm{NE}}(x)), \\ \dot{r} = r(1-r)\left[\pi_s^*(x) - \pi_c^*(x)\right]. \end{cases} \tag{5.44}$$

By using (5.43), the total harvesting becomes

$$h(x) = N\left[r\frac{a_s q_s}{2\gamma}x + (1-r)\frac{a_c q_c}{2\gamma}x\right], \tag{5.45}$$

and $\Delta\pi$, that is the difference between profits at the equilibrium harvestings, is

$$\Delta\pi := \pi_s^*(x) - \pi_c^*(x) = \frac{a_s^2 q_s - a_c^2 q_c}{4\gamma}x + \xi. \tag{5.46}$$

By (5.45) and (5.46), (5.44) can be written as follows:

$$\begin{cases} \dot{x} = x(\alpha - \beta x) - N\left[r\frac{a_s q_s}{2\gamma}x + (1-r)\frac{a_c q_c}{2\gamma}x\right], \\ \dot{r} = r(1-r)\left[\frac{a_s^2 q_s - a_c^2 q_c}{4\gamma}x + \xi\right], \end{cases} \tag{5.47}$$

where $\xi = c_c - c_s$ can be regarded as a policy parameter, since it includes taxes imposed in order to obtain the predominance of one technology over the other. In the following we consider $\xi < 0$, assuming that the fixed costs for the more intensive harvesting method are higher, due to a more sophisticated technology and higher taxes, or equivalently due to government subsidies for agents adopting the more ecological fishing methods.

The steady states of (5.47) are given by

$$E_{0,0} = (0,0) \quad \text{and} \quad E_{0,1} = (0,1).$$

The point $E_{0,0}$ is the outcome where the natural resource is extinct and all fishermen employ the standard technology. This steady state always exists. The local stability analysis shows that $E_{0,0}$ is either stable if $a_c q_c > 2\gamma\alpha/N$ or unstable (saddle point) if $a_c q_c < 2\gamma\alpha/N$. The point $E_{0,1}$ represents the outcome where the natural resource is extinct and all fishermen employ the clean technology. This equilibrium is always unstable.

Other steady states of (5.47) entail that the natural resource reaches a *modified* carrying capacity level and the employment of only one of the two available technologies:

$$\widehat{E}_{\widehat{x},0} = (\widehat{x}, 0) = \left(\frac{\alpha}{\beta} - \frac{N a_c q_c}{2\beta\gamma}, 0\right) \quad \text{and} \quad \overline{E}_{\overline{x},1} = (\overline{x}, 1) = \left(\frac{\alpha}{\beta} - \frac{N a_s q_s}{2\beta\gamma}, 1\right). \tag{5.48}$$

Notice that $\widehat{E}_{\widehat{x},0}$ is feasible, in the sense that biomass is positive, whenever $a_c q_c < 2\alpha\gamma/N$, i.e., if and only if $E_{0,0}$ is unstable along the r axis. Similarly, $\overline{E}_{\overline{x},1}$ is feasible provided that $a_s q_s < 2\alpha\gamma/N$, i.e., if and only if $E_{0,1}$ is unstable along the invariant line $r = 1$.

Interestingly, $\widehat{E}_{\widehat{x},0}$ and $\overline{E}_{\overline{x},1}$ can be both feasible; this occurs for parameters such that $2\alpha\gamma/N > \max\{a_c q_c, a_s q_s\}$. Moreover, $\widehat{E}_{\widehat{x},0}$ is a stable node for $\xi < \widehat{\xi}$ or a saddle point for $\xi > \widehat{\xi}$, whereas $\overline{E}_{\overline{x},1}$ is a stable node for $\xi > \overline{\xi}$ or a saddle point for $\xi < \overline{\xi}$, where

$$\widehat{\xi} = \frac{(a_c^2 q_c - a_s^2 q_s)(2\alpha\gamma - N a_c q_c)}{8\beta\gamma^2} \quad \text{and} \quad \overline{\xi} = \frac{(a_c^2 q_c - a_s^2 q_s)(2\alpha\gamma - N a_s q_s)}{8\beta\gamma^2}.$$

This means that these two boundary equilibria can coexist and they can be both stable. This occurs when the difference in the fixed costs of the two technologies $\xi = c_c - c_s$ is such that $\overline{\xi} < \xi < \widehat{\xi}$. If the system is in this state, by subsidizing the clean technology or taxing the traditional technology, it is possible for the clean technology to prevail without reducing the stock size of the target resource. This occurs by reducing the difference in the fixed costs between the two technologies so that $\xi < \overline{\xi} < \widehat{\xi}$. Hence, the two boundary fixed points do not change their position in the state space (their values do not depend on ξ), but the equilibrium $\overline{E}_{\overline{x},1}$ is now unstable, and $\widehat{E}_{\widehat{x},0}$ is stable and is the only attractor, as shown below.

In addition to the previous fixed points, there also exists a unique equilibrium that involves the employment of both technologies:

$$E^* = (x^*, r^*) = \left(\frac{4\xi\gamma}{a_c^2 q_c - a_s^2 q_s}, \frac{a_c q_c N - 2\alpha\gamma + \frac{8\xi\beta\gamma^2}{a_c^2 q_c - a_s^2 q_s}}{N(a_c q_c - a_s q_s)}\right).$$

In [22] it is shown that E^* is always a stable point (either spiral or node) when $a_c q_c \geq a_s q_s$ or when $a_c q_c < a_s q_s$ with $a_c^2 q_c < a_s^2 q_s$ provided that

$$\widehat{\xi} < \xi < \overline{\xi}. \tag{5.49}$$

In the case $a_c q_c < a_s q_s$ with $a_c^2 q_c > a_s^2 q_s$, E^* is a saddle point when

$$\overline{\xi} < \xi < \widehat{\xi}. \tag{5.50}$$

When conditions (5.49) or (5.50) do not hold, i.e., when the difference in the fixed costs of the two technologies ξ is sufficiently large, E^* is not meaningful. For further details on the dynamics and economic considerations, we refer the reader to [22].

5.6.2 *The Case of Discrete Switching and the Hybrid Model*

In a discrete time scale, the natural resource grows in non-overlapping generations according to the discrete-time logistic equation (5.40) where $h(x)$ is given in (5.45). The fraction r of agents employing the standard technology at time t evolves with the same time scale. For technical reasons, [2] adopts an exponential replicator dynamics (see, e.g., [10, 19]) of the form

$$r' = \frac{r e^{\beta \pi_s^*(x)}}{r e^{\beta \pi_s^*(x)} + (1-r) e^{\beta \pi_c^*(x)}} = \frac{r}{r + (1-r) e^{-\beta \Delta \pi}}, \tag{5.51}$$

where $\beta \in [0, +\infty)$ is the so called intensity of choice parameter which measures the reactiveness of agents to adopt the more profitable strategy, and $\Delta \pi$ is given in (5.46). The monotone transformation $\pi_i^*(x) \rightarrow e^{\theta \pi_i^*(x)}$ is employed to obtain fitness measures with strictly positive values (see [19, 30] for details).

All in all, the dynamic model can be written in the form of an iterated map of the plane $T : (x, r) \rightarrow (x', r')$ with

$$T : \begin{cases} x' = x + \alpha x - \beta x^2 - N \left[r \frac{a_s q_s}{2\gamma} x + (1-r) \frac{a_c q_c}{2\gamma} x \right], \\ r' = \frac{r}{r + (1-r) \exp\left[\beta \left(\frac{a_c^2 q_c - a_s^2 q_s}{4\gamma} x - \xi \right) \right]}, \end{cases} \tag{5.52}$$

where the dynamic variables represent feasible states of the system if $x \geq 0$ and $0 \leq r \leq 1$. In [2] the analysis of (5.52) is started by considering its dynamics along the following invariant sets: $x = 0$, characterized by resource extinction, and the two lines of pure strategies, $r = 0$ (all agents use the clean fishing technology) and $r = 1$ (all agents use the standard fishing technology). Being invariant sets, whenever the initial condition of the system is on each of these lines, the dynamics never leave that line. In this way, it is possible to characterize the behavior of the system in simpler yet useful cases. The existence of these invariant lines, that bound the two-dimensional phase space of the dynamical system is important in order to characterize its global dynamical properties. Moreover, the knowledge of the kind of dynamic motion occurring along the two lines where a single pure strategy exists, tells us what will happen in the long run when one of the two strategies becomes dominant in terms of profits, so that it will prevail due to evolutionary pressures. The latter problem may be equivalently stated by asking when the one-dimensional attractors of the restrictions along the invariant lines $r = 0$ and $r = 1$ are also attractors of the two-dimensional dynamical system. This depends on the transverse stability as well as on the existence of attractors internal to the phase space, i.e., characterized by $r \in (0, 1)$. These are the questions examined, analytically and numerically, in [2].

With respect to the dynamical properties of the bidimensional map (5.52), it is worth noticing here that its fixed points must satisfy exactly the same conditions considered for the model in continuous time (5.47). Hence, the fixed points of (5.52) are the same as of model (5.47). However, discrete time dynamics determine different stability properties of these fixed points under different time scales. We refer the reader to [2] for the details.

If the natural resource grows with non-overlapping generations, the most appropriate way of modeling the resource dynamics is in continuous time. However, dynamics in continuous time cannot describe properly the fishermen behavior. Therefore, [22] reformulates the model as a hybrid system, where both discrete and continuous time scales are present.[17]

In detail, [22] assumes that there exists a minimum time interval Δl after which switching technology may take place. The time interval Δl can be interpreted as the time of a single fishing operation. Thus, at the end of each time period of length Δl, a representative agent employing technology $i \in \{s, c\}$ measures $\overline{\pi}_i^* (x(t))$, the net performance of his/her current harvesting strategy, as the exponentially decaying weighted average of historical profits over a time interval $[t - \Delta t, t]$, i.e.,

$$\overline{\pi}_i^* (x(t)) = \frac{\delta}{1 - e^{-\delta \Delta t}} \int_{t-l}^{t} e^{-\delta (t-\tau)} \pi_i^{\mathrm{NE}} (x(\tau))\, d\tau, \quad i \in \{s, c\}, \tag{5.53}$$

where time delay $\Delta t \in (0, \infty)$ represents the profits history used to calculate the average past profits of the two harvesting strategies, and $\delta > 0$ is a decay rate, assumed equal for all agents. For sake of simplicity, [22] considers the case $\Delta t = \Delta l$. The magnitude of $\overline{\pi}_i^* (x(t))$ represents a fitness measure of playing strategy i (see [30]). Again letting $r(t) = n_s/N$, the replicator dynamics can be expressed by a continuous-time growth equation for the biomass and a discrete (or pulse) strategy switching (a discrete decision-driven time). Under a *synchronous updating* for the adopted technologies, the model can be written as

$$\begin{cases} \dot{x}(t) = x(t)\,(\alpha - \beta x(t)) - N\left(r(t) h_s^{\mathrm{NE}} (x(t)) + (1 - r(t)) h_c^{\mathrm{NE}} (x(t))\right), \\ r(t) = \begin{cases} \frac{r(t-\Delta l) e^{\beta \overline{\pi}_s^* (x(t))}}{r(t-\Delta l) e^{\beta \overline{\pi}_s^* (x(t))} + (1 - r(t-\Delta l)) e^{\beta \overline{\pi}_c^* (x(t))}} & \text{if } \frac{t}{\Delta l} = \left\lfloor \frac{t}{\Delta l} \right\rfloor, \\ r\left(\left\lfloor \frac{t}{\Delta l} \right\rfloor \Delta l\right) & \text{otherwise,} \end{cases} \end{cases} \tag{5.54}$$

[17] Related models with multispecies interactions are analyzed in [5, 6], while [7] studies the modeling of a fishery with different time scales and hybrid modeling as well.

where $\lfloor x \rfloor$ is the largest integer not greater than x (i.e., the floor of x), and $h_i^{\mathrm{NE}}(x(t))$, $\overline{\pi}_i^*(x(t))$, $i = 1, 2$, are given, respectively, in (5.43) and (5.53).

The dynamical model in (5.54) is a hybrid system because it combines the population growth model (5.39) in continuous-time together with the discrete time evolution of the fraction of agents adopting the clean technology. We refer the interested reader to [22] for analysis and simulations of the model (5.54).

References

1. Baiardi, C.L., Lamantia, F., Radi, D.: Evolutionary competition between boundedly rational behavioral rules in oligopoly games. Chaos Solitons Fractals **79**, 204–225 (2015)
2. Bischi, G., Cerboni-Baiardi, L., Radi, D.: On a discrete-time model with replicator dynamics in renewable resource exploitation. J. Differ. Equ. Appl. **21**(10), 954–973 (2015). http://www.tandfonline.com/eprint/3vXI7RBRkR2NGkCHrXCc/full
3. Bischi, G.I., Lamantia, F., Sbragia, L.: Competition and cooperation in natural resources exploitation: an evolutionary game approach. In: Carraro, C., Fragnelli, V. (eds.) Game Practice and the Environment. Edward Elgar publishing (2009)
4. Bischi, G.I., Lamantia, F., Sbragia, L.: Strategic interaction and imitation dynamics in patch differentiated exploitation of fisheries. Ecol. Complex. **6**, 353–362 (2009)
5. Bischi, G.I., Lamantia, F., Radi, D.: Multi-species exploitation with evolutionary switching of harvesting strategies. Nat. Resour. Model. **26**(4), 546–571 (2013)
6. Bischi, G.I., Lamantia, F., Radi, D.: A prey-predator model with endogenous harvesting strategy switching. Appl. Math. Comput. **219**(20), 10123–10142 (2013)
7. Bischi, G.I., Lamantia, F., Tramontana, F.: Sliding and oscillations in fisheries with on-off harvesting and different switching times. Commun. Nonlinear Sci. Numer. Simul. **19**(1), 216–229 (2014)
8. Bischi, G., Lamantia, F., Radi, D.: An evolutionary Cournot model with limited market knowledge. J. Econ. Behav. Organ. **116**, 219–238 (2015)
9. Bischi, G.I., Lamantia, F.: Harvesting dynamics with protected and unprotected areas. J. Econ. Behav. Organ. **62**, 348–370 (2009)
10. Cabrales, A., Sobel, J.: On the limit points of discrete selection dynamics. J. Econ. Theory **57**(2), 407–419 (1992)
11. Clark, C.W.: Mathematical Bioeconomics: The Optimal Management of Renewable Resources, 2nd edn. Wiley-Intersciences, New-York (1990)
12. Clark, C., Munro, G.: The economics of fishing and modern capital theory: a simplified approach. J. Environ. Econ. Manag. **2**, 92–106 (1975)
13. Conrad, J.M., Smith, M.: Nonspatial and spatial models in bioeconomics. Nat. Resour. Model. **25**(1), 52–92 (2012)
14. Dockner, E., Feichtinger, G., Mehlmann, A.: Non-cooperative solutions for a differential game model of fishery. J. Econ. Dyn. Control **13**, 1–20 (1989)
15. Droste, E., Hommes, C.H., Tuinstra, J.: Endogenous fluctuations under evolutionary pressure in Cournot competition. Games Econ. Behav. **40**(2), 232–269 (2002)
16. Dockner, E., Jørgensen, S., Long, N.V., Sorger, G.: Differential Games in Economics and Management Science. Cambridge University Press (2000)
17. Gordon, H.S.: The economic theory of a common property resource: the fishery. J. Polit. Econ. **62**, 124–142 (1954)
18. Hardin, G.: The tragedy of the commons. Science **162**, 1243–1248 (1968)
19. Hofbauer, J., Sigmund, K.: Evolutionary game dynamics. Bull. (New Series) Am. Math. Soc. **40**(4), 479–519 (2003)

20. Hofbauer, J., Sigmund, K.: Evolutionary Games and Population Dynamics. Cambridge University Press, Beverly Hills CA (1998)
21. Kopel, M., Lamantia, F., Szidarovszky, F.: Evolutionary competition in a mixed market with socially concerned firms. J. Econ. Dyn. Control **48**, 394–409 (2014)
22. Lamantia, F., Radi, D.: Exploitation of renewable resources with differentiated technologies: an evolutionary analysis. Math. Comput. Simul. **108**, 155–174 (2015)
23. Levhari, D., Mirman, L.J.: The great fish war: an example using a dynamic cournot-nash solution. Bell J. Econ. **11**, 322–334 (1980)
24. Okuguchi, K.: Long-run fish stock and imperfectly competitive international commercial fishing. Keio Econ. Stud. **35**(1), 9–17 (1998)
25. Schaeffer, M.B.: Some aspects of the dynamics of populations important to the management of the commercial marine fisheries. Bull. Math. Biol. **53**(1/2), 253–279 (1991)
26. Sethi, R., Somanathan, E.: The evolution of social norms in common property resource use. Am. Econ. Rev. **86**, 766–788 (1996)
27. Smith, V.L.: On models of commercial fishing. J. Polit. Econ. **77**(2), 181–198 (1969)
28. Szidarovszky, F., Okuguchi, K.: An oligopoly model of commercial fishing. Seoul J. Econ. **11**(3), 321–330 (1998)
29. Verhulst, P.F.: Notice sur la loi que la populations suit dans son accroissement. Corresp. Math. et Phys. **X**, 113–121 (1838)
30. Weibull, J.: Evolutionary Game Theory. The MIT Press, Cambridge (1995)

Chapter 6
Dynamic Models of Financial Markets with Heterogeneous Agents

Fabio Tramontana

Abstract In this Chapter the qualitative theory of discrete time dynamical systems is applied to financial markets populated by heterogeneous and boundedly rational traders. It is shown as by using these assumptions some well known stylized facts of financial markets can be replicated even by using small scale models.

6.1 Introduction

Among the numerous applied contexts in which the qualitative theory of dynamical systems is used, financial markets represent one of the most impressive examples of its utility. Asset price dynamics is typically characterized by some phenomena that can hardly be explained by the mainstream theory (namely, the so-called *Efficient Markets Hypothesis*, EMH henceforth, see [3, 5]). According to this theory, asset prices should move in time in a manner that can be attributed to a random walk, that is a discrete stochastic process [4]. In other words, there should be no place for deterministic processes in explaining price movements in the stock market. Nevertheless, actual asset prices do not always behave as a random walk, and a huge amount of attempts to connect these movements to a more complicate stochastic process seem to fail in explaining a lot of stylized facts.

One of these stylized facts of financial markets is represented by the occurrence of long periods of time where an asset price increases without any connection with the right value of the represented asset (also called *fundamental value*). This is what is popularly known as a financial bubble. Typically, this bubble grows for a long period and then suddenly explodes, leading back the price to its fundamental value (or even lower than it) in a short time. The understanding of such a phenomenon is extremely important because financial bubbles are not so rare, and the may have

F. Tramontana (✉)
Department of Mathematical Sciences, Mathematical Finance and Econometrics, Catholic University, 9 Via Necchi, 20123 Milan, MI, Italy
e-mail: fabio.tramontana@unicatt.it

G.I. Bischi et al. (eds.), *Qualitative Theory of Dynamical Systems, Tools and Applications for Economic Modelling*, Springer Proceedings in Complexity,
DOI 10.1007/978-3-319-33276-5_6

important consequences, among which they can originate a recession for one or more countries, as it occurred for the recent real estate financial bubble.

Other relevant stylized facts of financial markets are the alternation of period of low volatility of asset prices with periods of high turbulence (something known in the literature as *volatility clusters*, that technically speaking means that the autocorrelation of the returns in absolute value is significantly positive for long delays) or the not temporary misalignments between the fundamental value and the asset price detected by the researchers working in the field of behavioral finance.[1]

Indeed, from behavioral finance comes one of the possible explanations of such puzzles. By relaxing the hypothesis of perfect rationality of the agents, it is possible to build financial market models where price dynamics is closer to real one than to a random walk. They start from the experimental observation of the decisional mechanisms adopted by professional traders, who are subject to biases as a consequence of the heuristics used to decide which asset to buy and which to sell.

As a first approximation, investors can be subdivided into two groups, according to the kind of behavioral rule they adopt (see [6, 7], for some survey studies). On the one hand there are the so-called *fundamentalists*, who tend to correct any mispricing between the current price and the fundamental. This kind of investor is similar to arbitrageurs of the EMH with the difference that now this behavior is no more necessarily a rational one, because also other groups of investors may operate in the market. In fact, on the other hand there is the second group of traders, called *chartists*, who look at the time-series of prices in order to detect some regularity or trend permitting to foresee the future price movements. The coexistence of these groups of traders may cause a shift from a purely stochastic explanation of price movements, to a (at least partially) deterministic one because now future prices are determined by past prices, that influence the behavior of (at least some) agents.

In such models, nonlinearities easily arise from behavioral rules or some other mechanism, and as a consequence deterministic chaos may characterize the dynamics of prices. It turns out that in some cases the chaotic motion of price is qualitatively similar to real motion and some of the stylized facts previously mentioned are replicated.

In this chapter we will see how to introduce nonlinearities in dynamical systems explaining the asset price motion, and how the output of such simple models can give a possible explanation to long standing financial markets puzzles.

6.2 A Linear Model with Heterogeneous Traders

In this section we will show how heterogeneity by itself is not enough to qualitatively replicate stylized facts of financial markets, even if it could be a good starting point.

Let us consider a market of only one asset. What are the forces behind the movements of the asset price? The price will raise if the amount of assets the traders want to buy exceeds the amount other traders want to sell. The opposite if the amount

[1] From a survey of the empirical studies that contradict the EMH see, among the others, [1].

traders want to sell is higher than the amount other traders aim at buying. This is the classical economic law of supply and demand. So we will use a variable called *excess demand* (denoted by D) denoting the difference between the demand and the supply for the considered asset. How much does the asset price move as a consequence of the non-negativity of the excess demand? There is not a simple answer to this question, a typical assumption is to consider the presence of an agent, called *market maker*, who observes the excess demand and then regulates the asset price with a rule like the following one:

$$P_{t+1} = P_t + \alpha D_t, \tag{6.1}$$

where the price movement is assumed to be proportional to the excess demand and α is the constant of proportionality (without loss of generality we will assume it equal to one).

Now, we need to build the excess demand by considering two kinds of traders: fundamentalists and chartists. So, at each time period the excess demand is made up by two components, as follows:

$$D_t = D_t^f + D_t^c, \tag{6.2}$$

where D_t^f and D_t^c denotes the excess demands of fundamentalists and chartists, respectively.

Starting with fundamentalists, they buy the asset when its price is undervalued (i.e. lower than the fundamental value), while they sell it when it is overvalued (i.e. priced over the fundamental value). In other words, they bet on a fast correction of the misalignment. By assuming that the fundamental value F is fixed, exogenously given and known by everyone, and by using a linear trading rule, we can characterize the excess demand of fundamentalists as follows:

$$D_t^f = f(F - P_t), \tag{6.3}$$

where the positive parameter f measures the reactivity of fundamentalists to the misalignment, and it is called *speed of reaction*.[2]

Chartists interpret the gap between price and fundamental value exactly at the opposite. When the price is high (i.e. over the fundamental value) they believe that optimism is characterizing such a market and they bet on a further increasing in the asset price, at least in the short-run. As a consequence they buy it today to sell it tomorrow at a higher price. The opposite occurs when the price is low (i.e. lower than the fundamental value) and pessimism prevails in the market. Their linear excess demand can be obtained by

$$D_t^c = c(P_t - F), \tag{6.4}$$

where $c > 0$ as the same interpretation as f.

---

[2]Actually f can also be interpreted as a measure of the numerosity of the fundamentalists. By considering that the more they are, the more they buy/sell, then the two interpretations become interchangeable.

By inserting (6.3), (6.4) into (6.2) and then in (6.1), we get a first order linear difference equation regulating the asset price dynamics

$$P_{t+1} = h(P_t) = P_t + f(F - P_t) + c(P_t - F) = P_t(1 + c - f) + F(f - c). \tag{6.5}$$

This map has a unique equilibrium point, given by

$$P^* = F \tag{6.6}$$

that corresponds to the correct price according to the EMH. Concerning its stability properties, in order to be globally stable the following condition, obtained by imposing $|h'(P^*)| < 1$, must hold:

$$f - 2 < c < f. \tag{6.7}$$

The interpretation of the global stability condition (6.7) is straightforward: the price will converge to the fundamental value provided that chartists are not more reactive (or numerous) than fundamentalists ($c < f$) but also provided that fundamentalists are not excessively reactive (or numerous) with respect to chartists ($f < c + 2$). While the first part of the interpretation is easy to understand, in fact, chartists typically play a destabilizing role in the market by betting against the foresight of the EMH, a little bit more complicated is the second part. It is required that fundamentalists, who apparently play the role of stabilizing force, do not loom excessively before chartists. That is because when they prevail the price moves in the direction of the fundamental value but it can go beyond it, at a value even more misaligned than before. This phenomenon is called *overshooting*.

This simple linear model permits to make two interesting conclusions:

1. In order to have an asset price correctly evaluated by the market, both kinds of investors are necessary, in the right proportion;
2. When the price does not converge to the fundamental value, it diverges to infinity.

While the first conclusion is a realistic one, the second is not good for our purposes. It is clearly a consequence of the linearity of the model that must be removed in order to obtain more realistic price movements.

6.3 A Nonlinear Model with Heterogeneous Traders

It is not so complicated to insert some meaningful nonlinearities in the price formation mechanism. Without introducing some more advanced assumptions (such as a switching mechanism between one trading strategy and a different one) we may just assume a nonlinear trading rule of one or both kinds of investors.

Chartists, for instance, can be assumed to be more prudent or cautious when the misalignment between price and fundamental value becomes larger. This behavior

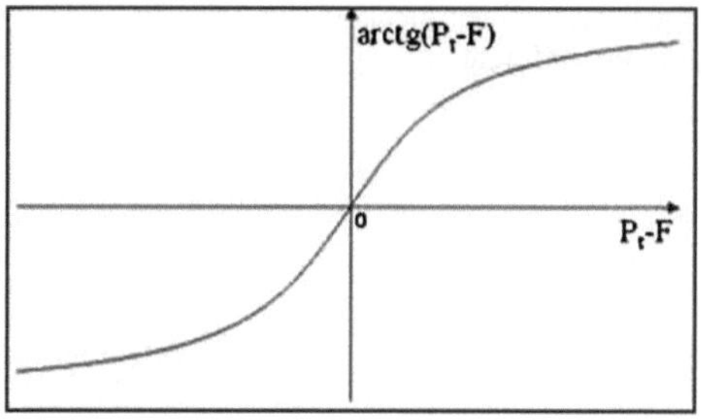

Fig. 6.1 Prudent chartists

can be captured by using a nonlinear trading rule like the following:

$$D_t^c = \arctan(P_t - F), \tag{6.8}$$

where the arctangent permits to obtain an excess demand like the one represented in Fig. 6.1, where the slope of the functions becomes lower and lower as the price deviates more and more from the fundamental value. This assumptions may lead to interesting price dynamics, but we would lose the opportunity to study analytically the local bifurcations that originate such price movements.

Another way of introducing a nonlinearity in the model consists in assuming, following [2], that fundamentalists become more and more aggressive (or numerous) as the price reaches values distant from the fundamental one. The explosion of such a bubble (or the end of such a pessimistic period) is perceived as extremely near, and they bet stronger on it. To obtain such an effect we can adopt the following formalization of the fundamentalists' trading rule:

$$D_t^f = f(F - P_t)^3, \tag{6.9}$$

where the cubic power makes more relevant high misalignments.

If we use (6.9) and (6.4) into (6.2) and then in (6.1) we get the following nonlinear map:

$$P_{t+1} = P_t + f(F - P_t)^3 + c(P_t - F),$$

that can be rewritten by using the auxiliary variable $x_t \equiv P_t - F$ as

$$x' = g(x) = x(1 + c) - fx^3, \tag{6.10}$$

where $'$ denotes the unit-time advancement operator.

Differently from the linear map (6.5) now the model is characterized by three equilibria:

$$x_0^* = 0, \quad x_{1,2}^* = \pm\sqrt{\frac{c}{f}}. \tag{6.11}$$

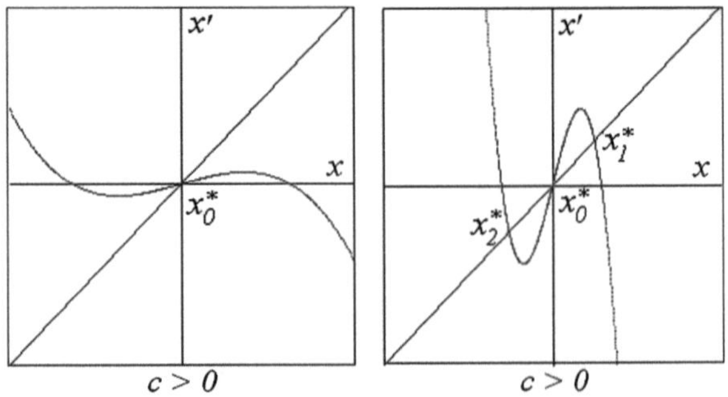

Fig. 6.2 A supercritical pitchfork bifurcation

While the equilibrium x_0^* (corresponding to the price equal to its fundamental value) always exists, the other two equilibria only exist for non-negative values of both speeds of reaction.[3] Equilibrium x_1^* characterizes a scenario where the asset price remains fixed at a value higher than the fundamental value, while in x_2^* the price is fixed at value lower than F.

In order to study the local stability of the three equilibria, we must compute the first derivative of $g(x)$:

$$g'(x) = 1 + c - 3fx^2 \tag{6.12}$$

and calculate it in correspondence of their values:

$$g'(x_0^*) = 1 + c, \quad g'(x_{1,2}^*) = 1 - 2c. \tag{6.13}$$

By applying the local stability condition $|g'(x^*)| < 1$ we obtain that the fundamental equilibrium is always unstable because of assumption that c is positive. The other two equilibria are locally stable provided that $c < 1$.

At the moment we can say that if we keep fixed the fundamentalists' speed of adjustment at a positive value, at $c = 0$ the equilibrium x_0^* undergoes a supercritical pitchfork bifurcation. In particular, at the bifurcation value the point x_0^* is nonhyperbolic and two new equilibria are generated. For $0 < c < 1$ the two new equilibria are separated from (and symmetric to) the fundamental equilibrium and are both locally stable, while x_0^* is unstable and separates the basins of attraction of x_1^* and x_2^*. Figure 6.2 shows geometrically the emergence of the two new equilibria.

At $c = 1$ a simultaneous flip bifurcation of both x_1^* and x_2^* occurs, and by further increasing the value of the parameter, a typical period-doubling bifurcation cascade leads to chaotic motion, as illustrated by the bifurcation diagram in Fig. 6.3. For a certain range of values of the parameter c, two attractors coexist: one located above the fundamental value (the so-called *bull region*), the other located below the fundamental value (the *bear region*). According to the initial condition, trajectories will converge to a fixed point (for $0 < c < 1$), to a periodic cycle or to a chaotic

[3]Mathematically they would also exist for negative values of both speeds of reaction, but they would lose any interpretability.

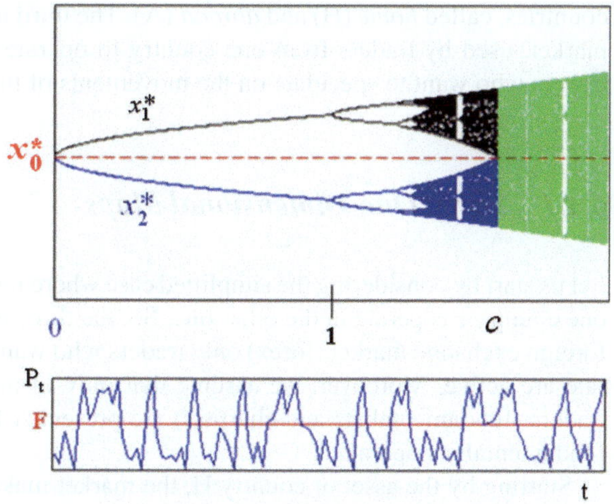

Fig. 6.3 Bifurcation diagram

attractor. In the green (rightmost) region of the bifurcation diagram, for large values of c, the two attractors merge forming a unique attractor (mostly chaotic but with the typical periodic windows) covering both bull and bear regions. This is the most interesting part of the diagram, where more realistic dynamics occur. One example is provided by the timeplot in Fig. 6.3. In this case the asset price alternate periods in the bull region and periods in the bear one, and the evidence of something qualitatively similar to those phenomena known as financial bubbles are clearly present.

6.4 From the Cause to the Consequences

Once we find some possible cause for the stylized facts we wanted to understand, then we can deepen the investigation to try explaining further phenomena. Doing so, we have found that the coexistence of a heterogeneity of investors using different trading strategies may be at the origin of phenomena like the emergence and the explosion of financial bubbles and more in general may cause fluctuations of the asset price that is not necessarily equal (or at least close) to its fundamental value, as predicted by the EMH.

By using the theoretical framework introduced in the previous section, we can investigate other issues. The one that we will deepen in this section is the following: what happens if asset markets where the EMH works become connected with other markets where turbulence and bubbles characterize it? Which one will infect the other? These are important questions especially nowadays when the globalization of financial markets has been realized.

To give an answer to these questions we present the works of [8, 9]. They consider three financial markets. Two of them characterize the market for an asset in different countries, called *home* (H) and *abroad* (A). The third market is the foreign exchange market, used by traders from one country to operate in the other one, but also by traders who want to speculate on the movements of the exchange rate.

6.4.1 Three One-Dimensional Maps

Let us start by considering the simplified case where it is not possible for traders from one country to operate in the other one. So, the three markets are isolated and in the foreign exchange market (forex) only traders who want to speculate on the exchange rate are active. Moreover, we assume that only in the forex market both kinds of traders (fundamentalists and chartists) are present, while in the two countries only fundamentalists operate.

Starting by the asset of country H, the market maker moves the asset price (P^{H}) as follows:

$$P_{t+1}^{\mathrm{H}} = P_t^{\mathrm{H}} + a^{\mathrm{H}} \left(D_t^{\mathrm{HH}} \right), \tag{6.14}$$

where $a^{\mathrm{H}} > 0$ is the constant of proportionality of price movements and D_t^{HH} is the excess demand of fundamentalists, given by

$$D_t^{\mathrm{HH}} = b^{\mathrm{H}}(F^{\mathrm{H}} - P_t^{\mathrm{H}}). \tag{6.15}$$

As usual, parameter $b^{\mathrm{H}} > 0$ is the constant of proportionality of price movements and F^{H} denotes the fundamental value of the asset traded in country H. So, the one-dimensional linear map regulating the dynamic of the asset price in country H is obtained by substituting (6.15) into (6.14):

$$P_{t+1}^{\mathrm{H}} = P_t^{\mathrm{H}} + a^{\mathrm{H}} b^{\mathrm{H}}(F^{\mathrm{H}} - P_t^{\mathrm{H}}). \tag{6.16}$$

The unique equilibrium ($P^{\mathrm{H}^*} = F^{\mathrm{H}}$) is globally stable provided that

$$a^{\mathrm{H}} b^{\mathrm{H}} < 2. \tag{6.17}$$

Similarly, for country A, where the market maker equation is the following:

$$P_{t+1}^{\mathrm{A}} = P_t^{\mathrm{A}} + a^{\mathrm{A}} \left(D_t^{\mathrm{AA}} \right) \tag{6.18}$$

with $a^{\mathrm{A}} > 0$ being reactivity of the market maker and the excess demand D_t^{AA} being a linear function of the mispricing between current price and the fundamental one (F^{A}), defined (according to the fundamentalists trading rule) as follows:

$$D_t^{\mathrm{AA}} = b^{\mathrm{A}}(F^{\mathrm{A}} - P_t^{\mathrm{A}}), \tag{6.19}$$

where $b^{\mathrm{A}} > 0$ is the speed of adjustment.

By using (6.19) into (6.18), we get

$$P_{t+1}^{\mathrm{A}} = P_t^{\mathrm{A}} + a^{\mathrm{A}} b^{\mathrm{A}} (F^{\mathrm{A}} - P_t^{\mathrm{A}}) \tag{6.20}$$

that regulates the dynamics of the asset traded in country A. Here, the unique equilibrium ($P^{\mathrm{A}^*} = F^{\mathrm{A}}$) is globally stable provided that:

$$a^{\mathrm{H}} b^{\mathrm{H}} < 2. \tag{6.21}$$

Finally, in the stock exchange market, the market maker adjusts the exchange rate (S) by summing up the excess demands of both fundamentalists (D_F^S) and chartists (D_C^S) according to the rule

$$S_{t+1} = S_t + d\left(D_{F,t}^S + D_{C,t}^S\right), \tag{6.22}$$

where $d > 0$ is the constant of proportionality of exchange rate movements.

If F^S is the exogenous fundamental value of the exchange rate, then the linear trading rule of chartists is the following:

$$D_{C,t}^S = e\left(S_t - F^S\right) \tag{6.23}$$

with $e > 0$ being their reactivity.

For fundamentalists, we use the nonlinear (cubic) trading rule seen in Sect. 6.3:

$$D_{F,t}^S = f\left(F^S - S_t\right)^3 \tag{6.24}$$

with the reactivity measured by $f > 0$ (cf. (6.9)).

By inserting (6.23) and (6.24) into (6.22), we obtain the nonlinear one-dimensional difference equation

$$S_{t+1} = S_t + de\left(S_t - F^S\right) + df\left(F^S - S_t\right)^3. \tag{6.25}$$

By using the auxiliary variable $z = S - F^S$, we can rewrite the map as follows:

$$z' = m(z) = z(1 + de) - dfz^3, \tag{6.26}$$

that is topologically equivalent to the cubic map (6.10). Thus, we expect to find qualitatively the same bifurcation structure.

By using the fixed point condition $z' = z = z^*$, we get the three equilibria. The fundamental one

$$z_0^* = 0 \tag{6.27}$$

and the two symmetric nonfundamental equilibria

$$z_{1,2}^{*} = \pm\sqrt{\frac{e}{f}} \tag{6.28}$$

that exist only for positive values of e and f, according to our assumptions.

To study the local stability properties of the equilibria we need the derivative

$$m'(z) = 1 + de - 3dfz^2, \tag{6.29}$$

that confirms the instability of the fundamental equilibrium (in fact, $m'(z_0^*) = 1 + de > 1$), while the other two equilibria are locally stable iff

$$de < 2. \tag{6.30}$$

At $de = 2$ the equilibria undergo a simultaneous flip bifurcation and by further increasing e (or d) a cascade of period doubling bifurcations followed by chaotic motion similar to the one displayed in the bifurcation diagram in Fig. 6.3 can be obtained.

Next, we connect the three markets to understand what happens to the dynamic of the three asset prices.

6.4.2 A Three-Dimensional Map

If we admit fundamentalists from country H to trade the asset of country A and vice versa, then the model changes.

First of all, we must add a new component to the excess demands of the two assets, namely D^{HA} and D^{AH}, denoting the excess demands of fundamentalists from country A who operate in country H and the opposite, respectively. They are given by

$$D_t^{\mathrm{HA}} = c^{\mathrm{H}}\left[\left(F^{\mathrm{H}} - P_t^{\mathrm{H}}\right) + \gamma^{\mathrm{H}}\left(F^S - S_t\right)\right] \tag{6.31}$$

and

$$D_t^{\mathrm{AH}} = c^{\mathrm{A}}\left[\left(F^{\mathrm{A}} - P_t^{\mathrm{A}}\right) + \gamma^{\mathrm{A}}\left(\frac{1}{F^S} - \frac{1}{S_t}\right)\right]. \tag{6.32}$$

The two equations above can be explained by considering that foreign fundamentalists may benefit from exchange rate movements, so, in their excess demand there is a term $\gamma^{\mathrm{H}}\left(F^S - S_t\right)$ and $\gamma^{\mathrm{A}}\left(1/F^S - 1/S_t\right)$ that is dependent on the mispricing in the foreign exchange market.[4]

[4]Note that traders from H to A consider the reciprocal values of the exchange rate and the fundamental value.

Moreover, in the dynamic equation regulating the exchange rate movements, now two more components must be added to the excess demand, namely,

$$P_t^{\mathrm{H}} D_t^{\mathrm{HA}}, \tag{6.33}$$

representing the demand for currency H generated by the orders from A to H, from which we must subtract the demand for currency A that are necessary for trading from H to A:

$$\frac{P_t^{\mathrm{A}}}{S_t} D_t^{\mathrm{AH}}, \tag{6.34}$$

where the reciprocal of the exchange rate permits to convert it into an amount in currency H.

By adding (6.31) and (6.32) to the excess demand in (6.16) and (6.20) and by updating the demand for currency, we get the following three-dimensional nonlinear dynamical system:

$$T: \begin{cases} x' = x - a^{\mathrm{H}} \left[\left(b^{\mathrm{H}} + c^{\mathrm{H}} \right) x + c^{\mathrm{H}} \gamma^{\mathrm{H}} z \right], \\ y' = y - a^{\mathrm{A}} \left[\left(b^{\mathrm{A}} + c^{\mathrm{A}} \right) y - c^{\mathrm{A}} \gamma^{\mathrm{A}} \frac{z}{F^S (z + F^S)} \right], \\ z' = z - d \left[c^{\mathrm{H}} \left(x + F^{\mathrm{H}} \right) \left(x + \gamma^{\mathrm{H}} z \right), \right. \\ \quad \left. + \, c^{\mathrm{A}} \frac{y + F^{\mathrm{A}}}{z + F^S} \left(\gamma^{\mathrm{A}} \frac{z}{F^S (z + F^S)} - y \right) - ez + f z^3 \right]. \end{cases} \tag{6.35}$$

Equilibrium points of the map (6.35) cannot be found analytically and, as a consequence, neither their stability properties. Nevertheless, numerical tools can be used to say something about the 3D map of interconnected markets. In particular, we can look at the following bifurcation diagrams in Figs. 6.4, 6.5 and 6.6, obtained by keeping fixed all parameters but the speed of reaction of chartists in the exchange rate market (e), that we know has a destabilizing role if it is increased.[5] This is confirmed by the bifurcation diagrams but even more interesting are the new findings with respect to the three separate markets:

- The flip bifurcations of the two nonfundamental equilibria are no more simultaneous. Even if it is not easy to see from the picture, the bifurcation of one equilibrium occurs at a lower value of the bifurcation parameter with respect to the loss of stability of the other one;
- Not only the exchange rate starts to oscillate periodically or chaotically when e is large enough, but also the prices of the asset traded in countries H and A. The timeplots in Figs. 6.4, 6.5 and 6.6 show the alternation of bubbles and crushes also in these markets and it was not possible before the markets were connected. So it proves that it can be sufficient, in a globalized and interconnected financial

[5] In particular, $a^{\mathrm{H}} = 0.41, b^{\mathrm{H}} = 0.11, c^{\mathrm{H}} = 0.83,\ \gamma^{\mathrm{H}} = 0.3, a^{\mathrm{A}} = 0.43,\ b^{\mathrm{A}} = 0.21,\ c^{\mathrm{A}} = 0.9,$ $\gamma^{\mathrm{H}} = 0.36, d = 0.35, f = 0.7$ and $F^S = 6.07$.

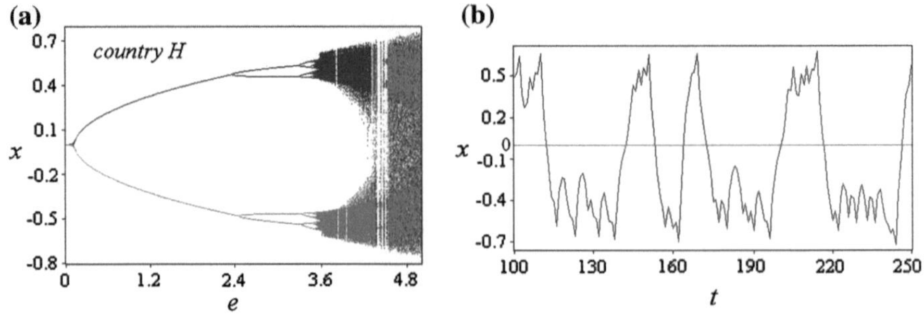

Fig. 6.4 **a** Bifurcation diagram where e varies between 0 and 5 and the asymptotic value of x is on the *vertical axis*. The two *different colors* (tonalities) denote two different initial conditions. **b** A timeplot obtained at $e = 4.86$

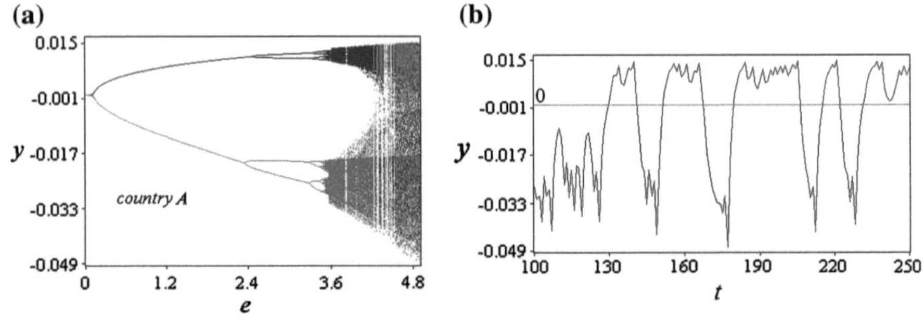

Fig. 6.5 **a** Bifurcation diagram where e varies between 0 and 5 and the asymptotic value of y is on the *vertical axis*. The two *different colors* (tonalities) denote two different initial conditions. **b** A timeplot obtained at $e = 4.86$

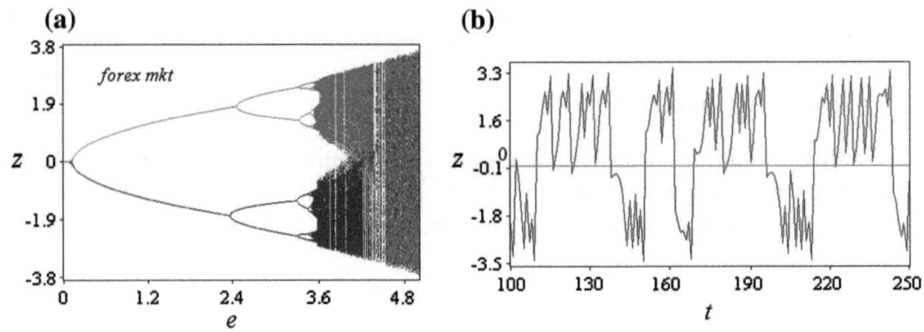

Fig. 6.6 **a** Bifurcation diagram where e varies between 0 and 5 and the asymptotic value of z is on the *vertical axis*. The two *different colors* (tonalities) denote two different initial conditions. **b** A timeplot obtained at $e = 4.86$

world, that one market which is unstable also infects other markets. This is a quite interesting result obtained by using a simple low-dimensional dynamical system characterized by nonlinearities.

6.5 An Alternative with Discontinuity

Results similar to those that can be obtained by introducing a nonlinearity in a financial market model, can also be obtained without renouncing to adopt linear trading rules for both kinds of traders. It is realistic to assume that the reactivity of traders when the asset is overvalued is not the same as the reactivity when it is undervalued.

So, we can specify the excess demand of fundamentalists as follows:

$$D_t^f = \begin{cases} f_1(F - P) + f_2 & \text{if } P \geq F, \\ f_3(F - P) + f_4 & \text{if } P < F. \end{cases} \tag{6.36}$$

Similarly, for chartists we have the following trading rule:

$$D_t^c = \begin{cases} c_1(P - F) + c_2 & \text{if } P \geq F, \\ c_3(P - F) + c_4 & \text{if } P < F. \end{cases} \tag{6.37}$$

The terms f_1 , f_3, c_1 and c_3 are the different (but all positive) reactivity parameters, while f_2, f_4, c_2 and c_4 are fixed components of the excess demand and can be positive or negative.

By using (6.36) and (6.37) into (6.1), we obtain a piecewise linear discontinuous map explaining the dynamics of the asset price

$$P' = \begin{cases} P + [c_1(P - F) + f_1(F - P) + f_2 + c_2] & \text{if } P \geq F, \\ P + [c_3(P - F) + f_3(F - P) + f_4 + c_4] & \text{if } P < F, \end{cases} \tag{6.38}$$

and in term of deviation from the fundamental value

$$x' = \begin{cases} x\,(1 + c_1 - f_1) + f_2 + c_2 & \text{if } x \geq 0, \\ x(1 + c_3 - f_3) + f_4 + c_4 & \text{if } x < 0. \end{cases} \tag{6.39}$$

Discontinuous maps are characterized by bifurcations called "border-collision bifurcations" and may lead to chaotic motion also when the map is piecewise linear. A numerical proof is obtained by using the following combination of parameters ($c_1 = 0.1$, $c_2 = 0.5$, $c_3 = 0.4$, $c_4 = 0$, $f_1 = 2.3$, $f_2 = 0.5$, $f_3 = 3.4$ and $f_4 = -0.5$) that permits to obtain the timeplot represented in Fig. 6.7. We can see also in this case bubbles and crushes phenomena similar to those obtained with a nonlinear map.

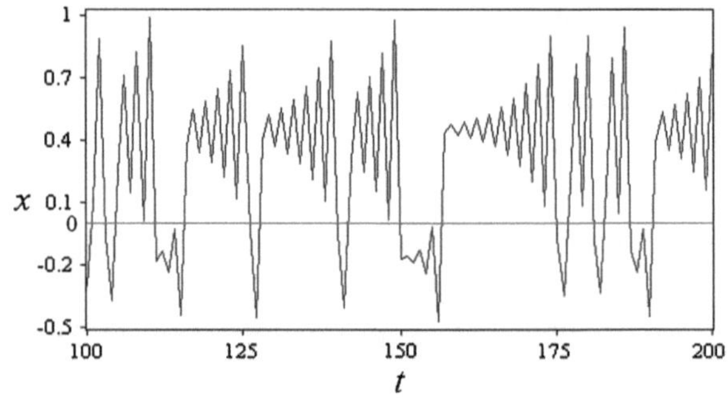

Fig. 6.7 Timeplot

6.6 Conclusions

We have shown how low-dimensional discrete time dynamical systems can be used to qualitatively replicate important stylized facts of financial markets such as the occurrence of bubbles and crushes and excess volatility. Behavioral assumptions permit to introduce in the deterministic system some nonlinearity (or discontinuity) that can be at the origin of the complex motion of actual asset prices. A mix of analytic and numerical tools is necessary to study these systems, explaining the passage from the convergence to the fundamental value, as predicted by the EMH, to complex and unpredictable motion.

References

1. Barberis, N.C., Thaler, R.K.: Chapter 18. A survey of behavioral finance. In: Constantinides, G.M., Harris, M., Stulz, R.M. (eds.) Financial Markets and Asset Pricing. Handbook of the Economics of Finance, vol. 1, Part B, pp. 1053–1128. Elsevier, North Holland (2003)
2. Day, R.H., Huang, W.: Bulls, bears and market sheep. J. Econ. Behav. Organ. **14**, 299–329 (1990)
3. Fama, E.F.: The behavior of stock market prices. J. Bus. **38**, 34–105 (1965a)
4. Fama, E.F.: Random walks in stock market prices. Finan. Anal. J. **21**, 55–59 (1965b)
5. Fama, E.F.: Efficient capital markets: a review of theory and empirical work. J. Finan. **25**(2), 383–417 (1970)
6. Menkhoff, L., Taylor, M.: The obstinate passion of foreign exchange professionals: technical analysis. J. Econ. Lit. **45**, 936–972 (2007)
7. Taylor, M., Allen, H.: The use of technical analysis in the foreign exchange market. J. Int. Money Finan. **11**, 304–314 (1992)
8. Tramontana, F., Gardini, L., Dieci, R., Westerhoff, F.: The emergence of "bull and bear" dynamics in a nonlinear model of intracting markets. Discrete Dyn. Nat. Soc., 310, 471 (2009)
9. Tramontana, F., Gardini, L., Dieci, R., Westerhoff, F.: Global bifurcations in a three-dimensional financial model of bull and bear interactions. In: Bischi, G.I., Chiarella, C., Gardini, L. (eds.) Nonlinear Dynamics in Economics, Finance and the Social Sciences, pp. 333–352. Springer, Heidelberg (2009b)

Chapter 7
A Dynamical Model of Proximal Development: Multiple Implementations

Ugo Merlone and Paul van Geert

Abstract Dynamical systems are quite important in psychological research as virtually all psychological processes occur in time. In this chapter we show how to implement a dynamical model of proximal development using a spreadsheet, R and C++. We discuss strengths and weakness of each approach. Using a spreadsheet or a statistical software such as R make these approaches palatable both for people with background in economics and psychology; on the other hand, using C++ provides better efficiency at the cost of requiring some more competencies. Last but not least, all the approaches we propose use free and open source software.

7.1 Introduction

There is a strong tradition in using dynamical systems in psychology. Starting from the seminal contribution [1] by Newell and Simon, in which a model of adaptive behavior and learning is analyzed as a system of differential equations and simulation in psychology is discussed, several other authors have used dynamical systems both descriptively and analytically. In fact, comparing the following two quotes—"[t]he observation that psychological processes occur in time is trite" [2, p. 231] and "anything that evolves over time can be thought of as a dynamical system" [3, p. 1]—the importance of dynamical systems in psychology is evident. Several books provide evidence of this interest: the reader may refer to [4] for an introduction to dynamical systems; to [5] for applications to social psychology; to [6] for an introduction to nonlinear dynamics in psychology; to [7] for an application to organi-

U. Merlone (✉)
Department of Psychology, Center for Cognitive Science,
University of Torino, via Verdi 10, 10124 Torino, Italy
e-mail: ugo.merlone@unito.it

P. van Geert
Heymans Institute, Grote Kruisstraat 2/1, 9712 TS Groningen, Netherlands
e-mail: p.l.c.van.geert@ppsw.rug.nl

G.I. Bischi et al. (eds.), *Qualitative Theory of Dynamical Systems, Tools and Applications for Economic Modelling*, Springer Proceedings in Complexity,
DOI 10.1007/978-3-319-33276-5_7

zational psychology; to [8] for applications to system analysis; to [9] for application to development psychology; and to [10] for a psychoanalytic theory application.

In this chapter we consider a well known model of educational dynamics and show how to implement it using different tools. The model we consider is a version of the model of proximal development presented in [9, 11, 12]. The multiple implementation technique we show can be applied to any model.

The chapter is organized as follows. In Sect. 7.2 we introduce the model. In Sect. 7.3 we discuss how multiple implementations can be useful when studying complex systems. In Sects. 7.4–7.6 we show how the proposed model can be implemented by using a spreadsheet, R and C++ respectively. Finally, in the last section, some suggestions about using multiple implementations in order to exploit the benefits of different simulation tools to study dynamical systems are provided.

7.2 The Mathematical Formalization

In order to arrive at a mathematical model of educational dynamics, we start from the assumption that

1. there exists an educational goal,
2. there is a person to be educated—for instance a child,
3. there is an educator who will actively pursue the process of education in the form of child-directed educational activity.

Education also requires active participation of the educated person, the child. Let us refer to this active participation by means of the term "learning", and use it as a general and overarching term for a wide variety of processes ranging from simple imitation to complex internal construction.

The educational goal can be conceived of as a stock of information, skills, cognitive complexity or whatever. This stock of information and skills may be vast but it is always limited. It can be characterized by a particular magnitude, K, and for reasons of simplicity, we set K to a dimensionless magnitude, represented by the number 1. Note that this is just a "form of" simplification because any other numerical representation will do. However, educational goals are not only characterized by a purely quantitative measure (how much is there to be learned), but also by a qualitative dimension. That is, the information, skills or whatever constitutes the content of the educational goal differ in terms of complexity or conditionality. That is to say, some things are easier to learn than others, or some things constitute elements that serve as building blocks for more complex elements and so forth. This means that the content of a particular educational goal stock can be ordered in terms of its "learnability" (some things must be learned before other things, or some things are much more easily learned than other things).

In a similar vein, the child can also be conceived of as a stock of information, skills, etc. In principle, this stock is virtually empty at the beginning. Take for instance the teaching and learning of arithmetic as an example. The educational goal is to teach arithmetic to the child, and what arithmetic means may be represented in the form of an informal or a formal curriculum, such as the primary school arithmetic curriculum. As the child enters primary school, his knowledge of arithmetic is highly elementary. This may vary from very simple number concepts to some form of emergent numeracy that the child has picked up spontaneously. Anyway, whatever the knowledge that is present at the beginning, there is still a very considerable distance between this initial state and the goal state. Let us call the current level of the child's stock of knowledge or skills the child's actual level of development, represented by $A\ (t)$.

Formally speaking, learning—in whatever form it occurs—can be defined as a flow from the educational goal stock to the stock representing an individual learner. Note that this is indeed a purely formal definition. It has no bearing on the nature of the processes that take place during learning or knowledge acquisition, in that these processes should not be equated with straightforward transmission of content from an educator to a child. That is, whatever the nature of the learning process, it can be formally defined as a flow from a limited goal stock to another limited stock, namely, the learning person. Thus, at any point in time, the system consisting of the educator and the educated person can be represented by two variables, namely, the child's actual level $A\ (t)$ characterizing the learner stock and $K - A\ (t)$ characterizing whatever needs to be appropriated by the child from the goal stock.

The simplest mathematical formalization of the learning process is that it amounts to a flow ΔA from the goal stock to the individual stock, and that this flow depends on

1. a flow parameter,
2. the current content of the goal stock,
3. the current content of the individual stock (the child's actual developmental level).

The flow parameter is in fact the rate of learning, rate of acquisition, rate of assimilation of the content to be learned, etc., and can be represented by the parameter r. Put differently, learning depends on what you already know (the actual level, i.e., the learner stock), on what you still have to learn (the goal stock) and on the rate of learning. This model can be represented in the form of a simple equation which is nothing else than the classical logistic equation first coined by the 19th century Belgian mathematician François Ferdinand Verhulst:

$$\Delta A = R \cdot A \cdot (K - A)\,.$$

The effect of the distance between K and A can also be represented in proportional terms, leading to the following form of the logistic equation

$$\Delta A = R \cdot A \cdot (1 - A/K)\,.$$

Thus, in order to learn, the learner must have access to the goal stock. In educational settings, the learner's access to the goal stock is guided by or organized by the educator. This can happen in the form of the educator giving examples, of the educator demonstrating a model of the goal behavior to be imitated, of giving assignments to the learner, giving instructions, and whatever else belongs to the myriad forms of interaction that count as education and teaching. In short, the educator is actively presenting the educated person with the goal stock in order to allow the educated person to learn, that is to say, in order to allow the flow from goal stock to individual stock to happen (recall that this notion of flow is only a formal way to represent the process of learning).

Vygotsky's great insight, which is basically the intuitive insight of all good educators, was that you cannot just present the learner with the goal stock as it is. What educators need to do to keep the flow going is to present the learner with a reduced and selective portion of the goal stock, namely, that portion that lies within a range of accessibility that depends on the learner's current state of learning, i.e., that depends on what the learner has already learned. Vygotsky defined this range of accessibility by what a learner can accomplish in terms of new knowledge, new skills, new insights, if that learner is given appropriate help. This range between the already consolidated things that the learner can do without help, and the new, not yet consolidated things that the learner can do if appropriate help is given, is Vygotsky's famous zone of proximal development. Given this help, the learner is capable of appropriating the new knowledge, skills or insights, that is to say, of making them his own. This process of appropriation is formally defined as a flow, slowly depleting the goal stock.

Defined in terms of our flow model, the flow will only happen if the goal stock K is presented in the form of a much reduced goal stock, more particularly, a reduced version that is close to the current individual stock, A. How "close" it must be depends on the learner. If it is too close, the learner will get bored and will not learn anything or only very slowly, if it is not close enough, the learner will get overwhelmed and confused and will not learn anything either. The optimum distance will differ between individual learners, and thus constitutes a typical person-specific parameter in the model. So, the educator's task is to present the learner with a goal stock that is sufficiently close to the learner's stock to make learning possible, i.e., to enable the flow from the goal stock to the individual stock. We shall call this reduced, adapted and learner-directed goal stock P, referring to the concept of proximal development. This reduced goal stock will now serve as the learner's attractor state. Given all this, we need to rewrite the equation for the learning process as follows:

$$\Delta A = R_a \cdot A \cdot (P - A)$$

or, alternatively,

$$\Delta A = R_a \cdot A \cdot (1 - A/P)\,.$$

An essential aspect of the concept of the zone of proximal development is that the rate of learning will be maximal (all other things being equal) if the distance between the current, individual stock level A and the current goal stock level P is optimal, that is to say, if the difference between P and A is equal to an optimal distance, represented by O_a.

As a consequence of presenting P to the learner, there will be a flow from P to A, and P will be used up until the point where $A = P$, after which no learning will take place anymore (the P stock is entirely depleted). Since the educator wants the learner to achieve the final goal state, i.e., to appropriate the content represented by K, P must be updated as the learner progresses, and this updating must be in the direction of the goal stock K. That is to say, the educator is monitoring a process that goes from an initial P, that is close to the learner's initial level A (t_0), and in that sense quite far away from the ultimate goal state K, to the final state that is ideally similar to the goal state, K.

In order to model this process of adaptation in the educator's learner-directed activity, we can use the same logic as we did for the process of learning. The educator's process of updating the learner-directed educational content depends on

1. an update rate,
2. the current level of the learner's proximal development P,
3. the content of the current goal stock, which is $K - P$.

Hence, the equation for the updating function of the educational content is

$$\Delta P = R_p \cdot P \cdot (K - P)$$

or, alternatively,

$$\Delta P = R_p \cdot P \cdot (1 - P/K).$$

We just stated that the rate of learning will be maximal if the distance between the level of proximal development, presented in the learner-directed activities of the educator, and the level of the learner's actual development is optimal. This distance can best be represented in proportional terms, that is to say, in function of the level already attained by the learner. In that case, there is some optimal proportion of P (potential development, i.e., the level expressed in the learner-directed educational activities of the educator) over A

$$O = P/A.$$

Hence, the rate of learning, R_a, is maximal, that is to say, equal to some learner-specific maximum rate of learning r_a, if $P/A - O = 0$. This means that we can express the education-dependent rate of learning R_a as follows:

$$R_a = r_a - |P/A - O| \cdot b$$

for b being a learner-specific parameter that moderates or dampens the effect of the distance between P/A and O. Since learning depends on the *distance* between P/A and O, we must use the absolute difference between P/A and O in our equation. That is to say, the more P moves away from the optimal P/A proportion, the slower the learning process (i.e., the slower the growth of A). Finally, we may assume that the effect of the distance between P and A also depends on the distance between the current level of development and the final goal level. The developmental level can be represented either by P (potential level) or by A (the actual level). The closer P or A approach the final goal level, the less will be the effect of the distance between P and A on the rate of learning.

Hence, the equation for updating the rate of learning R can be written either as

$$R_a = r_a - |P/A - O_a| \cdot b_a \cdot (1 - P/K),$$

or as

$$R_a = r_a - |P/A - O_a| \cdot b_a \cdot (1 - A/K).$$

(The second equation counts as the more preferable alternative, since the learner is primarily characterized by the actual level A).

A comparable line of reasoning applies to the change in the potential level, which is the level governed by the educator's learner-directed activities. The change in P should be so that the proportion P/A is always as close as possible to what the educator conceives to be the optimal proportional distance between P and A, expressed by O_p. The optimal distance preferred by the educator may be different from the real optimal distance O_a, as it functions for the learner. It is clear that the most optimal process of learning and educating is one in which the distance between the learner's and the educator's optimal P/A proportion is arbitrarily small. If the P/A proportion lies above the optimum O_p the rate of change of P must slow down and eventually become negative until P/A is close enough to O_p.

If the P/A proportion lies below the optimum O_p, the change in P must speed up to reach a point where P/A is arbitrarily close to O_p. Hence, the rate of change of P, as expressed in the form of the educator's learner-directed activities, is at its maximum if the *difference* between P/A and O_p approaches zero, namely,

$$R_p = r_p - \left(P/A - O_p\right) \cdot b_p$$

for b_p a damping parameter. Finally, the effect of the difference between P/A and O_p can be moderated by the distance between P and the final goal level K as follows:

$$R_p = r_p - \left(P/A - O_p\right) \cdot b_p \cdot (1 - P/K).$$

We can now combine all the equations to arrive at a complete mathematical model of Vygotsky's model of actual and potential development in the interaction between a learner and an educator/teacher in the form of four coupled equations:

$$\begin{cases} \Delta A = R_a \cdot A \cdot (1 - A/P) \\ \Delta P = R_p \cdot P \cdot (1 - P/K) \\ R_a \ = r_a - |P/A - O_a| \cdot b_a \cdot (1 - A/K) \\ R_p \ = r_p - \left(P/A - O_p\right) \cdot b_p \cdot (1 - P/K). \end{cases} \tag{7.1}$$

In order to study the dynamics, we rewrite (7.1) as

$$\begin{cases} A(t+1) = A(t)\left[1 + R_a(t)\left(1 - \frac{A(t)}{P(t)}\right)\right] \\ P(t+1) = P(t)\left[1 + R_p(t)\left(1 - \frac{P(t)}{K}\right)\right], \end{cases} \tag{7.2}$$

where

$$\begin{cases} R_a(t) = r_a - \left|\frac{P(t)}{A(t)} - O_a\right| b_a\left(1 - \frac{A(t)}{K}\right) \\ R_p(t) = r_p - \left(\frac{P(t)}{A(t)} - O_p\right) b_p\left(1 - \frac{P(t)}{K}\right) \end{cases} \tag{7.3}$$

and

- r_a and r_p: constant growth rates,
- O_a and O_p: optimality parameters,
- $b_a = d$ and $b_p = e$: damping parameters,
- K is a constant, fixed to 1.

7.3 Multiple Implementation

Becker [13] claims:

> As applications become more complex, their resource management requirements become more complex, and despite our best efforts, our designs often have holes in them, or we apply our designs incorrectly, or we make coding errors [13].

This applies also when modeling complex systems with Agent Based Modeling or even with numerical simulation. For example, [14] illustrates how, even in simple computations, floating point errors may cause serious discrepancies.

One possible way to avoid, or at least to be aware of, these problems may be to use different simulating environments to model the same complex phenomena as illustrated in [15]. Besides helping to detect coding errors, this approach can be helpful when dealing with complex systems which are sensitive to initial conditions and, therefore, tend to be extremely sensitive to implementation details.

For these reasons we will show how the model we consider can be implemented using three different simulation environments; namely, a spreadsheet, R and C++. After illustrating how the model is implemented, we will discuss strengths and weaknesses of each of the considered approaches.

7.4 Spreadsheet

As illustrated in [9], even absolute novices may start using a spreadsheet to implement dynamical systems. Furthermore, spreadsheets are available on every computer and even on mobile devices. In particular, we will use LibreOffice Calc which produces documents compatible to Microsoft Excel. For a first introduction to LibreOffice, the reader may refer to [16].

The model is implemented step by step; for each step we provide a picture illustrating the result. The final document is available at the link [17].

Let us start from an empty spreadsheet as the one we find when we start Libreoffice Calc, illustrated in Fig. 7.1. The next step is to prepare some cells for the map parameters as illustrated in Fig. 7.2. Parameter names are placed in the first row cells and their values in the cells below.

In row 3 we prepare some columns for time and state variables, and in row 4 we put initial conditions as illustrated in Fig. 7.3. We do not fill values for $R_a(0)$ and $R_p(0)$ as they are computed by (7.3). In fact, in cell `D4`, $R_a(0)$ is computed by = B\$2 − ABS(C4/B4 − C\$2) ∗ D\$2 ∗ (1 − B4/\$A\$2) as illustrated in Fig. 7.3, and in cell `E4`, $R_p(0)$ is computed by = E\$2 − (C4/B4 − F\$2) ∗ G\$2 ∗ (1 − C4/ \$A\$2) as illustrated in Fig. 7.4.

In cell `A5` we update time with formula = A4 + 1 as illustrated in Fig. 7.5. Now it is turn to code each of the difference equations of the map (7.2). Difference equation

$$A(t+1) = A(t)\left[1 + R_a(t)\left(1 - \frac{A(t)}{P(t)}\right)\right]$$

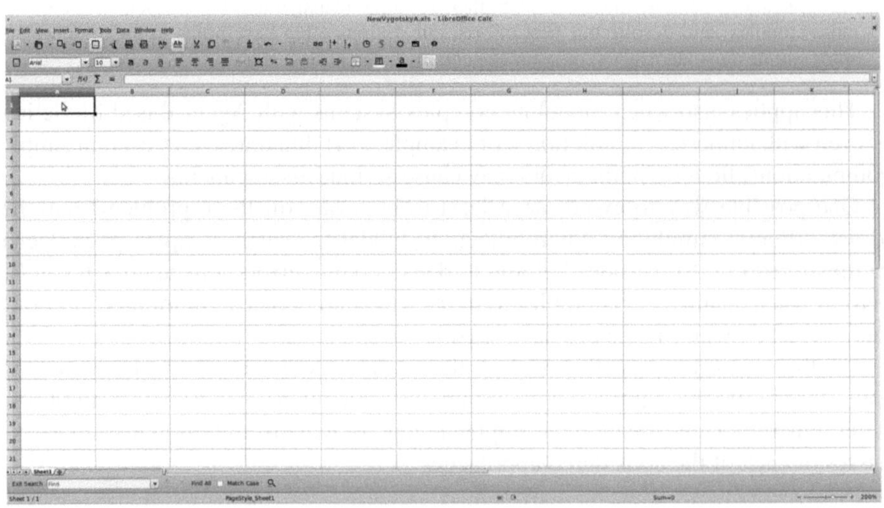

Fig. 7.1 An empty spreadsheet

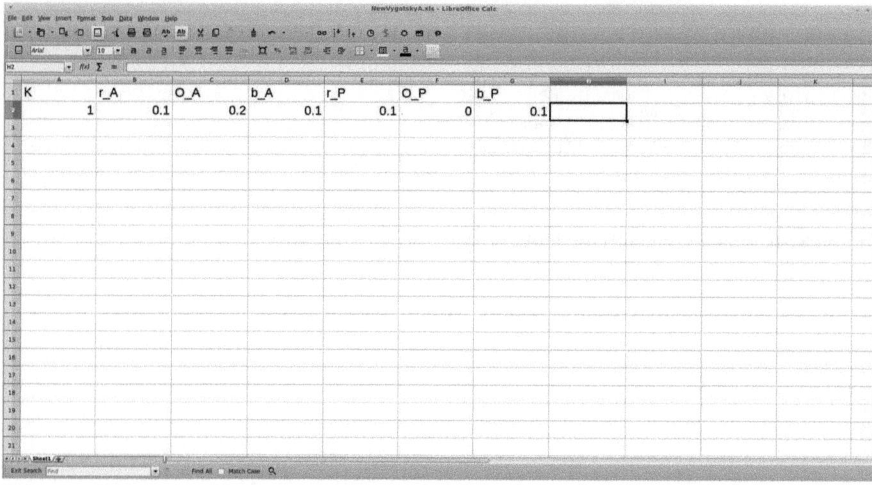

Fig. 7.2 The map parameters

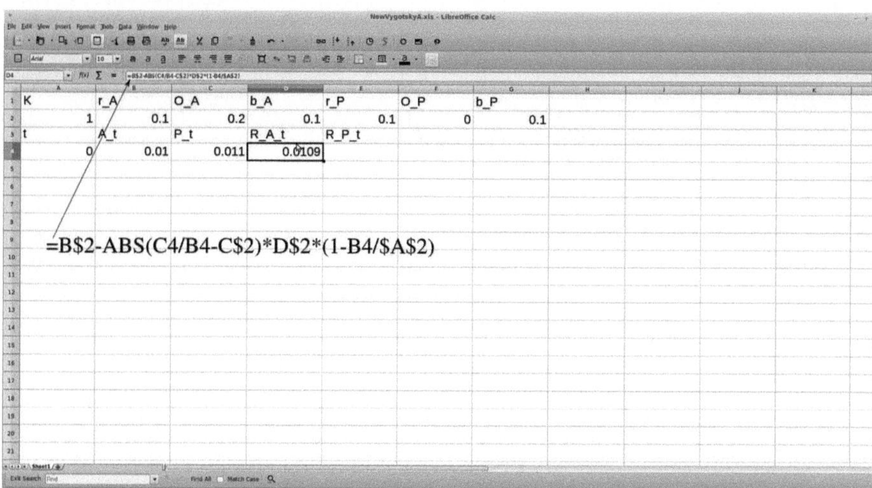

Fig. 7.3 $R_a(0)$ as defined in (7.3)

is coded as $= \mathrm{B4} * (1 + \mathrm{D4} * (1 - \mathrm{B4}/\mathrm{C4}))$ in cell `B5` as illustrated in Fig. 7.6. Then,

$$P\,(t+1) = P\,(t)\left[1 + R_p\,(t)\left(1 - \frac{P\,(t)}{K}\right)\right]$$

is coded $= \mathrm{C4} * (1 + \mathrm{E4} * (1 - \mathrm{C4}/\$\mathrm{A}\$2))$ in cell `C5` as illustrated in Fig. 7.7.

Finally, in cell `D5` and `E5` we update $R_a\,(1)$ and $R_p\,(1)$ copying and pasting the formulas from the respective cells above as illustrated in Fig. 7.8.

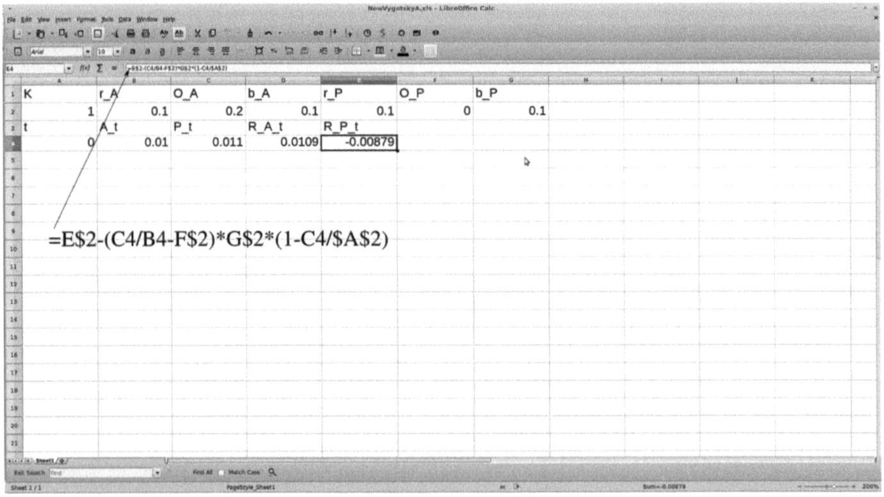

Fig. 7.4 $R_p(0)$ as defined in (7.3)

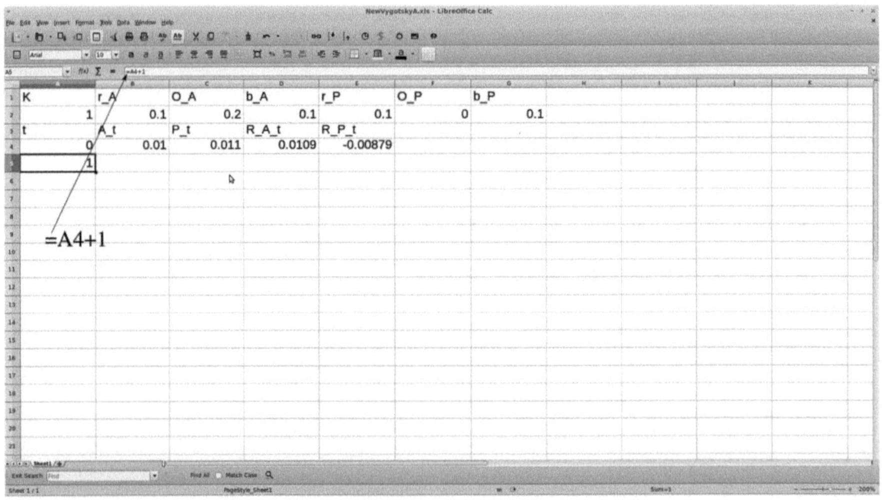

Fig. 7.5 Coding $t+1$

To have the whole trajectory, or at least the trajectory as long as we are interested in, we copy row 5 as many times as we want as illustrated in Fig. 7.9.

Now we are able to modify either the values of the parameters or the initial conditions or both to assess how the dynamics changes. Let us discuss the strengths and weaknesses of using a spreadsheet to model a dynamical system. This approach can be used almost on any computer as spreadsheets are common and can be used on different platforms; almost all computers have spreadsheets and there are even free and

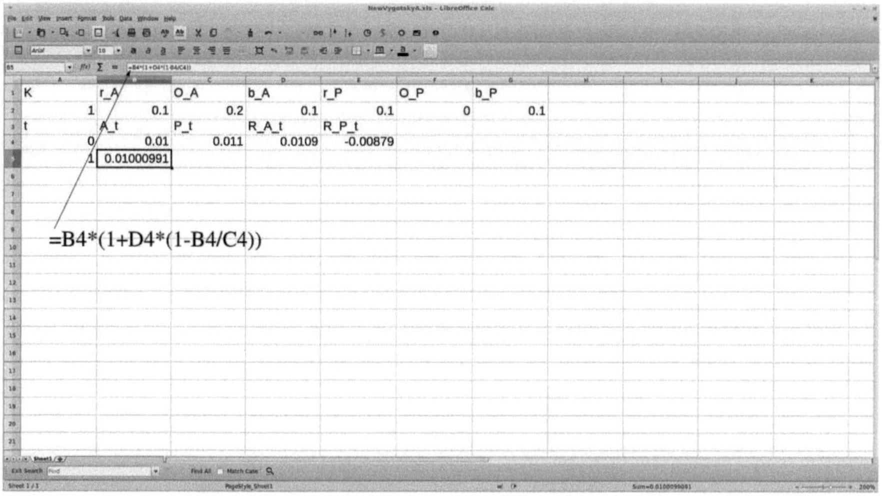

Fig. 7.6 Coding $A(t+1) = A(t)\left[1 + R_a(t)\left(1 - A(t)/P(t)\right)\right]$

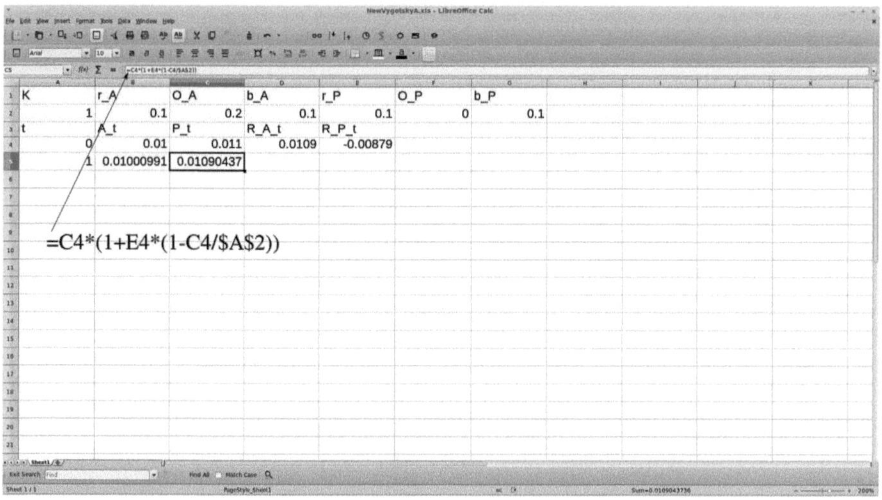

Fig. 7.7 Coding $P(t+1) = P(t)\left[1 + R_p(t)\left(1 - P(t)/K\right)\right]$

open source software such as the one we used. Finally, using spreadsheets involves almost no programming and is relatively fast. When considering weaknesses, equations become difficult to read, the graphical representation may be a little cumbersome, at least for those unfamiliar with making graphs with spreadsheets; finally, spreadsheets allow limited analysis only.

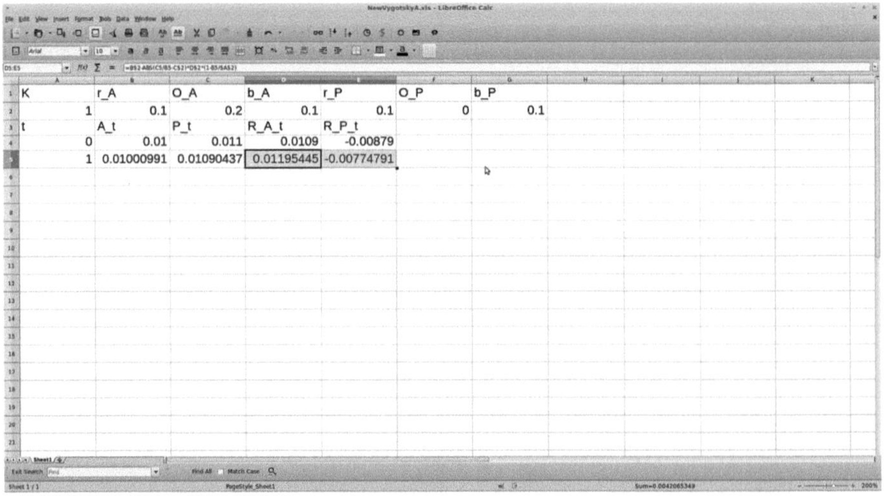

K	r_A	O_A	b_A	r_P	O_P	b_P
1	0.1	0.2	0.1	0.1	0	0.1
t	A_t	P_t	R_A_t	R_P_t		
0	0.01	0.011	0.0109	-0.00879		
1	0.01000991	0.01090437	0.01195445	-0.00774791		

Fig. 7.8 R_a (1) and R_p (1) are obtained by coping and pasting the cells in the row above

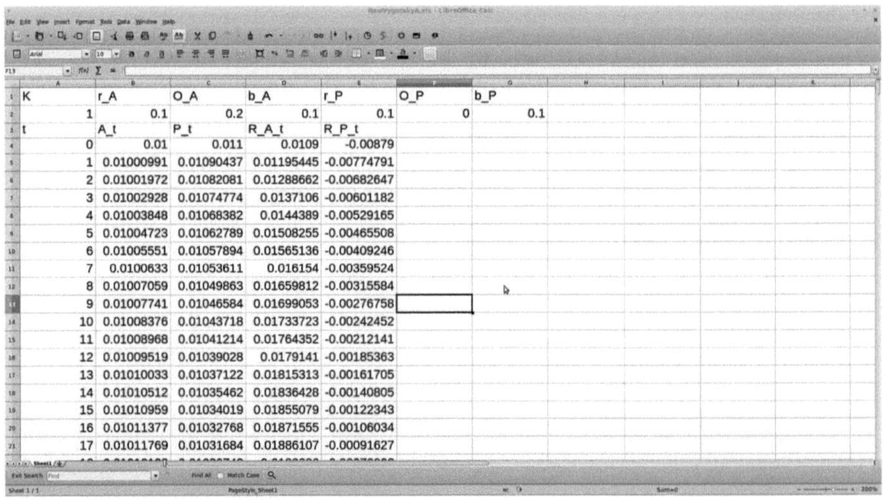

K	r_A	O_A	b_A	r_P	O_P	b_P
1	0.1	0.2	0.1	0.1	0	0.1
t	A_t	P_t	R_A_t	R_P_t		
0	0.01	0.011	0.0109	-0.00879		
1	0.01000991	0.01090437	0.01195445	-0.00774791		
2	0.01001972	0.01082081	0.01288662	-0.00682647		
3	0.01002928	0.01074774	0.0137106	-0.00601182		
4	0.01003848	0.01068382	0.0144389	-0.00529165		
5	0.01004723	0.01062789	0.01508255	-0.00465508		
6	0.01005551	0.01057894	0.01565136	-0.00409246		
7	0.0100633	0.01053611	0.016154	-0.00359524		
8	0.01007059	0.01049863	0.01659812	-0.00315584		
9	0.01007741	0.01046584	0.01699053	-0.00276758		
10	0.01008376	0.01043718	0.01733723	-0.00242452		
11	0.01008968	0.01041214	0.01764352	-0.00212141		
12	0.01009519	0.01039028	0.0179141	-0.00185363		
13	0.01010033	0.01037122	0.01815313	-0.00161705		
14	0.01010512	0.01035462	0.01836428	-0.00140805		
15	0.01010959	0.01034019	0.01855079	-0.00122343		
16	0.01011377	0.01032768	0.01871555	-0.00106034		
17	0.01011769	0.01031684	0.01886107	-0.00091627		

Fig. 7.9 The complete spreadsheet

7.5 R

R is a powerful and free statistical software and is rapidly becoming the standard setting for quantitative analysis and graphics. Although there are several guides to use R, see for instance [18], and books devoted to special applications, to the best of our knowledge there are no books explaining how to use R for analyzing dynamical systems. Nevertheless, the reader may find useful some blogs and other online material such as [19, 20]. Finally, there is a package (`deSolve`) implementing some functions to solve initial value problems of a system of first-order ordinary differential equations, of partial differential equations, differential algebraic equations and of delay differential equations [21, 22].

In this section we will not rely on this material, rather we will use basic instructions in R. In Listing 7.1 we provide some R code for computing a trajectory.

```
1  rm(list=ls())
2
3  N<- 4000   # number of iterations
4
5  # preparing variables
6  time <- c(1:N)
7  A     <-  rep(0, length(t))
8  P     <-  rep(0, length(t))
9  rA    <-  rep(0, length(t))
10 rP    <-  rep(0, length(t))
11
12 # parameters
13 K    <- 1.000000
14 r_A  <- 0.120000
15 o_A  <- 0.150000
16 b_A  <- 0.100000
17 r_P  <- 0.100000
18 o_P  <- 0.000000
19 b_P  <- 0.100000
20
21
22 # Initial condition
23 A0  <-   0.010000
24 P0  <-   0.011000
25 rA0  <- r_A-abs(P0/A0-o_A)*b_A*(1.0-A0/K)
26 rP0  <- r_P-(P0/A0-o_P)*b_P*(1.0-P0/K)
27
28
29 # first Iteration
30 A[1]  <- A0*(1.0+rA0*(1.0-A0/P0))
31 P[1]  <- P0*(1.0+rP0*(1.0-P0/K))
32 rA[1]  <- r_A-abs(P[1]/A[1]-o_A)*b_A*(1.0-A[1]/K)
33 rP[1]  <- r_P-(P[1]/A[1]-o_P)*b_P*(1.0-P[1]/K)
34
35 # remaining iterations
36  for(i in 1:(N-1)){
37     A[i+1]  <- A[i]*(1.0+rA[i]*(1.0-A[i]/P[i]))
38     P[i+1]  <- P[i]*(1.0+rP[i]*(1.0-P[i]/K))
39     rA[i+1]  <- r_A-abs(P[i+1]/A[i+1]-o_A)*b_A*(1.0-A[i+1]/K)
40     rP[i+1]  <- r_P-(P[i+1]/A[i+1]-o_P)*b_P*(1.0-P[i+1]/K)
41 }
42
43 # a glimpse of trajectory
44 A[1:100]
45 P[1:100]
```

Listing 7.1 R code for computing a trajectory

The code is commented; in R the sign # comments the rest of the line. Nevertheless, some clarifications will help the reader following the code. In line 1 we remove all the variables from the workspace, i.e., we start from a "clean state". In line 3 we set the number of iterations equal to 4000. In lines 6–10 we prepare some variables to store values. Parameters are defined in lines 13–19, and initial condition is selected in lines 23–24. In lines 25–26, we compute $R_a(0)$ and $R_p(0)$ following (7.3). Then, in lines 30–33 the first iteration is computed; finally, with a for cycle all the remaining iterations are computed in lines 36–41. Lines 30–33 are the analogous of Figs. 7.6, 7.7 and 7.8. Lines 36–41 are analogous of copying the spreadsheet lines. Finally, we obtain the trajectory with lines 44–45.

Using `ggplot2`, which is a powerful data visualization package for R, we can display some trajectories. We do not discuss the details here; we just mention that, with line 46 in Listing 7.2, the trajectory is coded as a *dataframe* that is a common object for data analysis in R. Line 49 loads package `ggplot2`; besides the online material [23] the reader may consult [24] to have further information about `ggplot2`.

```
46  df <- data.frame(x=rep(time,2), y=c(A, P), level=c(rep("A", length(time)),
    rep("P", length(time))))
47
48
49  library(ggplot2)
50
51  PatternA <- ggplot(df, aes(x=x, y=y, color=level)) + geom_point()+    xlab("t") +
    ylab("level") +  geom_line() +  geom_point( size=4, shape=21, fill="white") +
52  theme(legend.text=element_text(size=30),legend.position="none")
53
54  PatternA
```

Listing 7.2 R code for graphical representation of a trajectory

Following [9], we consider different parameters and initial conditions constellations. Figures 7.10 illustrate different behaviors of the system; they replicate some of examples reported in [9, p.268]. For example, Fig. 7.10a shows a sudden jump in both the help and the actual developmental level; parameter values are $r_a = 0.120$, $O_a = 0.150$, $b_a = 0.100$, $r_p = 0.100$, $O_p = 0.000$, $b_p = 0.100$ and initial condition is $A(0) = 0.010$, $P(0) = 0.011$. Figure 7.10b is obtained by parameter values $r_a = 0.000$, $O_a = 0.010$, $b_a = 0.100$, $r_p = 0.000$, $O_p = 0.010$, $b_p = 0.100$, initial condition $A(0) = P(0) = 0.010$ and represents a process where both help and actual developmental level reach a maximum being far below the potential maximum, which is 1. Figure 7.10c shows another process where both help and actual developmental level reach a maximum being far below the potential maximum. In this case parameters values are $r_A = b_a = r_p = o_p = b_p = 0.100$, $o_A = 0.010115$ and initial condition is $A(0) = 0.010$, $P(0) = 0.0101$. Finally, Fig. 7.10d represents a process in which both help and actual developmental level drop to zero. For this case parameters values are $r_a = O_a = b_a = r_p = b_p = 0.100$, $O_p = 0.120$ and initial condition is $A(0) = 0.010$, $P(0) = 0.0101$.

When comparing strengths and weakness of using R for studying dynamical systems, some of them boil down to those of R itself. First of all, R is free and open source software, allowing anyone to use and, importantly, to modify it. In fact, R is licensed under the GNU General Public License, with copyright held by The R Foundation for Statistical Computing. R is cross-platform: it runs on many operating systems and different hardware. It is popularly used on GNU/Linux, Macintosh, and Microsoft Windows, running on both 32 and 64 bit processors. Furthermore, the equations can be coded in a form close to the actual formulae. The graphical capabilities of R are outstanding and can also produce graphics output in PDF, JPG,

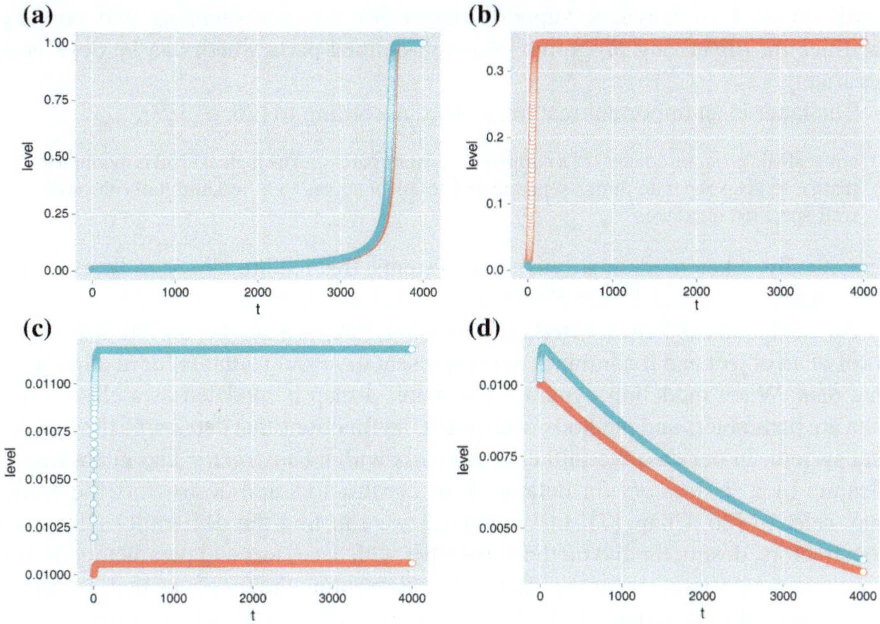

Fig. 7.10 Qualitatively different growth curves—$A(t)$ in *red* and $P(t)$ in *blue*—resulting from parameters configurations similar to those reported in [9]

PNG, and SVG formats, and table output for LATEX and HTML. Finally, R plays well with many other tools, importing data, for example, from CSV files, or directly from Microsoft Excel among the others.

In terms of weaknesses, R has a steep learning curve and it does take a while to get used to the power of R. Furthermore, installing R can be tricky, at least for those unfamiliar with installing software on their computer.

Finally, script must be run and many R commands give little thought to memory management; therefore it may be unfeasible to perform long simulations.

For a thorough discussion about R advantages and disadvantages when doing statistical analysis and data mining the reader may refer to [25].

7.6 C++

C++ is a general-purpose programming language. In the preface of the first edition of his guide, C++ creator states:

> The key concept in C++ is class. A class is a user-defined type. Classes provide data hiding, guaranteed initialization of data, implicit type conversion for user-defined types, dynamic typing, user-controlled memory management, and mechanisms for overloading operators [26, p. xiii].

Furthermore, C++ language supports *object-oriented programming* and provides facilities for organizing programs into well-defined parts which can be developed separately.

The latter is an important feature. In fact, according to [26, p. 389]:

> any realistic program consists of a number of separate parts. ...The logical ideal is modularity, that is, to keep separate things separate and to allow access to a 'module' only through a well-specified interface.

Specifically, when modeling dynamical systems, this feature allows to keep a common structure and use different maps as separate modules.

As it concerns *object-oriented programming*, in C++ a *class* is used to specify the form of an object and it combines data representation and methods for manipulating that data. When modeling dynamical systems, a map is modeled as a class where data are parameters and methods implement the functions that determine the dynamical system. In our case we initialize the class with a *constructor* and guarantee its cleanup by a *destructor*; for details about constructors and destructors the reader may refer to [26, Chap. 17]. In Listing 7.3 we can find the declarations for class constructors, destructor and methods together with accessors and parameters. In line 7 we find the declaration for the methods implementing the functions that determine the dynamical system, namely (7.2). Also in lines 10–26 we find an example of the *accessor functions*, i.e., the part of code which allows us to examine and change the parameter values (see [26, p. 541]).

```
1  class Map {
2    public:
3      Map();
4      Map(double _k,double _r_A,double _o_A,double _b_A,
5          double _r_P,double _o_P,double _b_P);
6      ~Map();
7
8      void Iteration(long double *_x);  // find x(t+1)
9
10   /////////////////////////////////////////
11   //  variables setter and getter
12   /////////////////////////////////////////
13   char* Name()
14   {
15     return name;
16   }
17
18   // map parameters
19   double K()
20   {
21     return k;
22   }
23   void SetK(double newK)
24   {
25     k=newK;
26   }
27   protected:
28   private:
29   // name of the map
30   char* name;
31   // parameters
32   double k;
```

```
    double r_A;
    double o_A;
    double b_A; // b_a
    double r_P;
    double o_P;
    double b_P;   // b_p
};
```

Listing 7.3 Header file for class Map

Listing 7.4 shows the `.cpp` file of the class Map, that is the file which is compiled. In particular, lines 34 and 35 compute $R_a(t)$ and $R_p(t)$ according to (7.3), while lines 36 and 37 compute $A(t+1)$ and $P(t+1)$ according to (7.2). To pass state variable values we use pointers as $R_a(t)$, $R_p(t)$, $A(t)$ and $P(t)$ are stored in an array; for details see [26, Chap. 7].

```
1  // Constructor performs initialization
2  Map::Map() :
3    name("Vygotsky"),
4    k(1),
5    r_A(0.1),
6    o_A(0.1),
7    b_A(0.1),
8    r_P(0.1),
9    o_P(0.1),
10   b_P(0.1)
11 {
12 }
13 //
14 // Overloaded Constructors
15 //
16 Map::Map(double _k,double _r_A,double _o_A,double _b_A,
17          double _r_P,double _o_P,double _b_P) :
18   name("Vygotsky"),
19   k(_k),
20   r_A(_r_A),
21   o_A(_o_A),
22   b_A(_b_A),
23   r_P(_r_P),
24   o_P(_o_P),
25   b_P(_b_P)
26 {
27 }
28 // Destructor performs cleanup.
29 Map::~Map()
30 {
31 }
32  void  Map::Iteration(long double *x){  // find x(t+1)
33  double newx[4];    // new value:  A_t+1, P_t+1, R_A_t+1, R_P_t+1
34  // compute x[2] and x[3]
35  x[2]=r_A-fabsl(x[1]/x[0]-o_A)*b_A*(1.0-x[0]/k);     // R_A_t
36  x[3]=r_P-(x[1]/x[0]-o_P)*b_P*(1.0-x[1]/k);     // R_P_t
37  newx[0]=x[0]*(1.0+x[2]*(1.0-x[0]/x[1]));     // A_t+1
38  newx[1]=x[1]*(1.0+x[3]*(1.0-x[1]/k));     // P_t+1
39  newx[2]=r_A-fabsl(x[1]/x[0]-o_A)*b_A*(1.0-x[0]/k);     // R_A_t+1
40  newx[3]=r_P-(x[1]/x[0]-o_P)*b_P*(1.0-x[1]/k);     // R_P_t+1
41  // update values to pass back
42  x[0]=newx[0];
43  x[1]=newx[1];
```

```
    x[2]=newx[2];
    x[3]=newx[3];
  };
```

Listing 7.4 Class Map .cpp file

The strengths of using C++ for simulating dynamical systems' dynamics are that C++ is flexible, cross-platform and, finally, really fast. On the other hand, C++ presents a steep learning curve, code needs to be compiled, graphical output is not immediate and also the programmers may need to be familiarized with a *rapid application development* framework in order to write and compile their code.

7.7 Conclusions

The three approaches we have presented in this chapter have both strengths and weaknesses; therefore there is no perfect approach.

Although C++ can become the preferred choice when speed is a requirement, the need for effective graphic representation calls for using other approaches or at least integrating several approaches. One option is to combine the different approaches in order to exploit their relative strengths. For example, it can be possible to perform the calculations in C++ and use R for representing graphics. To this purpose it is useful to use a format such as Comma-Separated Values (CSV) files, as data exchange for moving data between programs. Although [27] proposes a specification for the CSV format, and this is the definition commonly used, there is not a single, well-defined format for CSV files. Nevertheless, it is easy to exchange data using CSV like files, as proposed in Fig. 7.11. Besides this aspect, probably the main benefit of using multiple implementation is given by debugging and controlling numerical artifacts. Concerning the first benefit, the fact that equations coded with a spreadsheet are rather distant from the coding C++ and R call for, makes using spreadsheets a good way to check for mistakes. As it concerns the second aspect, this is quite important when studying chaotic dynamical systems, as one of the ingredient of a chaotic map

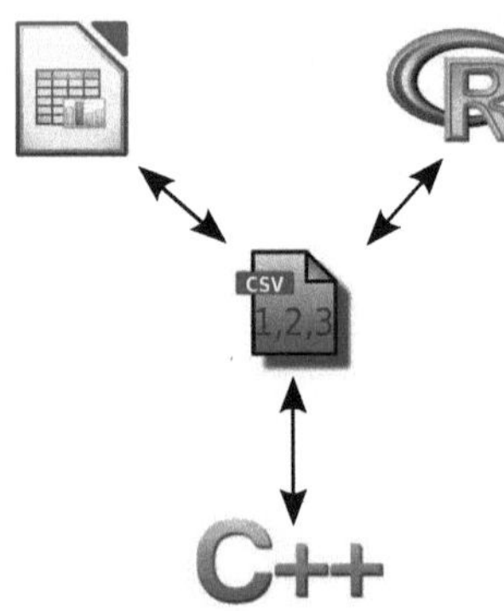

Fig. 7.11 Moving data between spreadsheet, R and C++ using CSV files

we have sensitive dependency on initial conditions [28] and, therefore, numerical approximations can play an important role.

Acknowledgments We are grateful to Arianna Dal Forno and Angelo Romano for helpful comments and suggestions. Usual caveats apply.

References

1. Newell, A., Simon, H.: Computers in psychology. In: Luce, R., Bush, R., Galanter, E. (eds.) Handbook of Mathematical Psychology, vol. 1, Chap. 7, pp. 361–428. Wiley, New York (1963)
2. Coombs, C.H., Dawes, R.D., Tversky, A.: Mathematical Psychology. Prentice-Hall, Englewood Cliffs (1970)
3. Scheinerman, E.R.: Invitation to Dynamical Systems. Prentice Hall, Upper Saddle River (1996)
4. Abraham, F.D., Abraham, R.H., Shaw, C.D.: A Visual Introduction to Dynamical Systems Theory for Psychology. Aerial Press, Santa Cruz (1990)
5. Nowak, A., Vallacher, R.R.: Dynamical Social Psychology. The Guildford Press, New York (1998)
6. Guastello, S.J., Koopmans, M., Pincus, D.: Chaos and Complexity in Psychology. Cambridge University Press, Cambridge, UK (2009)
7. Dal Forno, A., Merlone, U.: Effort dynamics in supervised work groups. J. Econ. Behav. Organ. **75**, 413–425 (2010)
8. Levine, R.L., Fitzgerald, H.E.: Analysis of Dynamic of Psychological Systems, vol. 1. Plenum Press, New York (1992)
9. van Geert, P.: Dynamic System of Development. Harvester Wheatsheaf, New York (1994)
10. Dal Forno, A., Merlone, U.: Nonlinear dynamics in work groups with bions basic assumptions. Nonlinear Dyn. Psychol. Life Sci. **17**(2), 295–315 (2013)
11. van Dijk, M., van Geert, P., Korecky-Kroll, K., Maillochon, I., Laaha, S., Dressler, W.U., Bassano, D.: Dynamic adaptation in child-adult language interactione. Lang. Learn. **63**(2), 243–270 (1982)
12. van Geert, P., Steenbeek, H.: Explaining after by before: basic aspects of a dynamic systems approach to the study of development. Dev. Rev. **25**(2), 408–442 (2005)
13. Becker, P.: Bad pointers. C/C++ Users J. **23**(9), 37–41 (2005)
14. Cerruti, U., Giacobini, M., Merlone, U.: A new framework to analyze evolutionary 2x2 symmetric games. In: IEEE Proceedings of "CIG'05: Symposium on Computational Intelligence and Games", April 4–6 2005. Essex University, Colchester, Essex, UK (2005)
15. Merlone, U., Sonnessa, M., Terna, P.: Horizontal and vertical multiple implementations in a model of industrial districts. J. Artif. Soc. Social Simul. **11**(2) (2008)
16. LibreOffice Documentation Team: LibreOffice Calc 4.1. Samurai Media Limited, Wickford, UK (2015)
17. Merlone, U.: Personal webpage. http://www.ugomerlone.net/documenti/merlonevangeert.zip. Accessed 24 Feb 2016
18. Crawley, M.J.: The R Book. Wiley, Chichester, UK (2007)
19. Chivers, C.: Dynamical systems: mapping chaos with R. http://bayesianbiologist.com/2012/07/13/dynamical-systems-mapping-chaos-with-r/. Accessed 24 Feb 2016
20. Fenu, L.: Shiny, desolve and ggplot play nicely together. http://grrrraphics.blogspot.it/2013/01/shiny-desolve-and-ggplot-play-nicely.html. Accessed 24 Feb 2016
21. Soetaert, K., Petzoldt, T., Setzer, R.W.: Solving differential equations in R. R J. **2**(2), 5–15 (2010)
22. Soetaert, K., Petzoldt, T., Setzer, R.W.: Solving differential equations in R: package deSolve. J. Statist. Softw. **33**(9), 1–25 (2010)

23. Wickham, H.: ggplot2. http://ggplot2.org/. Accessed 24 Feb 2016
24. Wickham, H., Sievert, C.: ggplot2. Springer, Dordrecht (2016)
25. Williams, G.: Data Mining with Rattle and R. Springer, Dordrecht (2011)
26. Stroustrup, B.: The C++ Progamming Language, 4th edn. Addison Wesley, Upper Saddle River (2013)
27. Shafranovich, Y.: Common format and MIME type for comma-separated values (CSV) files. RFC 4180 (Informational) (2005)
28. Devaney, R.L.: Chaotic Dynamical Systems, 2nd edn. Perseus Books, MA (1989)

Index

G.I. Bischi et al. (eds.), *Qualitative Theory of Dynamical Systems, Tools and Applications for Economic Modelling*, Springer Proceedings in Complexity,
DOI 10.1007/978-3-319-33276-5

MIX
Papier aus verantwortungsvollen Quellen
Paper from responsible sources
FSC® C105338

If you have any concerns about our products,
you can contact us on
ProductSafety@springernature.com

In case Publisher is established outside the EU,
the EU authorized representative is:
Springer Nature Customer Service Center GmbH
Europaplatz 3, 69115 Heidelberg, Germany

Printed by Libri Plureos GmbH
in Hamburg, Germany